AF361881

Energy Efficiency in Electric Motors, Drives, Power Converters and Related Systems

Energy Efficiency in Electric Motors, Drives, Power Converters and Related Systems

Special Issue Editor

Mario Marchesoni

MDPI • Basel • Beijing • Wuhan • Barcelona • Belgrade • Manchester • Tokyo • Cluj • Tianjin

Special Issue Editor
Mario Marchesoni
University of Genova
Italy

Editorial Office
MDPI
St. Alban-Anlage 66
4052 Basel, Switzerland

This is a reprint of articles from the Special Issue published online in the open access journal *Energies* (ISSN 1996-1073) (available at: https://www.mdpi.com/journal/energies/special_issues/efficiency_electric).

For citation purposes, cite each article independently as indicated on the article page online and as indicated below:

LastName, A.A.; LastName, B.B.; LastName, C.C. Article Title. *Journal Name* **Year**, *Article Number*, Page Range.

ISBN 978-3-03936-390-2 (Pbk)
ISBN 978-3-03936-391-9 (PDF)

Contents

About the Special Issue Editor

Mario Marchesoni received his Ph.D. in electrical engineering in power electronics in 1990 from the University of Genova, Italy. Today he is a Full Professor at the University of Genova, where he teaches Electrical Drive Control. He has more than 30 years of professional experience. He is the Head of the PETRA Group (Power Electronics, TRansportation and Automation) of the University of Genova, which has been active in research activity in the field of power electronics for decades, with particular focus on industrial automation, road, rail, naval and aerospace transportation. His technical and scientific activity, as evidenced by about 200 papers presented at international conferences and published in international journals, has been carried out within research contracts and in cooperation with national and international companies. In addition, he has worked on several research contracts funded by the Italian Ministry of University and Scientific and Technological Research, by the Italian National Research Council and by the European Commission.

Preface to "Energy Efficiency in Electric Motors, Drives, Power Converters and Related Systems"

The promise of sustainable growth through the use of renewable energy has been attracting increased attention around the world. With the same goal of sustainable growth in mind, the potential for obtaining transformative results (that have an immediate, short-term impact) through increasing energy efficiency should not be ignored. As an example, the European Union has set itself a 20% energy savings target by 2020, which is roughly equivalent to turning off a few hundred power stations. Today, 20% of all final energy consumption in the EU is electrical energy, but this is predicted to grow significantly over the next few decades. Within this scenario, power electronics is a key enabling technology that allows not only the efficient generation, use and distribution of electrical energy, but also the implementation of energy saving applications at reasonable costs. The widespread diffusion of electric motor drives has also been enabled by power electronics.

Power electronics is a transversal technology, which covers a very high power range, from the order of mW required for mobile phone operation to the order of GW for applications in the field of energy transmission. Advanced power electronics can achieve very substantial energy savings. There are many market segments that can potentially benefit from the use of this technology: home and office applications; heating, ventilation and air conditioning; consumer digital products; communications; factory automation and electric drives; electric traction; the automotive industry; and renewable energies.

Power electronics is the key technology that allows the flow of electricity from the source to the load to be controlled with extreme precision, satisfying the load specifications. It is responsible for the reliability and stability of the whole electrical grid, including sources, transmission, and distribution of energy to a very wide variety of applications in industry, transport systems and almost infinite domestic and office applications. As a technology that allows for the efficient use, distribution and generation of electricity, it enables significant energy savings. There are many fields of application worth mentioning.

The connection of renewable energy sources to the electricity grid would not be possible without power electronics: power electronic converters optimize the efficiency of photovoltaic panels and allow making the best use of the energy produced by wind turbines.

Electric motor drives use 50%–60% of all electricity consumed in industry. By using power electronics, it is possible to achieve a reduction in energy consumption of about 20%–30%.

In domestic applications, electronic thermostats for refrigerators and freezers can lead to savings of around 20%, while another 20% can be saved by using power electronics to control the compressor motor. Further energy can be saved if the motor is built with permanent magnets.

Advanced power electronics can, for example, achieve savings of more than 50% of energy losses in the conversion of mains or battery voltage to that used in electronic equipment.

New technologies in power supplies can increase efficiency by around 2%–4%, reducing absorption in conditions of low power requirement or standby, reducing losses from 14% to 30%. Digital control can further reduce energy consumption.

In automotive applications, electric or hybrid drives are only possible using intelligent power electronics. The concept of a "drive by wire" fully electric vehicle can allow savings of more than 20% thanks to power electronics.

Despite the strategic importance of power electronics, there is a lack of awareness of its role in modern industrial society, even among the generally well-informed public. The impact of power electronic technology, capable of providing reliable and precisely controlled electric power sources in all areas of human life, is not known at the public level. Energy savings, which would be achievable in the short term using the electronic power technologies already available today, are not implemented in many fields of application. The enormous energy saving potential available through the development of new technologies that affect the entire supply chain, including consumer equipment, remains to be understood and implemented. Furthermore, from both public and political points of view, it is clear that power electronics does not have the same appeal compared to, for example, microelectronics or nanotechnology, with a negative impact on the attractiveness for students and on the destination of research funds.

The publication of this collection of scientific articles, dedicated to the topic of energy efficiency using power electronics and electrical drives, is our contribution to better disseminate this information to people.

Mario Marchesoni
Special Issue Editor

Article

Robust Design Optimization with Penalty Function for Electric Oil Pumps with BLDC Motors

Keun-Young Yoon [1] and Soo-Whang Baek [2,*]

[1]   Department of Electrical Engineering, Honam University, 417 Eodeung-daero, Gwangsan-gu, Gwangju 62399, Korea; ky.yoon@honam.ac.kr
[2]   Department of Automotive Engineering, Honam University, 417 Eodeung-daero, Gwangsan-gu, Gwangju 62399, Korea
*   Correspondence: swbaek@honam.ac.kr; Tel.: +82-062-940-5408

Received: 29 November 2018; Accepted: 27 December 2018; Published: 2 January 2019

Abstract: In this paper, we propose and evaluate a robust design optimization (RDO) algorithm for the shape of a brushless DC (BLDC) motor used in an electric oil pump (EOP). The components of the EOP system and the control block diagram for driving the BLDC motor are described. Although the conventional deterministic design optimization (DDO) method derives an appropriate combination of design goals and target performance, DDO does not allow free searching of the entire design space because it is confined to preset experimental combinations of parameter levels. To solve this problem, we propose an efficient RDO method that improves the torque characteristics of BLDC motors by considering design variable uncertainties. The dimensions of the stator and the rotor were selected as the design variables for the optimal design and a penalty function was applied to address the disadvantages of the conventional Taguchi method. The optimal design results obtained through the proposed RDO algorithm were confirmed by finite element analysis, and the improvement in torque and output performance was confirmed through experimental dynamometer tests of a BLDC motor fabricated according to the optimization results.

Keywords: optimal design; oil pump; brushless DC; motor; robust; vehicles

---

1. Introduction

A hybrid vehicle combines two or more different power sources to obtain a driving force. In most cases, the power sources are an internal combustion engine that uses fuel and an electric motor driven by battery power. Hybrid vehicles drastically reduce fuel consumption and harmful gas emissions compared to conventional vehicles. In recent years, research on hybrid vehicles has been actively pursued in response to the demand for improved fuel efficiency and more environmentally friendly products [1–5].

The oil pump is an actuator between the engine and the electric motor and supplies the hydraulic fluid necessary for the transmission [6,7]. However, conventional oil pumps fail to deliver the required flow rate under operating conditions, thereby decreasing the vehicle's fuel efficiency [8]. To address this problem, an electric oil pump (EOP) has been developed that can supply a suitable flow rate under operating conditions [9,10]. Hybrid vehicles with conventional internal combustion engines use a mechanical oil pump to improve fuel economy and an EOP as an auxiliary; the EOP operates only when the mechanical oil pump stops, when the vehicle stops or travels at low speed, or when the transmission fluid flow is insufficient in the electric-vehicle mode. Thus, the EOP reduces mechanical power loss by operating only when the transmission requires oil pressure [11,12]. The EOP supplies the required flow rate to the engine clutch and the transmission by variably controlling the rotational speed of the EOP's drive motor according to the driving state of the vehicle and operator [13]. In addition to the oil pump installed inside the transmission, the hybrid vehicle also has an external oil pump attached to the outside of the transmission to supply sufficient operating fluid to the engine clutch [14]. The EOP is controlled by a brushless DC (BLDC) motor in low- and high-speed modes in consideration of system efficiency and operation responsiveness according to the driving state of the vehicle and driver [15].

Recently, as competition in the international industrial market has become more intense, new design methods are required to satisfy increasingly stringent product performance and quality requirements [16]. In the conventional motor design process, deterministic design optimization (DDO) methods have been used to optimize design, assuming constant design variable values to maximize performance [17]. However, all motors exhibit inherent uncertainties in their material properties, manufacturing tolerances, and operating conditions. Conventional design techniques consider product uncertainties by implementing safety factors based on design experience [18]. However, these safety factors do not accurately quantify the limitations and uncertainties, and insufficient safety factor estimates can degrade motor performance and service life, whereas excessive safety factors increase production costs. Therefore, a robust design optimization (RDO) technique that can systematically consider stochastic uncertainties is required [19]. In RDO, unlike conventional design methods, design variables are defined as randomly distributed according to specific dispersion characteristics centered on mean values, and the relative probabilities are considered [20]. RDO approaches have double-loop optimization structures in which the probabilistic constraint added to the conventional optimal design process is optimized separately from the objective function. Therefore, RDO approaches, such as the Monte Carlo and worst-case simulations that have recently been applied to this problem, significantly increase the design time and computational costs, compared with the conventional optimal design method [21–23]. To address these problems, the Taguchi method has been proposed, whereby the main design factors influencing the product characteristics are selected and the optimal design combination is obtained with a minimum number of experiments. However, the Taguchi method the preset level values in the design space are limited to the available experimental combinations, and unrestricted searches of the entire design space are thus not permitted [24,25]. In addition, various constraints related to motor performance cannot be easily investigated in the given design time.

In this paper, with the objective of refining the Taguchi method to develop an evaluation technique that offers excellent convergence stability and calculation efficiency, we introduce an optimal level search method that incorporates the inequality constraint by adding a corresponding penalty function to the loss function. The optimal level search method guarantees the free movement of repeated design points in the design space by automatically adjusting the intervals of the set design parameter levels to identify successive experimental combinations. To verify feasibility, the proposed RDO method was applied to the design of a 150 W BLDC motor with an EOP, and the method's performance was compared with conventional DDO results obtained using finite element analysis. Finally, prototype BLDC motors were fabricated and dynamometer tests were performed to confirm the suitability of the proposed RDO in a practical application.

2. EOP System and BLDC Motor

2.1. EOP System

Conventional oil pumps are driven by the driving force of the engine, which causes power loss from the engine. As the number of engine revolutions increases, the oil flow rate increases more than necessary. Therefore, unnecessary power consumption occurs, and the engine output and fuel efficiency are reduced. To solve this problem, a (motor-driven) EOP that is mechanically independent of the engine speed has been developed. EOPs provide the fluid pressure and flow rate required for the transmission and are especially useful in vehicles equipped with the "Idle Stop & Go" system function, where the EOPs reduce the shift shock that can occur with the automatic transmission because of lack of flow after stopping or starting the vehicle, thereby improving fuel economy and reducing exhaust gas. The EOP system consists of a motor, a motor driver, and a pump assembly for generating hydraulic fluid, as shown in Figure 1. The motor is a three-phase BLDC motor with a Hall sensor, and the motor driver uses a three-phase full bridge circuit consisting of an N-channel metal-oxide field-effect transistor (MOSFET). The motor driver controls the hydraulic pressure of the transmission to maintain suitable low-speed, reverse, or stop conditions.

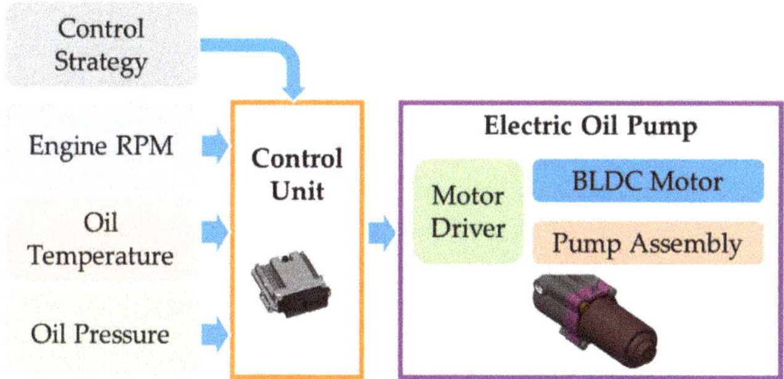

Figure 1. Configuration of the electric oil pump (EOP) system.

Figure 2 is a block diagram of the EOP's control unit, which includes a power supply unit that receives the vehicle battery power and supplies the required voltage to each device; a microcontroller unit (MCU) that receives the target rotational speed (rpm) of the EOP motor and outputs a pulse width modulation (PWM) control signal to the motor driver; a motor control unit; and a motor driving unit that supplies current to the motor by switching MOSFETs. The motor driving unit consists of six MOSFET switches that are sequentially turned on and off according to the signal provided by the motor control unit, which includes logic for turning the six MOSFETs on and off, as well as a Hall sensor input for detecting the position of the rotor, a function for diagnosing disconnection and short circuit of the motor, and a gate drive circuit for turning the MOSFETs on and off.

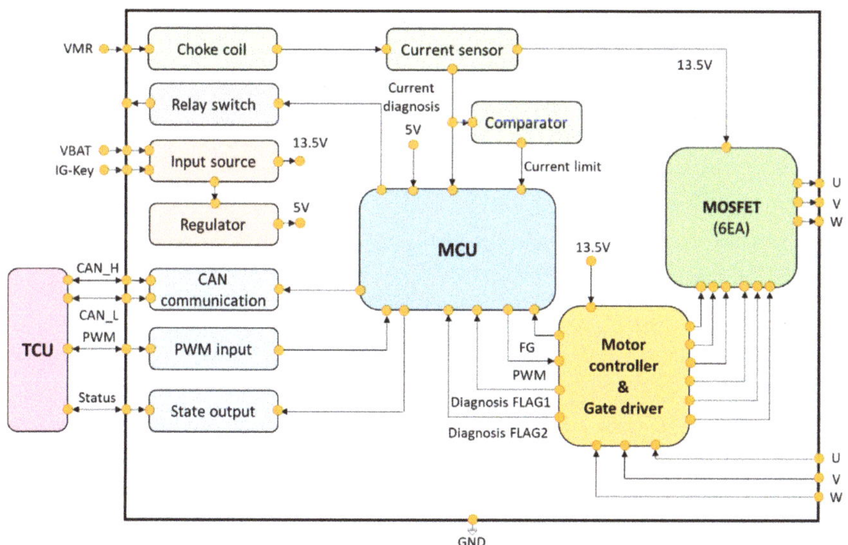

Figure 2. Block diagram of the EOP control unit.

The power supply uses three sources, as shown in Figure 2. The constant power source (VBAT) is supplied from the vehicle battery. The isolated ground (IG) power (IG-Key) is supplied only when the vehicle is started. The relay power supply (VMR) is entirely dedicated to the motor driver FET. The power unit receives the constant power supply and outputs 13.5 V and 5 V to the motor control unit and to the MCU and Hall sensor of the BLDC motor, respectively. When the IG power is turned on with constant power available, the MCU operates for a certain period of time using the constant power even after the IG power is turned off. The MCU is a key component for driving the oil pump and manages all operating procedures. When IG power is on, the MCU operates and controller area network communication is activated; the MCU diagnoses the internal condition and informs the transmission control unit of the abnormality via controller area network communication. When the transmission control unit receives an instruction to drive the motor, the transmission control unit turns on the external relay to supply power via the VMR and drive the motor. In addition, a PWM waveform is supplied to the motor control section to control the speed of the BLDC motor. The current value flowing through the motor is sensed by the current sensor, and the maximum current value is limited during overcurrent. To protect the system, when the maximum current continues to flow for a certain period of time, the MCU enters the power save mode and limits the current value to a lower level. The set BLDC motor speed is provided continuously from the transmission control unit to the MCU via controller area network communication or the PWM value, which acts as a backup for when the controller area network communication is disconnected. The MCU also provides the internal control status of the EOP control unit to the transmission control unit as a status update in preparation for controller area network disconnection.

2.2. BLDC Motor

Figure 3 shows the shape of a conventional BLDC motor used for an EOP, the specifications of which are presented in Table 1.

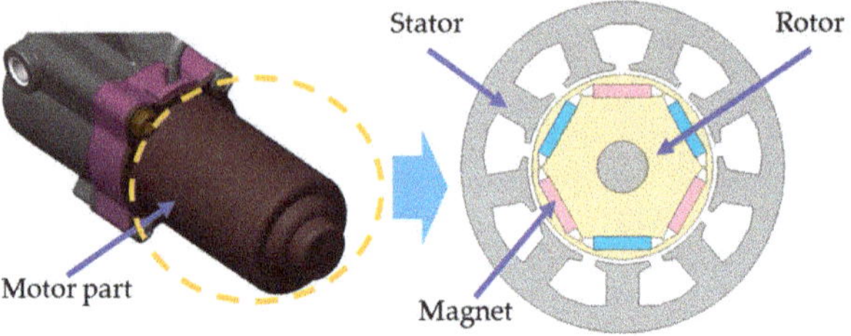

Figure 3. Shape of conventional brushless DC (BLDC) motor for an EOP.

The conventional BLDC motor selected for this study had a six-pole nine-slot structure, of the concentrated-winding interior permanent magnet (IPM) type, considering vibration and noise. An initial design was established to satisfy the given specifications. The current density of the stator windings was designed to be less than 6 A_{rms}/mm^2 at maximum speed. In this study, a previously reported torque per unit rotor volume (TRV) was used to establish the dimensions of the rotor [26]. The rotor diameter can be estimated using maximum torque, TRV, and stack length (rotor and stator axial length) and can be expressed as

$$D_r = \sqrt{\frac{4 \cdot T}{TRV \cdot \pi \cdot L_{stack}}} \tag{1}$$

where D_r is the diameter of the rotor, T is the maximum torque of the BLDC motor and L_{stack} is the stack length.

Table 1. Specifications of the conventional BLDC motor.

	Items	Value	Unit
Stator	Outer/Inner diameter	51/30	mm
	Slots	9	-
	Number of turns	14	turns
Rotor	Outer/Inner diameter	29/8	mm
	Pole	6	-
	Magnet grade	N42EH	NdFeB
Rated	Speed	4000	rpm
	Output power	150	W
	Efficiency	70	%
	Torque	0.358	Nm
	Stack length	35	mm
	Air-gap length	0.5	mm
	Maximum speed	6000	rpm
	Cogging torque (peak to peak)	0.037	Nm
	Torque ripple (peak to peak)	0.06	Nm

3. Methods

3.1. RDO Process

Figure 4 summarizes the RDO concept. If there is a design variable that varies with any distribution of the other system design variables, the response of the system to the first design variable should also appear as a distribution rather than as a single value. In such cases, RDO is executed to narrow the variation range of the objective function with respect to the variation of the design variables, applying constraints to ensure that the optimal design point is feasible.

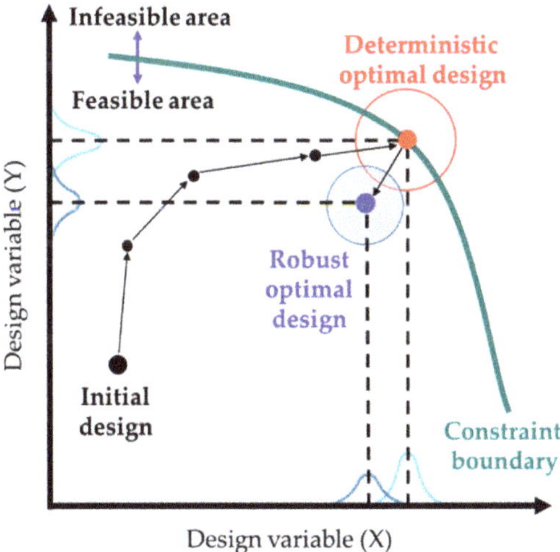

Figure 4. Robust design optimization (RDO) concept.

RDO thus minimizes a specific objective function at a set constraint. In this paper, the Taguchi method is modified as follows to enhance efficiency. In the Taguchi method, the loss that occurs as the product performance moves away from the target value is expressed as a loss function, defined based on the signal-to-noise ratio (SNR) to target a specific value, a maximum value, or a minimum value [27,28]. In this study, we use the loss function shown in Equation (2), which targets a minimum but non-negative value.

$$SNR = -\log\left[\frac{1}{n}\sum_{i=1}^{n}y_i^2\right] \tag{2}$$

Here, y_i is the performance value of the performance function h derived from the ith experimental combination, and n is the total number of experimental combinations.

The conventional Taguchi method derives the optimal combination of experiments using the SNR only for the specific target performance of the design object. However, a general RDO must consider multiple targets simultaneously, as well as other performance-related constraints, and we therefore introduce a penalty function that is added to the performance value y_i for the performance function h, as in Equation (3), to account for the constraints. That is, the penalty function increases in value according to the degree of violation of the constraint condition.

$$y_i = h(X) + W_pP(X) \tag{3}$$

Here, the performance value derived from the re-defined experimental combination includes the penalty function: W_p is the weight of the penalty function, and $P(X)$ is the penalty function considering the constraint. The definition of the penalty function used in this paper is expressed as Equation (4).

$$P(X) = \sum_{j=1}^{np}\left[maximum\{0, C(X)\}^2\right] \tag{4}$$

where C is a constraint. If all constraints are satisfied, the penalty function value becomes zero. However, if the constraint condition is violated, the square of the constraint condition value is applied to the penalty function.

To find the optimal solution by continuously searching the designated design space using the Taguchi method, the level of each design factor used to construct each experimental combination should be automatically changed. That is, when the design point moves in response to the optimal combination result derived from the current experimental combination, a new combination of experiments is constructed at the new design point. At this time, if a new level value is determined for each design factor, a more advanced design point search is possible. In this study, three design factor levels were defined: the interval between levels 1 and 2, Δ_1, and the interval between levels 2 and 3, Δ_2. The values of the current levels 1, 2, and 3 were set as X_{1prev}, X_{2prev}, and X_{3prev}, respectively. The new level values (X_1, X_2, X_3) for the subsequent experiment combination were then defined as shown in Table 2 according to the optimal level value derived from the current experimental combination.

Table 2. Definition of new level values.

	Items	Description
Case #1	When X_1 is determined as the optimal level	$X_1 = X_{1prev} - \Delta_1$ $X_2 = X_{1prev}$ $X_3 = X_{2prev}$
Case #2	When X_2 is determined as the optimal level	$X_1 = X_{2prev} - \Delta_1/2$ $X_2 = X_{2prev}$ $X_3 = X_{2prev} + \Delta_2/2$
Case #3	When X_3 is determined as the optimal level	$X_1 = X_{2prev}$ $X_2 = X_{3prev}$ $X_3 = X_{2prev} + \Delta_2$

Figure 5 shows the proposed RDO algorithm with the penalty function and the optimal level search method.

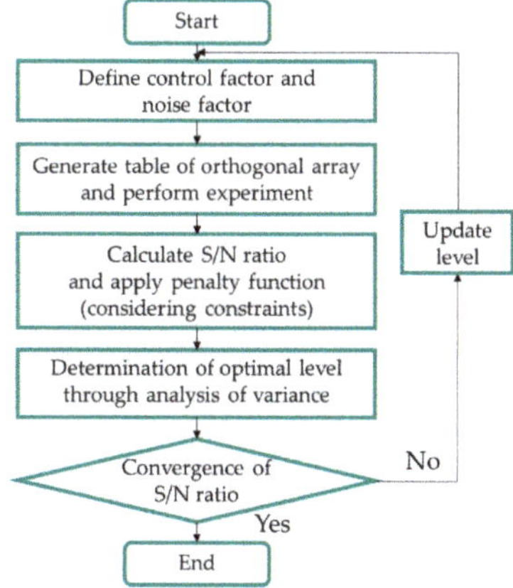

Figure 5. Proposed RDO algorithm.

The iterative design process of the proposed Taguchi method modification is as follows.

(1) Define noise factors caused by control factors and design variable uncertainties.
(2) Configure experimental combinations according to an orthogonal array table and perform experiments.
(3) Calculate the SNR to which the penalty function is added considering the constraint condition.
(4) Determine the optimal experimental combination based on the design factor levels estimated through analysis of variance.
(5) If the SNR does not converge to the set value, repeat steps (2) through (4) using the new levels identified in the optimal level search.
(6) When the SNR converges to the set value, the design process is complete.

To verify the feasibility of the proposed RDO, the method was applied to the BLDC motor described in Table 1, which drives an EOP with a rated output of 150 W, rated torque of 0.358 Nm, and rated speed of 4000 rpm. In BLDC motors, cogging torque is generated by a permanent magnet, flux barrier, and stator slot structure, which causes vibration and noise. Therefore, in this study, to minimize the cogging torque of the conceptually designed motor, five major design variables were set as shown in Figure 6.

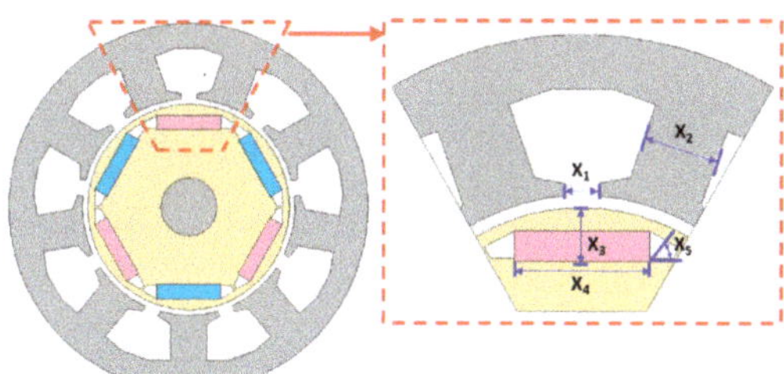

Figure 6. Cross-section of the BLDC motor and design variables. where X_1 is the length of slot opening, X_2 is the width of stator tooth, X_3 is the position of permanent magnet from rotor outer diameter, X_4 is the length of permanent magnet and X_5 is the angle between permanent magnet and magnetic flux barrier.

ANSYS Electromagnetics Suite 18.0, a commercial electromagnetic analysis tool based on finite element analysis, was used to simulate the experimental combinations of the Taguchi method. For comparison, the conventional DDO and the proposed RDO techniques were each applied to the motor cogging torque reduction design.

3.2. DDO Process

The maximum cogging torque of the BLDC motor to be designed is defined as an objective function f. In this case, the objective function, given by Equation (5), is minimized by considering the following two performance constraints: C_1, whereby the rated torque of an optimized motor shall be greater than 0.358 Nm, and C_2, whereby the cogging torque of the optimized motor should be less than 24.85% of the torque ripple of the conventional motor.

$$Minimize\ f(X)$$

$$Subject\ to\ C_1(X) = 0.358 - T_{rated} \leq 0,\ C_2(X) = \frac{T_{max} - T_{min}}{T_{rated}} \times 100 - 24.85 \leq 0$$

$$X_L \leq X \leq X_U \tag{5}$$

Here, T_{rated} is the rated torque; T_{max} and T_{min} are the maximum and minimum torque ripple, respectively; X_L and X_U are the set lower and upper limits of each design variable, the values of which are shown in Table 3. The optimal solution of the design problem given by Equation (5), excluding the fluctuation of design variables due to uncertainty, was explored by applying sequential quadratic programming.

Table 3. Design and noise factors of the BLDC motor.

Items	Parameters	Level 1	Level 2	Level 3	Unit	X_L Lower	X_U Upper
	X_1	2.1	2.2	2.3	mm	1.5	2.5
	X_2	5.0	5.1	5.2	mm	4.5	5.5
Design	X_3	3.44	3.54	3.64	mm	3.0	4.0
factor	X_4	8.9	9.0	9.1	mm	8.0	10.0
	X_5	18.9	19.0	19.1	degree	10	60
Noise factor	-	Level 1	Level 1	Level 1	-	-	-
Tolerance	-	−0.1	0	0.1	-	-	-

3.3. RDO Definition

To formulate the RDO problem using the proposed method, the penalty function $P(X)$ was constructed using the average torque C_1 and the torque ripple C_2, which were the Equation (5) constraints, and a new performance value was defined, as shown in Equation (6).

$$Minimize\ T_{cog}\ :\ f = -10\log\left[\frac{1}{n}\sum_{i=1}^{n} y_i'^2\right]$$

$$y_i' = h(X) + W_p P(X)$$

$$P(X) = \sum_{j=1}^{np}\left[maximum\{0, C_j(X)\}^2\right] \tag{6}$$

Here, T_{cog} is the cogging torque, and the value of the weight W_p of the penalty function is set to 1.

When an experimental combination violated a constraint, the penalty function increased the performance value such that it was excluded from the optimal combination selection. All design factors were assigned three levels, and experimental combinations were constructed using an L18(3^5) orthogonal array of five factors with three levels each, without considering alternation. The variation of the objective function caused by a tolerance of ± 0.1 was calculated through analysis of variance by setting the uncertainty caused by the manufacturing tolerance of a design factor as a noise factor.

4. Verification

4.1. Simulation Results

Figure 7 compares the shapes derived from the DDO and the proposed RDO methods, based on the design parameter values of the conventional model. The optimal design models provided by both methods had the same stator and rotor dimensions, winding specifications and permanent magnet amounts as the conventional model.

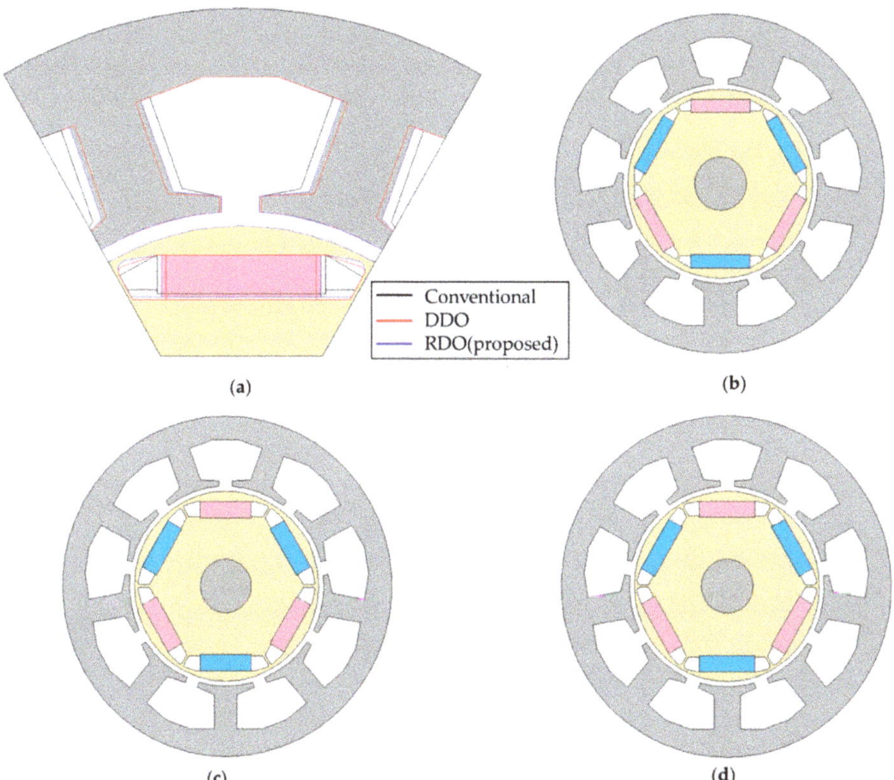

Figure 7. Shape comparison: (**a**) Comparison; (**b**) Conventional; (**c**) Deterministic design optimization (DDO); (**d**) RDO (proposed).

Figure 8 shows the characteristics of the cogging torque. One cycle of the cogging torque had a mechanical angle of 20 degrees per motor revolution, based on the least common multiple of 18 of the nine slots and six poles. The cogging torque characteristics represent the difference between the maximum and minimum values, which were 0.037, 0.019, and 0.011 Nm according to the conventional, DDO, and proposed RDO models, respectively. These results confirmed analytically that the value of cogging torque decreases with optimization, which was attributed to the reduced magnetic reluctance in the air gap between the BLDC's stator and rotor due to their optimized shapes.

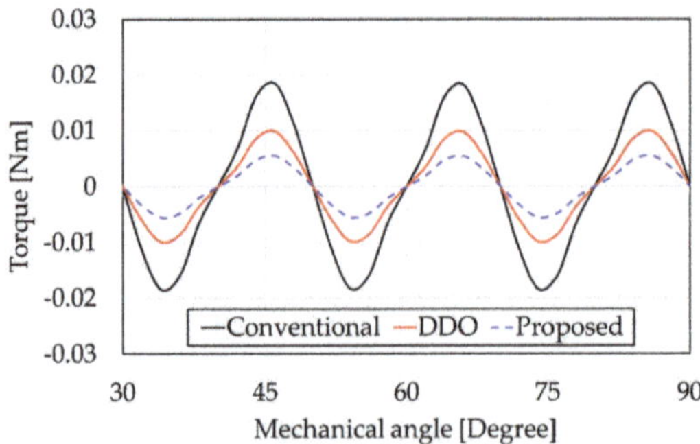

Figure 8. Cogging torque characteristic comparison.

Figure 9 compares the characteristics of the torque ripple, which was 0.087, 0.052, and 0.017 Nm according to the conventional, DDO, and proposed RDO models, respectively. These results confirmed that the proposed RDO method reduces the torque ripple characteristic compared to the conventional and DDO models.

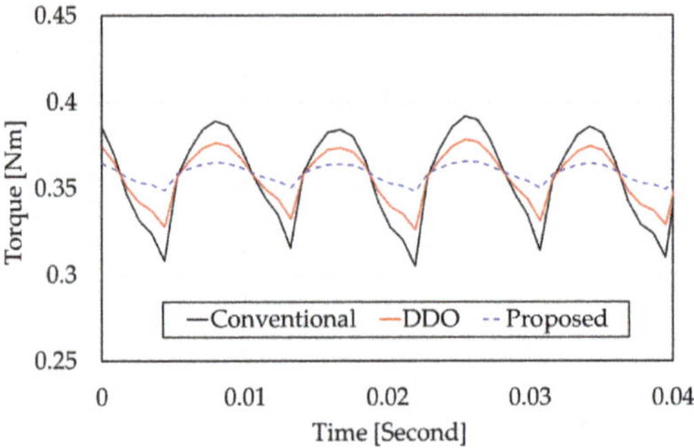

Figure 9. Torque ripple characteristic comparison.

Table 4 compares the design values derived from the DDO and proposed RDO methods, based on the design parameter values of the conventional model. When tolerances of ±0.1 were included in the design parameters, the cogging torque (T_{cog}), torque ripple (T_{ripple}), and standard deviation of the maximum cogging torque (σ_{cog}) appeared in each of the three motor designs. Both the DDO and proposed RDO results satisfied the constraints on rated torque and torque ripple. Table 4 also shows the number of iterations and simulations that the DDO and proposed RDO methods required to derive the optimal design values for the BLDC motor.

Table 4. Design values and characteristics.

Items	Conventional	DDO	RDO	Unit
X_1	2.2	2.2	2.0	mm
X_2	5.1	3.5	4.1	mm
X_3	3.54	3.84	3.7	mm
X_4	9.0	8.0	8.5	mm
X_5	19.0	33.0	42.0	degree
T_{cog}	0.037	0.019	0.011	Nm
T_{ripple}	0.087	0.052	0.017	Nm
σ_{cog}	-	0.0484	0.0315	-
Iterative designs	-	9	15	-
Required simulations	-	124	336	-

4.2. Experimental Results

To verify the proposed RDO method, a prototype of the optimal BLDC motor was fabricated. The prototype BLDC motor assembly, stator, and rotor are shown in Figure 10. The stator had a split core structure to increase the winding fill factor in the slot. The rotor used NdFeB magnets rated at N42EH with high output density and excellent high-temperature demagnetization characteristics, in consideration of the environment in which the EOP would be mounted on the vehicle.

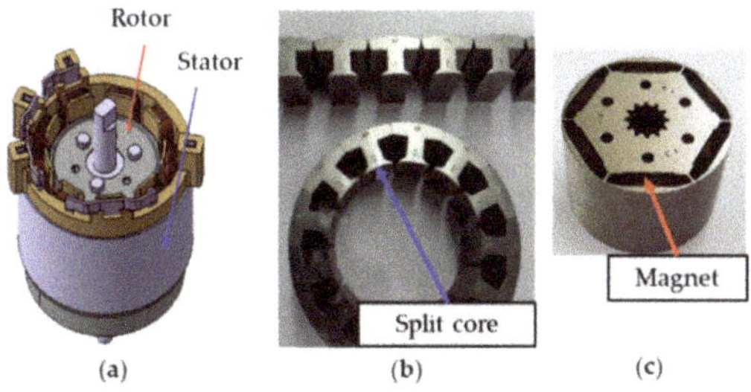

Figure 10. Prototype of optimal design: (**a**) Assembly; (**b**) Stator; (**c**) Rotor.

To measure the performance of the BLDC motors thus designed, a motor dynamometer test environment was constructed as shown in Figure 11.

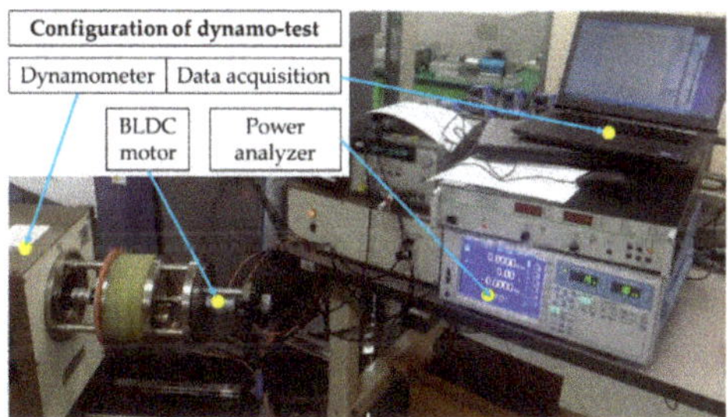

Figure 11. Experimental setup.

Figure 12 shows the experimentally determined cogging torque characteristics. The experimental results showed that the cogging torque characteristics were 0.071, 0.046, and 0.014 Nm for the conventional, DDO, and proposed RDO models, respectively. The experimental characteristics of the cogging torque were consistent with the decreasing tendency of the analytical results, despite some value differences due to the inertia of the test apparatus. In particular, the characteristics of the model identified by the proposed RDO method were clearly the most reduced.

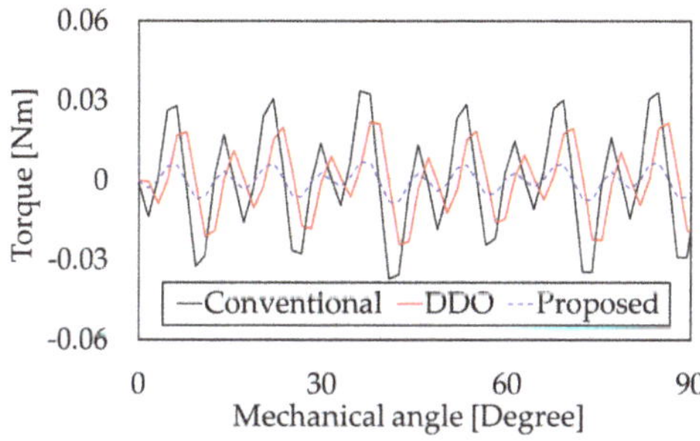

Figure 12. Comparison of experimentally determined cogging torque characteristics.

The BLDC motor performance was tested using the conventional, DDO, and proposed RDO models, and the results are shown in Figure 13.

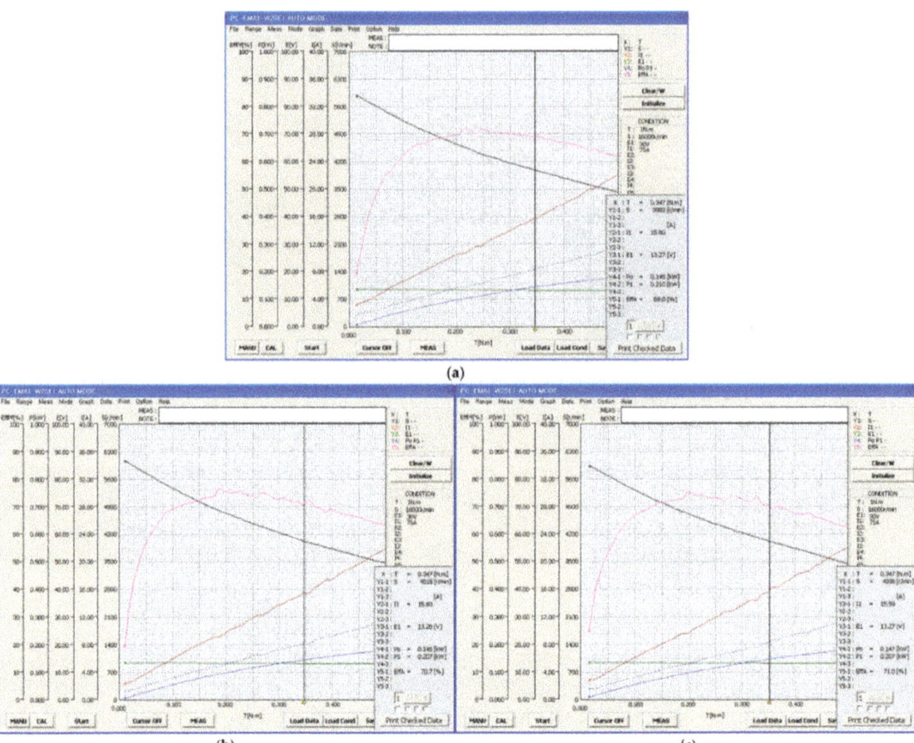

Figure 13. Comparison of experimental motor dynamometer test results: (**a**) Conventional; (**b**) DDO; (**c**) Proposed.

The measurement targets were speed and rated efficiency in response to increased load. The input voltage was 13.5 V DC. To compare the operating characteristics, the BLDC motor was started at a no-load speed of 6000 rpm and the load torque was continuously increased during the experiment. The experimental results showed that the rated efficiency of the conventional model was 69.0%, that of the DDO model was 70.7%, and that of the RDO model was 71.0%, under the rated operating conditions (rotational speed of 4000 rpm and output of 150 W). The rated efficiency of the test results of the conventional model was reduced by approximately 1.0% compared to the analysis results because the long lead wire used in the experiment increased the terminal resistance, and the magnitude of the resistance can have a large effect on the terminal voltage. The voltage dropped as the terminal resistance increased, and the copper loss was increased, resulting in a difference in voltage and efficiency characteristics. These results show that the rated efficiency of the proposed RDO model exceeds that of the conventional and DDO models. Table 5 summarizes the simulation and experimental results from each model.

Table 5. Simulation and experimental results.

	Items	Conventional	DDO	RDO	Unit
	Cogging torque	0.037	0.019	0.011	Nm
Simulation	Torque ripple	0.087	0.052	0.017	Nm
	Rated efficiency	70.0	71.2	72.2	%
	Cogging torque	0.071	0.046	0.014	Nm
Experiment	Torque ripple	0.155	0.097	0.055	Nm
	Rated efficiency	69.0	70.7	71.0	%

5. Conclusions

In this paper, we optimized the torque characteristics of a BLDC motor used in an EOP system. The design domain was set to have a normal distribution, and an orthogonal array table was applied to minimize the analysis in the design domain. To address inequality constraints, Taguchi proposed a penalty function method for the limited optimization problem, whereby the optimal level is determined from experimental results and parameter levels are changed accordingly. In this study, this constraint was applied to the design method by adding a penalty function associated with the constraint to the loss function and adjusting the parameter level intervals by applying an optimal level search method. Based on the performance of existing BLDC motors used in EOPs, we compared the DDO and RDO results by performing finite element analysis and experimental dynamometer testing. With the proposed RDO method, the cogging torque is reduced to 19%, the fluctuation of torque ripple is reduced to 35%, and the rated efficiency characteristic is improved by 2% compared with the conventional model. The proposed RDO improved the Taguchi method and was shown to be robust in comparison with DDO, although the number of iterations required was large. The cumulative results confirm that the proposed RDO algorithm improves the cogging torque, torque ripple, and rating efficiency characteristics of BLDC motors and is expected to improve the flow control stability when implemented in an EOP system in a test vehicle in the future.

Author Contributions: Conceptualization, S.-W.B.; Methodology, K.-Y.Y. and S.-W.B.; Software, K.-Y.Y. and S.-W.B.; Validation, K.-Y.Y. and S.-W.B.; Writing—original draft preparation, K.-Y.Y.; Writing—review and editing, S.-W.B.; Supervision, S.-W.B.; Funding acquisition, S.-W.B.

Acknowledgments: This study was supported by the National Research Foundation of Korea (NRF) grant funded by the Korea government (MSIT) (No. 2017R1C1B5075525); this study was also supported by a 2018 research fund from Honam University.

References

1. He, H.; Liu, Z.; Zhu, L.; Liu, X. Dynamic Coordinated Shifting Control of Automated Mechanical Transmissions without a Clutch in a Plug-In Hybrid Electric Vehicle. *Energies* **2012**, *5*, 3094–3109. [CrossRef]
2. Chen, J.-S. Energy Efficiency Comparison between Hydraulic Hybrid and Hybrid Electric Vehicles. *Energies* **2015**, *8*, 4697–4723. [CrossRef]
3. Chen, Z.; Xiong, R.; Wang, K.; Jiao, B. Optimal Energy Management Strategy of a Plug-in Hybrid Electric Vehicle Based on a Particle Swarm Optimization Algorithm. *Energies* **2015**, *8*, 3661–3678. [CrossRef]
4. Xia, C.; DU, Z.; Zhang, C. A Single-Degree-of-Freedom Energy Optimization Strategy for Power-Split Hybrid Electric Vehicles. *Energies* **2017**, *10*, 896. [CrossRef]
5. Capata, R. Urban and Extra-Urban Hybrid Vehicles: A Technological Review. *Energies* **2018**, *11*, 2924. [CrossRef]
6. Doikin, A.; Habib Zadeh, E.; Campean, F.; Priest, M.; Brown, A.; Sherratt, A. Impact of Duty Cycle on Wear Progression in Variable-displacement Vane Oil Pumps. *Procedia Manuf.* **2018**, *16*, 115–122. [CrossRef]
7. Rostek, E.; Babiak, M.; Wroblewski, E. The Influence of Oil Pressure in the Engine Lubrication System on Friction Losses. *Procedia Eng.* **2017**, *192*, 771–776. [CrossRef]

8.   Podevin, P.; Clenci, A.; Decombes, G. Influence of the lubricating oil pressure and temperature on the performance at low speeds of a centrifugal compressor for an automotive engine. *Appl. Therm. Eng.* **2011**, *31*, 194–201. [CrossRef]

9.   Miyachi, E.; Ishiguro, M.; Mizumoto, K. *Development of Electric Oil Pump*; SAE Technical Paper; SAE International: Warrendale, PA, USA, 2006. [CrossRef]

10. Kim, Y.; Lee, J.; Jo, C.; Kim, Y.; Song, M.; Kim, J.; Kim, H. Development and Control of an Electric Oil Pump for Automatic Transmission-Based Hybrid Electric Vehicle. *IEEE Trans. Veh. Technol.* **2011**, *60*, 1981–1990. [CrossRef]

11. Kim, H.; Wi, J.; Yoo, J.; Son, H.; Park, C.; Kim, H. A Study on the Fuel Economy Potential of Parallel and Power Split Type Hybrid Electric Vehicles. *Energies* **2018**, *11*, 2103. [CrossRef]

12. Zhang, X.; Li, C.-T.; Kum, D.; Peng, H. Prius+ and Volt−: Configuration Analysis of Power-Split Hybrid Vehicles with a Single Planetary Gear. *IEEE Trans. Veh. Technol.* **2012**, *61*, 3544–3552. [CrossRef]

13. Zhao, Z.; He, L.; Yang, Y.; Wu, C.; Li, X.; Karl Hedrick, J. Estimation of torque transmitted by clutch during shifting process for dry dual clutch transmission. *Mech. Syst. Signal Process.* **2016**, *75*, 413–433. [CrossRef]

14. Della Gatta, A.; Iannelli, L.; Pisaturo, M.; Senatore, A.; Vasca, F. A survey on modeling and engagement control for automotive dry clutch. *Mechatronics* **2018**, *55*, 63–75. [CrossRef]

15. Song, M.; Oh, J.; Kim, J.; Kim, Y.; Yi, J.; Kim, Y.; Kim, H. Development of an electric oil pump control algorithm for an automatic-transmission-based hybrid electric vehicle considering the gear shift characteristics. *Proc. Inst. Mech. Eng. Part D J. Automob. Eng.* **2013**, *228*, 21–36. [CrossRef]

16. Tu, J.; Choi, K.; Park, Y.H. A New Study on Reliability-Based Design Optimization. *J. Mech. Des.* **1999**, *121*, 557–564. [CrossRef]

17. Kim, N.; Kim, D.; Kim, D.; Kim, H.; Lowther, D.; Sykulski, J. Robust Optimization Utilizing the Second-Order Design Sensitivity Information. *IEEE Trans. Magn.* **2010**, *46*, 3117–3120. [CrossRef]

18. Lagaros, N.D.; Papadrakakis, M. Robust seismic design optimization of steel structures. *Struct. Multidisc. Optim.* **2007**, *33*, 457–469. [CrossRef]

19. Liu, C.; Lei, G.; Ma, B.; Guo, Y.; Zhu, J. Robust Design of a Low-Cost Permanent Magnet Motor with Soft Magnetic Composite Cores Considering the Manufacturing Process and Tolerances. *Energies* **2018**, *11*, 2025. [CrossRef]

20. Jun, C.-S.; Kwon, B.-I.; Kwon, O. Tolerance Sensitivity Analysis and Robust Optimal Design Method of a Surface-Mounted Permanent Magnet Motor by Using a Hybrid Response Surface Method Considering Manufacturing Tolerances. *Energies* **2018**, *11*, 1159. [CrossRef]

21. Zhan, S.; Li, Z.; Hu, J.; Liang, Y.; Zhang, G. Model Order Identification for Cable Force Estimation Using a Markov Chain Monte Carlo-Based Bayesian Approach. *Sensors* **2018**, *18*, 4187. [CrossRef]

22. Sun, Y.; Zhu, F.; Chen, J.; Li, J. Risk Analysis for Reservoir Real-Time Optimal Operation Using the Scenario Tree-Based Stochastic Optimization Method. *Water* **2018**, *10*, 606. [CrossRef]

23. Choi, S.-H.; Hussain, A.; Kim, H.-M. Adaptive Robust Optimization-Based Optimal Operation of Microgrids Considering Uncertainties in Arrival and Departure Times of Electric Vehicles. *Energies* **2018**, *11*, 2646. [CrossRef]

24. Pervez, M.N.; Shafiq, F.; Sarwar, Z.; Jilani, M.M.; Cai, Y. Multi-Response Optimization of Resin Finishing by Using a Taguchi-Based Grey Relational Analysis. *Materials* **2018**, *11*, 426. [CrossRef] [PubMed]

25. Hong, G.; Wei, T.; Ding, X.; Duan, C. Multi-Objective Optimal Design of Electro-Hydrostatic Actuator Driving Motors for Low Temperature Rise and High Power Weight Ratio. *Energies* **2018**, *11*, 1173. [CrossRef]

26. Hendershot, J.R.; Miller, T.J.E. *Design of Brushless Permanent-Magnet Machines*, 2nd ed.; Motor Design Books LLC: Venice, FL, USA, 2010.

27. Tseng, K.-H.; Shiao, Y.-F.; Chang, R.-F.; Yeh, Y.-T. Optimization of Microwave-Based Heating of Cellulosic Biomass Using Taguchi Method. *Materials* **2013**, *6*, 3404–3419. [CrossRef]

28. Solehati, N.; Bae, J.; Sasmito, A.P. Optimization of Wavy-Channel Micromixer Geometry Using Taguchi Method. *Micromachines* **2018**, *9*, 70. [CrossRef] [PubMed]

Article

Analytical Analysis of a Novel Brushless Hybrid Excited Adjustable Speed Eddy Current Coupling

Yibo Li [1], Heyun Lin [1,*], Hai Huang [2], Hui Yang [1], Qiancheng Tao [1] and Shuhua Fang [1]

[1] School of Electrical Engineering, Southeast University, Nanjing 210096, China; epolee@163.com (Y.L.); huiyang@seu.edu.cn (H.Y.); 220162293@seu.edu.cn (Q.T.); shfang@seu.edu.cn (S.F.)

[2] Jiangsu Magnet Valley Technologies Co., Ltd., Zhenjiang 212009, China; 13871110369@163.com

* Correspondence: hyling@seu.edu.cn; Tel.: +86-025-8379-4169 (ext. 805)

Received: 19 December 2018; Accepted: 17 January 2019; Published: 19 January 2019

Abstract: A novel brushless hybrid excited adjustable speed eddy current coupling is proposed for saving energy in the drive systems of pumps and fans. The topology and operation principle of the coupling are presented. Based on the real flux paths, the fluxes excited by permanent magnet (PM) and field current are analyzed separately. A magnetic circuit equivalent (MEC) model is established to efficiently compute the no-load magnetic field of the coupling. The eddy current and torque are calculated based on the proposed MEC model, Faraday's law, and Ampere's law. The resultant magnetic fields, eddy currents, and torques versus slip speeds under different field currents are studied by the MEC-based analytical method and verified by finite element analysis (FEA). The copper loss, core loss, and efficiency were investigated by FEA. The analytically predicted results agree well with the FEA, and the analysis results illustrate that a good speed regulation performance can be achieved by the proposed hybrid excited control.

Keywords: eddy current coupling; hybrid excited; magnetic equivalent circuit; magnetic field analysis; torque-slip characteristic

---

1. Introduction

The eddy current coupling forms a basic variable speed drive, which is widely applied to regulate the flows from large pumps and fans for a remarkable energy-saving effect [1]. With the development of permanent magnet (PM) material, a high torque density adjustable speed PM eddy current coupling (AS-PMECC) (Oregon State University, Corvallis, OR, USA) has been developed and applied in the field of variable speed drive [2]. Compared with other adjustable speed devices, such as gearboxes [3] and variable frequency drives [4], AS-PMECC has many advantages, such as lower sensitivity to environmental conditions, more reliable overload protection, and better energy-saving performance [5,6].

Based on the magnetized directions of PMs, AS-PMECC can be generally classified as axial flux [7] and radial flux couplings [8]. It is usually installed between prime motor and load, which commonly consists of a PM rotor (PMR), a conductor rotor (CR), and an additional mechanical manipulator (AMM) [6]. When a relative rotation happens between the two rotors, eddy currents can be induced in the copper sheet (CS) mounted on the iron core of the CR, which yields an electromagnetic torque from the interaction with the primary PM magnetic field [9–12]. For a given load torque, the speed of AS-PMECC can be adjusted by changing either the air-gap length or coupling area between the two rotors by the AMM [6,10].

In essence, the mechanical-based solutions usually realize speed regulation by adjusting the air-gap flux between the PMR and the CR. However, it is relatively hard to regulate the axial displacement between the two rotors, which requires a relative complicated design of AMM. Further, the complicated AMM is not reliable enough and takes more axial space, which is urgent to be simplified and improved for the energy-saving reconstruction of aged pump and fan systems. In addition, the local overheat problem produced by shifting a rotor is very serious in the conventional radial flux AS-PMECC [8,13]. Therefore, this paper proposes a novel brushless hybrid excited adjustable speed eddy current coupling (HE-ASECC) (Southeast University, Nanjing, China) based on a hybrid excited concept to solve the above problem [14,15].

The novelty of the proposed coupling lies on the brushless hybrid excited (HE) control by an additional field excitation stator (AFES), which can provide a flexible air-gap field adjustment. The AFES just occupies some radial space rather than axial space, which is very suitable for the energy-saving reconstruction of aged pump and fan systems. Due to the absence of brush, slip ring, and complicated AMM, the magnetic field control of the HE-ASECC becomes highly simple and reliable. For illustrating the advantages of the HE-ASECC more clearly, it is compared with valve and baffling vane (VB), AS-PMECC, and variable frequency drive (VFD) in Appendix A. Due to the special geometry of the coupling, three-dimensional (3D) finite element analysis (FEA) is usually required to accurately compute its electromagnetic characteristics, but it is very time-consuming. For improving computational efficiency, an analytical model based on the magnetic equivalent circuit (MEC) method is proposed to analyze the electromagnetic characteristics of the coupling. The eddy current and torque are calculated and analyzed based on Ampere's laws under an asymmetric magnetic field. Finally, the analytically predicted results are presented and compared with the FEA.

2. Structure and Principle

Figure 1a shows the structure of the studied HE-ASECC. It consists of an AFES, a PMR, and a CR, all of which are coaxial. Different from the AMM used in the conventional AS-PMECC, the AFES embedded with a toroidal field winding is located outside the PMR. Usually, the PMR and the CR are connected with the load mover and the prime motor, respectively. The PMR is characterized by two suits of axially parallel consequent-poles mounted on the inner surface of the PMR core that consists of two annular iron cores fixed together with an axial distance. The PMs mounted on the two annular iron cores are radial magnetized in opposite directions, respectively. The copper sheet (CS) is tightly mounted on the surface of iron core of the CR to provide paths for the induced eddy currents.

The field control principle can be explained by the magnetic fluxes passing through the PM and iron poles as shown in Figure 1b–d under different field currents. The PMs serve as constant magnetomotive force (MMF) sources, while the field winding acts as a changeable MMF source to perform a flexible air-gap flux adjustment. As a result, the fluxes produced by the field current can enhance or weaken the air-gap fluxes produced by PMs alone based on the directions and amplitudes of the applied field currents. Consequently, the slip speed between the PMR and the CR under a given load torque can be adjusted with the aid from HE control.

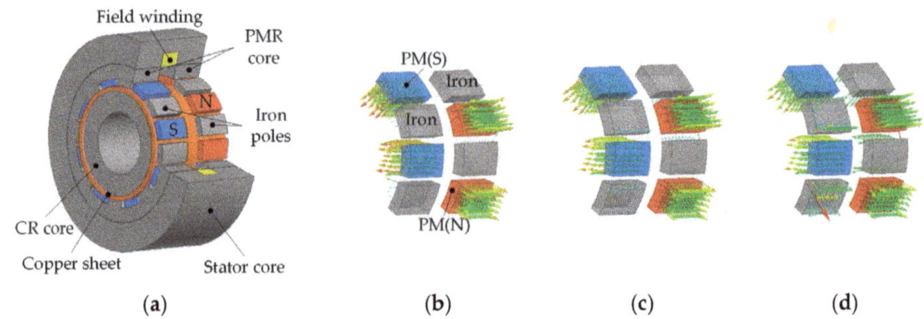

Figure 1. The magnetic field control principle of the coupling. (a) The topology. (b) $I_f = -10$ A. (c) $I_f = 0$ A. (d) $I_f = 10$ A.

3. Analytical Model

In the design process of an electromagnetic device, a reliable and effective theoretical model is desirable for evaluating its electromagnetic characteristics. Generally, the theoretical models based on the magnetic vector potential [9,16] and the MEC method [17–21] are very popular for the design and analysis of electromagnetic devices. Considering the special 3D structure of the proposed HE-ASECC, the latter is adopted in this paper.

3.1. No-Load Magnetic Field Calculation

Based on the real fluxes shown in Figure 1b–d and the reluctance elements of MEC, the fluxes created by the PMs and the field current can be analyzed separately. In addition, it should be noted that the CR core and the stator core are communal for the MEC branches excited by the PMs and the field current. Figure 2a,b give the MEC models of the coupling excited by both the PMs and the field current when the slip speed between the two rotors is 0 rpm, respectively.

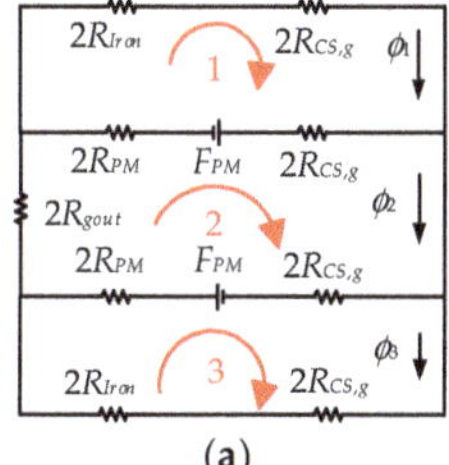
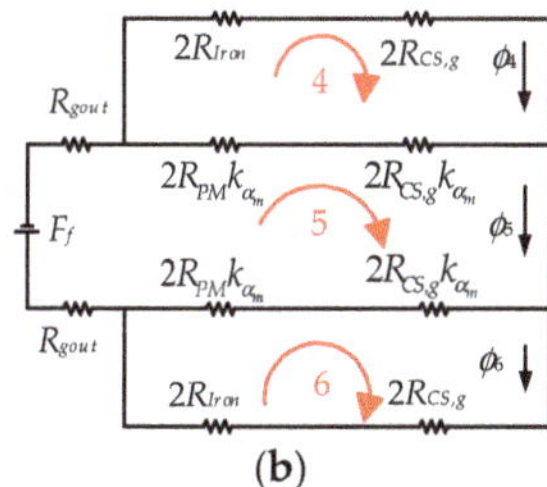

Figure 2. The magnetic circuit equivalent (MEC) model of the hybrid excited adjustable speed eddy current coupling (HE-ASECC). (**a**) permanent magnet (PM) excited. (**b**) I_f excited.

As shown in Figure 2a, the PM fluxes can be divided into two parts. One passes through the adjacent PM poles located on the two annular iron cores, the inner and outer air-gaps and the CS, and loops with the CR core and the stator core. The other begins with the PM poles and passes through the adjacent iron poles located on the same annular iron core, the inner air-gap and the CS, and loops with the PMR core and the CR core. As shown in Figure 2b, the fluxes created by the field winding pass through the outer air-gap, the PMR cores, the iron pole, and the PM-air region, including the PM poles and the air region between PM and iron poles located on the same annular iron core, the inner air-gap and the CS, and loop with the CR core. All elements of the MEC model are calculated in the cylindrical coordinate to permit higher accuracy. In addition, the symbols describing the structure parameters of the coupling are presented in its axial section and axial cross section shown in Figure 3a,b, respectively.

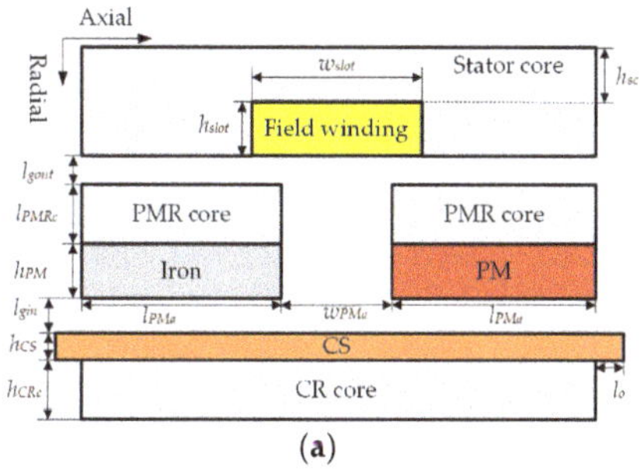
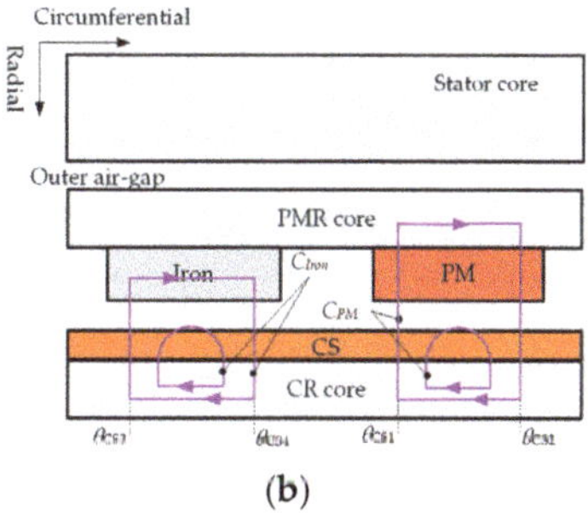

Figure 3. The 2D view of the HE-ASECC. (**a**) Axial section. (**b**) Axial cross section.

Assuming the iron parts of the coupling are unsaturated, the reluctances of the stator core, the PMR core, the iron poles, and the CR core can be ignored for simplifying calculation. The MMF sources are constituted by the PM poles and the field winding located in the AFES, which can be respectively expressed as

$$F_{PM} = H_c h_{PM} \tag{1}$$

$$F_f = I_f N_f \tag{2}$$

where I_f, N_f, H_c, and h_{PM} represent the field current, the turns of the field winding, the PM coercive force, and the thicknesses of the PM poles, respectively. Applying Kirchhoff's voltage (KVL) law to Loops 1–6 of the MEC model shown in Figure 2 yields [17,18]

$$\begin{bmatrix} R_{11} & -R_{12} & 0 \\ -R_{21} & R_{22} & -R_{23} \\ 0 & -R_{32} & R_{33} \end{bmatrix} \times \begin{bmatrix} \phi_1 \\ \phi_2 \\ \phi_3 \end{bmatrix} = \begin{bmatrix} -F_{PM} \\ 2F_{PM} \\ -F_{PM} \end{bmatrix} \tag{3}$$

$$\begin{bmatrix} R_{44} & -R_{45} & 0 \\ -R_{54} & R_{55} & -R_{56} \\ 0 & -R_{65} & R_{66} \end{bmatrix} \times \begin{bmatrix} \phi_4 \\ \phi_5 \\ \phi_6 \end{bmatrix} = \begin{bmatrix} 0 \\ -F_f \\ 0 \end{bmatrix} \tag{4}$$

where ϕ_1 to ϕ_6 are the fluxes flowing along Loops 1–6 in the MEC model.

$$\begin{cases} R_{11} = R_{33} = 2\left(R_{PM} + 2R_{CS,g} + R_{Iron}\right) \\ R_{22} = 4\left(R_{PM} + R_{CS,g} + R_{gout}\right) \\ R_{12} = R_{21} = R_{23} = R_{32} = 2\left(R_{PM} + R_{CS,g}\right) \end{cases} \tag{5}$$

$$\begin{cases} R_{44} = R_{66} = 2\left(k_{\alpha_m} R_{PM} + 2k_{\alpha_m} R_{CS,g} + R_{Iron}\right) \\ R_{55} = 4\left(k_{\alpha_m} R_{PM} + k_{\alpha_m} R_{CS,g} + R_{gout}\right) \\ R_{45} = R_{54} = R_{56} = R_{65} = 2\left(k_{\alpha_m} R_{PM} + k_{\alpha_m} R_{CS,g}\right) \end{cases} \tag{6}$$

where R_{PM}, R_{Iron}, $R_{CS,g}$, R_{gout}, and $k_{\alpha m}$ denote the reluctances of the PM pole, the iron pole, the copper sheet and the inner air-gap, the outer air-gap, and the coefficient to modify R_{PM} and $R_{CS,g}$ under the field current excitation alone, respectively. Assuming the permeabilities of the CS and PM are equal to μ_0, the above parameters can be calculated as [17,18]

$$\begin{cases} R_{PM} = \ln\left[1 + h_{PM}/\left(r_{CSo} + l_{gin}\right)\right] / \left(\mu_0 \alpha_m \tau_p l_{PMa}\right) \\ R_{CS} = \ln\left(1 + h_{CS}/r_{CSi}\right) / \left(\mu_0 \alpha_m \tau_p l_{PMa}\right) \\ R_{gin} = \ln\left(1 + l_{gin}/r_{CSo}\right) / \left(\mu_0 \alpha_m \tau_p l_{PMa}\right) \\ R_{gout} = 2\ln\left(1 + l_{gout}/r_{PMRco}\right) / \left[\mu_0 \alpha_m \tau_p \left(2l_{PMa} + w_{PMa} - w_{slot}\right)\right] \\ R_{CS,g} = R_{CS} + R_{gin} \\ k_{\alpha_m} = (2 - \alpha_m)/\alpha_m \end{cases} \tag{7}$$

where r_{CSi}, r_{CSo}, r_{PMRco}, α_m, τ_p, l_{PMa}, and μ_0 is the inner and outer radii of the CS, the outer radius of the PMR core, the pole-arc coefficients of PM and iron poles, the pole pitch in the circumferential direction, the axial length of PM poles, and the vacuum permeability, respectively.

Based on the analysis of the flux paths and reluctance parameters of the coupling, the flux density distributions can be obtained by a superposition of the fields produced by the PMs and the field current. Based on the MEC model in Figure 2 and the flux density model in Figure 4, the no-load air-gap flux density of the coupling can be expressed as [17,18,20]

$$B_{PM}(\theta) = \begin{cases} k_t B_{PM\_PM}(\theta - \theta_2)/(\theta_2 - \theta_1) + B_{PM\_I_f}; & \theta_1 \leq \theta < \theta_2 \\ k_t B_{PM\_PM} + B_{PM\_I_f}; & \theta_2 \leq \theta \leq \theta_3 \\ k_t B_{PM\_PM}(\theta - \theta_4)/(\theta_3 - \theta_4) + B_{PM\_I_f}; & \theta_3 < \theta \leq \theta_4 \end{cases} \tag{8}$$

$$
B_{Iron}(\theta) = \begin{cases} k_t B_{Iron\_PM}(\theta - \theta_2)/(\theta_2 - \theta_1) + B_{PM\_I_f}; & \theta_1 \leq \theta < \theta_2 \\ k_t B_{Iron\_PM} + B_{Iron\_I_f}; & \theta_2 \leq \theta \leq \theta_3 \\ k_t B_{Iron\_PM}(\theta - \theta_4)/(\theta_3 - \theta_4) + B_{PM\_I_f}; & \theta_3 < \theta \leq \theta_4 \end{cases} \tag{9}
$$

where k_t denotes the coefficient to modify the trapezoid flux density waveforms shown in Figure 4a,b. θ_1 to θ_4 and k_t can be obtained as follows:

$$
\begin{cases} \theta_1 = -\theta_4 = -\tau_p/2 \\ \theta_2 = -\theta_3 = -\tau_p \alpha_m/2 \\ k_t = 2\alpha_m/(1 + \alpha_m) \end{cases} \tag{10}
$$

In addition, B_{PM_PM}, B_{Iron_PM}, $B_{PM\_I_f}$, and $B_{Iron\_I_f}$ denote the average flux densities facing the PM and iron poles produced by the PM and the field current, respectively, which can be derived as

$$
\begin{cases} B_{PM\_PM} = 2(\phi_2 - \phi_1)/\left[\tau_p \alpha_m (r_{CSo} + l_g/2) l_{PM}\right] \\ B_{Iron\_PM} = 2\phi_1/\left[\tau_p \alpha_m (r_{CSo} + l_g/2) l_{PM}\right] \\ B_{PM\_I_f} = 2(\phi_5 - \phi_4)/\left[\tau_p (2 - \alpha_m)(r_{CSo} + l_g/2) l_{PM}\right] \\ B_{Iron\_I_f} = 2\phi_4/\left[\tau_p \alpha_m (r_{CSo} + l_g/2) l_{PM}\right] \end{cases} \tag{11}
$$

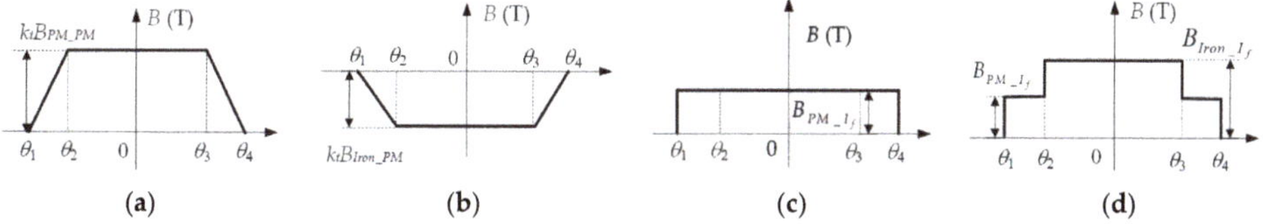

Figure 4. The model of air-gap flux density. (**a**) B_{PM} excited by PMs. (**b**) B_{Iron} excited by PMs. (**c**) B_{PM} excited by I_f. (**d**) B_{Iron} excited by I_f.

3.2. Eddy Current Field Calculation

Thanks to FEA, the distributions of eddy currents under different field currents can be easily calculated and shown in Figure 5 when the slip speed is 150 rpm. It can be seen that the directions of eddy currents do not change, while the direction of the field current is reversed. In addition, the eddy current densities rise with the increase in field current.

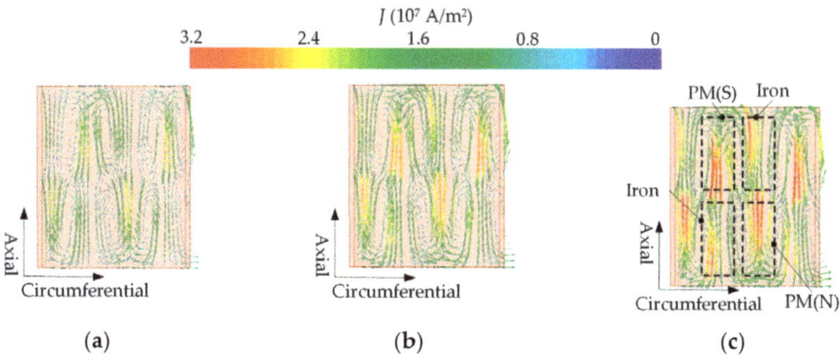

Figure 5. The distributions of eddy currents. (**a**) $I_f = -10$ A. (**b**) $I_f = 0$ A. (**c**) $I_f = 10$ A.

As a relative rotation between the PMR and the CR happens, the eddy current will be induced in the CS. Since the air-gap flux densities facing the PM and iron poles are different, the eddy current densities can be calculated separately, which can be respectively given as [17,18,20]

$$J_{PM}(r,\theta) = r\sigma\omega B_{r\_PM}(\theta) \tag{12}$$

$$J_{Iron}(r,\theta) = r\sigma\omega B_{r\_Iron}(\theta) \tag{13}$$

where σ and ω denote the conductivity of the CS and relative angular velocity, respectively. $B_{r\_PM}(\theta)$ and $B_{r\_Iron}(\theta)$ are the radial components of the resultant flux density in the inner air-gap facing the PM and iron poles, respectively. In fact, the axial components of the eddy current densities facing the PM and iron poles in the same loop are the same, i.e., the average eddy current densities of J_{PM} and J_{Iron} are equal, which can be expressed as

$$J_{av}(r,\theta) = (J_{PM}(r,\theta) - J_{Iron}(r,\theta))/2 = r\sigma\omega(B_{r\_PM}(\theta) - B_{r\_Iron}(\theta))/2. \tag{14}$$

In fact, the PM magnetic field is interacted by the magnetic field produced by the eddy current in the CS. Thus, an iterative process should be taken in the air-gap field calculation of the coupling. The air-gap flux density is first determined using the no-load MEC model. The eddy current density in the CS is then calculated by means of Faraday's law under the different slip speeds between the two rotors. Afterward, the impact of the magnetic field yielded by the eddy currents on the original air-gap flux density is taken into account through Ampere's law. The radial components of the resultant flux densities in the air-gap can be expressed as [17,18,20]

$$B_{r\_PM}^{(n)}(\theta) = B_{PM}(\theta) + B_{CS\_PM}^{(n-1)}(\theta) \tag{15}$$

$$B_{r\_Iron}^{(n)}(\theta) = B_{Iron}(\theta) + B_{CS\_Iron}^{(n-1)}(\theta) \tag{16}$$

where the positive integer n represents the iteration number, and $B_{CS\_PM}(\theta)$ and $B_{CS\_Iron}(\theta)$ denote the flux densities produced by the eddy current in the CS facing the PM and iron poles, respectively. The eddy current field can be calculated based on Ampere's laws expressed as [17,18,20]

$$\int_{C_{PM}} Hdl = \int_{\theta_{CS1}}^{\theta_{CS2}} \int_{r_{CSi}}^{r_{CSo}} J_{PM}(r,\theta)\, rdrd\theta \tag{17}$$

$$\int_{C_{Iron}} Hdl = \int_{\theta_{CS3}}^{\theta_{CS4}} \int_{r_{CSi}}^{r_{CSo}} J_{Iron}(r,\theta)\, rdrd\theta \tag{18}$$

where the right terms are the total currents enclosed in flux paths C_{PM} and C_{Iron}. θ_{CS1} to θ_{CS4} are the edges of the eddy current field flux paths shown in Figure 3b.

Based on Equations (17) and (18), the eddy currents under the iron and PM poles are integrated along the different lengths of the flux paths (C_{PM} and C_{Iron}) shown in Figure 3b. For simplifying calculation, the magnetic reluctances of the iron materials are ignored. Thus, the average lengths of the flux paths facing the PM and iron poles can be expressed as $l_{e\_PM} = \pi(\tau_p/4)$ and $l_{e\_Iron} = 2(l_{gin} + h_{CS})$, respectively [20]. Based on Equations (14), (17), and (18) and the selections of paths C_{PM} and C_{Iron} in Figure 3b, the flux densities $B_{CS\_Iron}(\theta)$ and $B_{CS\_PM}(\theta)$ can be respectively obtained as [17,18,20]

$$B_{CS\_Iron} = -k_{le}B_{CS\_Iron} \tag{19}$$

$$B_{CS\_PM}(\theta) = \begin{cases} B_{CS1} = k_1 e^{m'\theta}; & \theta_1 \le \theta < \theta_2 \\ B_{CS2} = k_2 e^{m'\theta} - B_m/(1+k_{le}); & \theta_2 \le \theta \le \theta_3 \\ B_{CS3} = k_3 e^{m'\theta}; & \theta_3 < \theta \le \theta_4 \end{cases} \tag{20}$$

where

$$B_m = \left(B_{PM\_PM} + B_{PM\_I_f} - B_{Iron\_PM} - B_{Iron\_I_f} \right)/2$$
$$m' = m_{PM}(1 + k_{le})/2$$
$$k_{le} = l_{e\_PM}/l_{e\_Iron}$$
$$m_{PM} = \mu_0 \sigma \omega \left(r_{CSo}^3 - r_{CSi}^3 \right)/(3l_{e\_PM}) \tag{21}$$

The following boundary conditions are taken as

$$B_{CS2}(\theta_0) = 0 \tag{22}$$

$$B_{CS1}(\theta = \theta_2) = B_{CS2}(\theta = \theta_2) \tag{23}$$

$$B_{CS2}(\theta = \theta_3) = B_{CS3}(\theta = \theta_3) \tag{24}$$

where Equation (22) is the main boundary condition of the problem referring to a particular point where the total currents enclosed in the intervals $[\theta_1, \theta_0]$ and $[\theta_0, \theta_4]$ are equal, and Equations (23) and (24) express the continuity of $B_{CS}(\theta)$. θ_0 can be determined by

$$\int_{\theta_1}^{\theta_0} \int_{r_{CSi}}^{r_{CSo}} J(r,\theta) r\, dr\, d\theta = \int_{\theta_0}^{\theta_4} \int_{r_{CSi}}^{r_{CSo}} J(r,\theta) r\, dr\, d\theta. \tag{25}$$

Thus,

$$\theta_0 = -\frac{1}{m'} \ln \left\{ \frac{1 + (1 - e^{\alpha_m}) \cosh\left[(1 - \alpha_m)m'\tau_p/2\right]}{(1 - e^{\alpha_m}) \cosh(m'\tau_p/2) + \cosh(m'\alpha_m \tau_p/2)} \right\}. \tag{26}$$

The coefficients k_1, k_2 and k_3 in (20) can then be expressed as [17,18]

$$\begin{cases} k_1 = B_m \left(e^{-m'\theta_0} - e^{-\alpha_m m'\theta_1} \right)/(1 + k_{le}) \\ k_2 = B_m e^{-m'\theta_0}/(1 + k_{le}) \\ k_3 = B_m \left(e^{-m'\theta_0} - e^{-\alpha_m m'\theta_4} \right)/(1 + k_{le}) \end{cases} \tag{27}$$

Thus, the radial components of the air-gap flux densities can be obtained by substituting Equations (19) and (20) into Equations (15) and (16), respectively.

3.3. Torque Calculation

The torque of the coupling can be finally determined by using the total ohmic losses dissipated in the CS [17–21]:

$$T = \frac{K_s p_{cu\_cs}}{\omega} = \frac{2K_s l_{PMa}}{\sigma \omega} \iint_{CS} |J(r,\theta)|^2 r\, dr\, d\theta = 2K_s l_{PMa} \pi \left(r_{CSo}^2 - r_{CSi}^2 \right) J_{av}^2/\sigma \omega \tag{28}$$

where K_s, p_{cu_CS}, and J_{av} denote the 3D coefficient, the copper loss in the CS, and the average eddy current density in the same loop. K_s can be derived as [22]

$$K_s = 1 - \frac{\tanh[pl_{PMa}/(2r_{av})]/[pl_{PMa}/(2r_{av})]}{1 + \tanh[pl_{PMa}/(2r_{av})]\tanh[pl_o/(2r_{av})]} \tag{29}$$

where $r_{av} = (r_{CSo} + r_{CSi})/2$, and l_o and p denote the overhang-length of the CS and the pole pair numbers of the PMR, respectively.

3.4. Efficiency

The output and input powers of the coupling can be respectively expressed as

$$P_{out} = T_{out}\Omega_{out} \tag{30}$$

$$P_{in} = P_{out} + p_{cu\_CS} + p_{cu\_fw} + p_{other} \tag{31}$$

where T_{out}, Ω_{out}, and p_{cu_fw} denote the output torque, the output angular speed, and the copper loss of the field winding, respectively. p_{other} denotes the other losses including core loss, stray loss, mechanical loss, and so on. The coupling efficiency can then be obtained as

$$\eta_C = P_{out}/P_{in} \times 100\%. \tag{32}$$

Ignoring other losses, the input torque is almost equal to the output torque. Thus, the input power of the coupling can be estimated as [20]

$$P_{in} = (1 - s)^2 P_N \tag{33}$$

where P_N and s denote the rated power of pump and fan loads, and the slip between the PMR and the CR, respectively. The efficiency of the coupling can be simplified as

$$\eta_C = (1 - s) \times 100\%. \tag{34}$$

With the increase in slip, P_{in} declines much faster than η_c. Based on [6], the energy-saving effect of the system can be still more pronounced when compared to the method through throttling valves and baffles.

4. Results and Discussions

The major design parameters of the HE-ASECC are tabulated in Table 1. The air-gap magnetic fields, eddy currents, and torque-slip characteristics are calculated by the proposed analytical method and verified by FEA. The loss and efficiency are subsequently investigated by FEA.

Table 1. Major design parameters of the coupling.

Item	Symbol	Value	Unit
Inner radical of CS	r_{CSi}	46	mm
CS thickness	h_{CS}	3	mm
CS conductivity	σ	5.77×10^7	S/m
Over hang length	l_o	5	mm
CR yoke thickness	h_{CRc}	10	mm
Air-gap length	l_{gout}/l_{gin}	1	mm
PM inner radical	r_{mi}	50	mm
Magnetic pole height	h_{PM}	6	mm
PM axial length	l_{PMa}	30	mm
PM coercive force	H_c	−890	kA/m
Pole-arc coefficient	α_m	0.8	
Pole pair numbers of the PMR	p	16	
Axial distance between the two annular iron cores	w_{PMa}	10	mm
PMR yoke thickness	h_{PMRc}	10	mm
Stator yoke thickness	h_{sc}	10	mm
Stator slot width	w_{slot}	40	mm
Stator slot height	h_{slot}	10	mm
Field coil turns number	N_f	300	turns

4.1. Air-Gap Magnetic Field

Figure 6 shows the radial components of the no-load inner air-gap flux densities between the PMR and the CR under different field currents. When I_f increases from −10 to 10 A, the amplitude of B_{PM} decreases from about 0.7 to 0.5 T (about 30%), while B_{Iron} decreases from 0.2 to −0.3 T. The variation of B_{PM} is smaller that of B_{Iron}, which can be explained by the fact that the reluctance of PM is much larger than that of iron poles. In addition, the difference between analytical and FEA methods is found, which is caused by the flux leakage between adjacent PM and iron poles, which is difficult to be considered in the analytical modeling. The above analysis illustrates a fairly good effect of the proposed approach on air-gap magnetic field regulation.

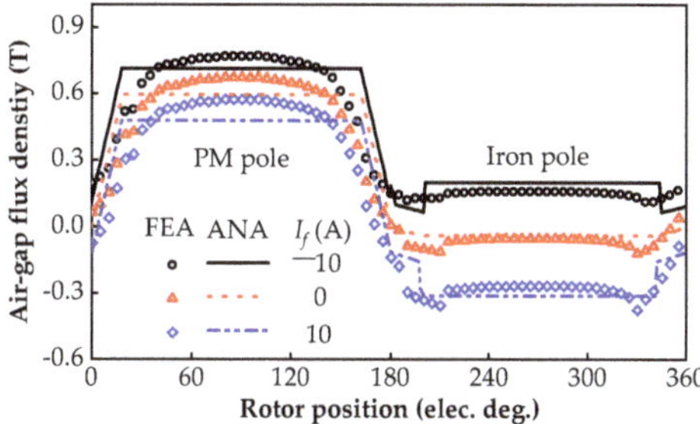

Figure 6. No-load air-gap magnetic flux density distributions.

Figure 7a,b show the average values of B_{r_PM} and B_{r_Iron} as functions of slip speed under different field currents. It is obvious that the average value of B_{r_PM} declines from about 0.62 to 0.41 T, while that of B_{r_Iron} increases from about -0.3 to 0.15 T when the slip speed is equal to 50 rpm. With the increase in slip speed, the average values of B_{r_PM} and B_{r_Iron} decline and rise gradually, respectively. Based on Equations (14) and (17), the eddy current field is enhanced with the increase in slip speed. Although the direction of the flux passing through iron pole is changed by the opposite field currents (e.g., $I_f = -10$ and 10 A), the direction of the eddy current field remains unchanged. The reason is that the direction of eddy current loop is mainly determined by J_{PM}, which is much larger than J_{Iron}.

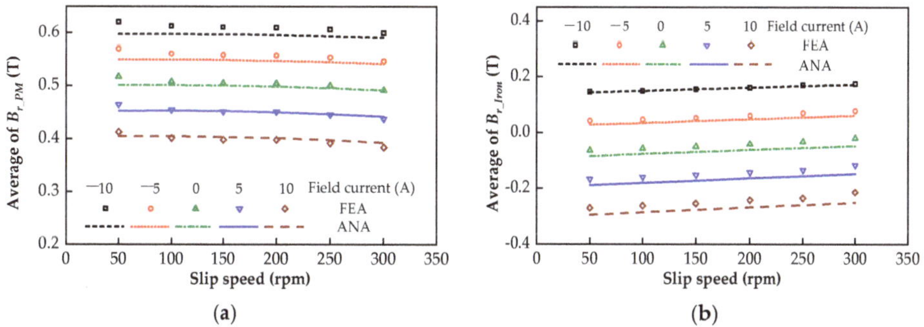

Figure 7. The average values of flux densities under different field currents. (**a**) The average values of B_{r_PM}. (**b**) The average values of B_{r_Iron}.

4.2. Eddy Current Density

As expected, the eddy current densities facing the PM and iron poles increase gradually with the increase in field current as shown in Figure 8. In addition, the eddy current versus rotor position distributions under the PM and iron poles are not uniform, which can be explained by Equations (12)–(14) and the air-gap flux density distributions shown in Figure 6. It should be noted that the directions of eddy currents facing the iron pole are unchanged when the field current changes from -10 to 10 A. The reason is that J_{PM} and J_{Iron} are respectively proportional to B_{r_PM} and B_{r_Iron}, while B_{r_PM} is much larger than B_{r_Iron}, which determines the flowing directions of eddy currents.

Figure 9 shows the variations in average eddy current densities facing the PM and iron poles of the proposed coupling under different slip speeds. With the increase in field current, both J_{PM} and J_{Iron} rise gradually, and they also increase with the increase in slip speed under a given field current. It should be noted that the average eddy current densities calculated by analytical and FEA methods agree well, which illustrates that the eddy current calculation model shown in Equation (14) is sufficiently reliable.

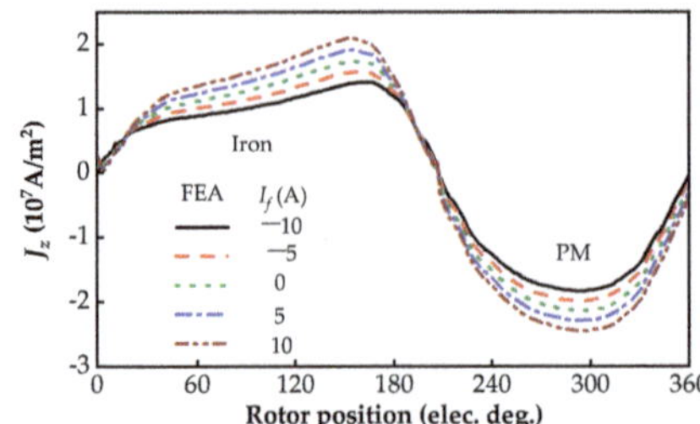

Figure 8. The eddy current densities versus the rotor position.

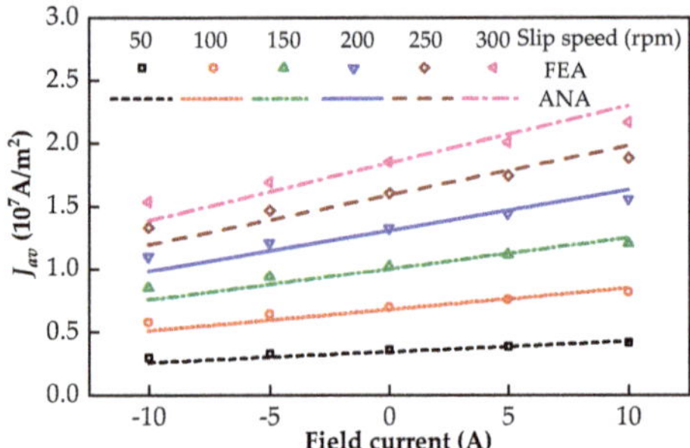

Figure 9. The average eddy current densities under different slip speeds.

4.3. Torque Characteristics

Figure 10 shows the torque-slip characteristics of the proposed coupling under different field currents. In the lower slip speed range, the torque presents an ascending trend, the reason of which is that the eddy current density increases proportionally with slip speed. Due to the skin effect, the torque increases more and more slowly. Moreover, it can be seen that the torque experiences a gradually reduction with the decrease in I_f from 10 to -10 A at a given slip speed. This implies that a wide torque regulation range can be achieved by field control. The above analysis confirms the validity of HE control for the load speed regulation by the proposed coupling.

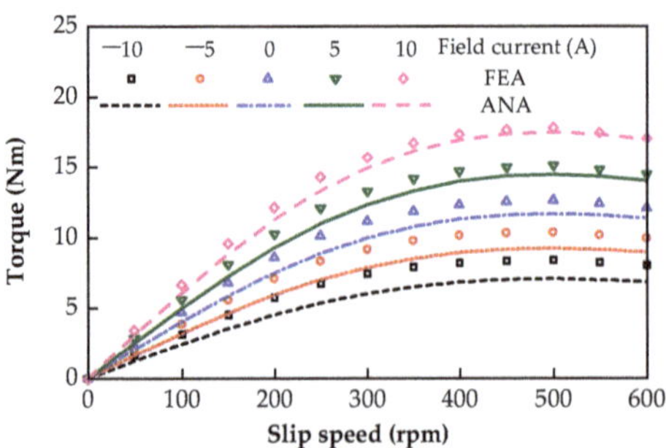

Figure 10. The torque–slip characteristics under different field currents.

4.4. Losses and Efficiency

Figure 11 shows the core losses of the coupling versus slip speed under different field currents. With the increase in slip speed, the core losses of the coupling gradually increase, and are proportional to the frequencies of the alternating magnetic fluxes passing through the cores. In fact, the magnetic fluxes passing through the PMR core and the stator core are almost not alternating, which results in low core losses in the PMR and stator cores. Thus, the core loss is mainly produced in the CR core. In addition, the core loss also increases with the increase in field current, which enhances the alternating magnetic fluxes passing through the CR core.

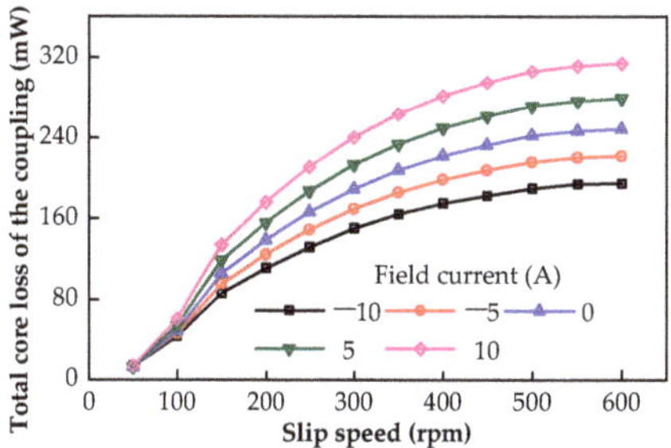

Figure 11. The core losses under different field currents.

Figure 12 shows the copper losses of the CS versus slip speed under different field currents. With the increase in slip speed and field current, the copper losses rise gradually. The copper loss of field winding is proportional to the field current, which can obtain the maximum value as 97.5 W when the magnitude of the field current is set to 10 A. It takes up a small percentage of the total copper loss of the coupling, which illustrates that the HE control is acceptable for the speed regulation application.

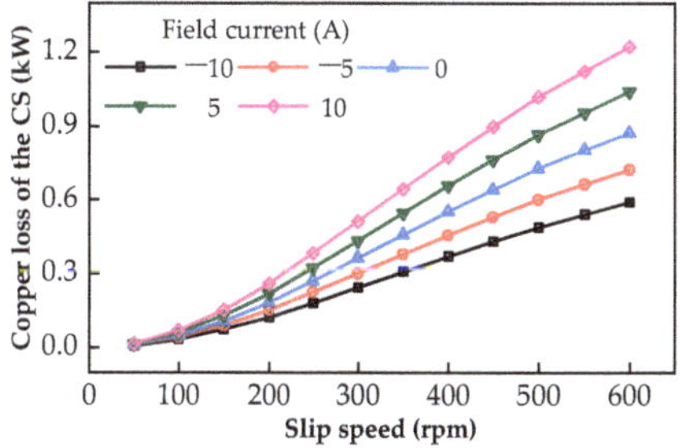

Figure 12. The copper losses of the copper sheet (CS) under different field currents.

Figure 13 shows the efficiencies of the coupling under different field currents when the speed of the prime motor is 1500 rpm. The efficiency of the coupling reaches the highest when I_f is zero, which is attributed to the fact that the copper loss of the field winding is the lowest. It decreases from 96.7 to 56.9% when the slip speed increases from 50 to 600 rpm. When I_f decreases from 10 to 0 A, the efficiency of the coupling slightly declines, while its reduction is enlarged with the further decrease in I_f from 0 to −10 A. This can be attributed to the fact that P_{out} decreases with the decrease in the field current. In addition, p_{cu_fw} declines when the field current decreases from 10 to 0 A, and it then rebounds to a higher value with the further decrease in the field current. In a lower slip speed range, the variations in efficiencies are obvious under different field currents. It is responsible for this that the copper loss of field winding occupies a large percentage of the relatively small output power when the torque of the coupling is small. Although the efficiency of the HE-ASECC declines with the increase in slip speed linearly, for a pump system, an energy-saving effect can still be achieved based on Equation (33) compared with the valves and baffles controls.

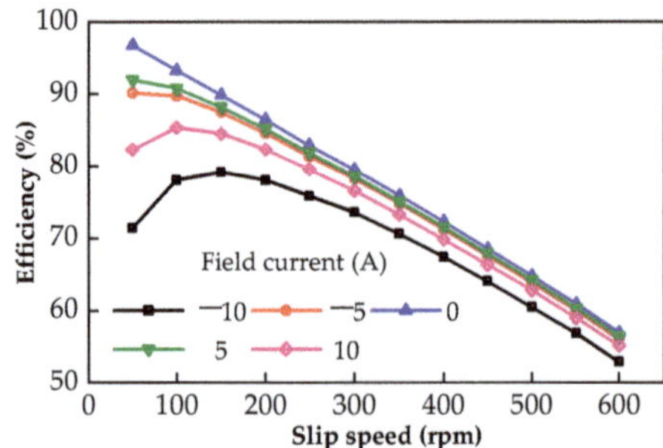

Figure 13. The efficiencies of the coupling under different field currents.

5. Conclusions

This paper proposes a novel brushless HE-ASECC with a simple AFES, which makes the online speed regulation reliably and effectively for the energy-saving of the aged pump and fan systems. The MEC-based analytical method is presented to calculate the magnetic field and torque of the proposed coupling efficiently. In addition, the analytical method provides an approach to solve the eddy current field problem under an asymmetric magnetic field of the PMR. Finally, the magnetic field and torque adjustment ability of the coupling illustrates that a wide speed regulation range can be achieved by the HE control. It also reveals that that the proposed speed regulation method can be applied in the field of variable speed drive for energy-saving reconstruction.

Author Contributions: Conceptualization: Y.L., H.L., and H.Y.; methodology: Y.L., Q.T., and H.L; software: Y.L. and Q.T.; validation: H.Y. and H.L.; formal analysis: Y.L. and H.Y.; investigation: Y.L., H.Y., and H.H.; resources: H.L.; data curation: H.L.; writing—original draft preparation: Y.L.; writing—review and editing: H.L., H.Y., and S.F; visualization: Y.L.; supervision: H.L. and H.H.; project administration: H.L.; funding acquisition: H.L.

Funding: This work was supported by the National Natural Science Foundation of China (51577026).

Conflicts of Interest: The authors declare no conflict of interest.

Appendix A

Table A1 shows a comparison of HE-ASECC with VB, AS-PMECC, and VFD in energy-saving effect, space, efficiency, power electronic converter capacity, reliability, and cost.

Table A1. Full comparison among different energy saving devices.

Item	VB	AS-PMECC	VFD	HE-ASECC
Energy-saving effect	★	★★★	★★★★★	★★★
Space occupation	★★★★★	★★★	★★	★★★★
Efficiency	★	★★★	★★★★★	★★★
Power electronic converter capacity	None	None	Large	None
Reliability	★★★★★	★★★	★★	★★★★
Cost	★★★★★	★★★★	★★	★★★★

References

1.   Bloxham, D.A.; Wright, M.T. Eddy-current coupling as an industrial variable-speed drive. *Proc. IEE* **1972**, *119*, 1149–1154. [CrossRef]

2. Wallace, A.; Wohlgemuth, C.; Lamb, K. A high efficiency, alignment and vibration tolerant, coupler using high energy-product permanent magnet. In Proceedings of the Seventh International Conference on Electrical Machines and Drives, Durham, UK, 11–13 September 1995; pp. 232–236.

3. Abdelaziz, E.A.; Saidur, R.; Mekhilef, S. A review on energy-saving strategies in industrial sector. *Renew. Sustain. Energy Rev.* **2011**, *15*, 150–168. [CrossRef]

4. Khalid, N. Efficient energy management: Is variable frequency drives the solution. *Procedia Soc. Behav. Sci.* **2014**, *145*, 371–376. [CrossRef]

5. Saidur, R.; Mekhilef, S.; Ali, M.B.; Safari, A.; Mohammed, H.A. Applications of variable speed drive (VSD) in electrical motors energy-savings. *Renew. Sustain. Energy Rev.* **2012**, *16*, 543–550. [CrossRef]

6. Wallace, A.; Jouanne, A. Industrial speed control: Are PM couplings an alternative to VFDs? *IEEE Trans. Ind. Appl.* **2001**, *7*, 57–63. [CrossRef]

7. Canova, A.; Vusini, B. Design of axial eddy-current couplers. *IEEE Trans. Ind. Appl.* **2003**, *39*, 725–733. [CrossRef]

8. Canova, A.; Vusini, B. Mathematic modeling of rotating eddy-current couplers. *IEEE Trans. Magn.* **2005**, *41*, 24–35. [CrossRef]

9. Wang, J.; Lin, H.; Fang, S.; Huang, Y. A general analytical model of permanent magnet eddy current couplings. *IEEE Trans. Magn.* **2014**, *50*, 1–9. [CrossRef]

10. Lubin, T.; Rezzoug, A. Steady-state and transient performance of axial-field eddy-current coupling. *IEEE Trans. Ind. Electron.* **2015**, *4*, 2287–2296. [CrossRef]

11. Wang, J.; Lin, H.; Fang, S. Analytical prediction of torque characteristics of eddy current couplings having a quasi-Halbach magnet structure. *IEEE Trans. Magn.* **2016**, *52*, 1–9. [CrossRef]

12. Dai, X.; Cao, J.Y.; Long, Y.J.; Liang, Q.H.; Mo, J.Q.; Wang, S.G. Analytical modeling of an eddy-current adjustable-speed coupling system with a 3-segment Halbach magnet array. *Electr. Power Comp. Syst.* **2015**, *43*, 1891–1901. [CrossRef]

13. Zheng, D.; Wang, D.; Li, S.; Zhang, H.; Yu, L.; Li, Z. Electromagnetic-thermal model for improved axial-flux eddy current couplings with combine rectangle-shaped magnets. *IEEE Access* **2018**, *6*, 26383–26390. [CrossRef]

14. Yang, H.; Zhu, Z.Q.; Lin, H.; Chu, W.Q. Flux adjustable permanent magnet machines: A technology status review. *Proc. CJEE* **2016**, *2*, 14–30.

15. Aydin, M.; Huang, S.; Lipo, T.A. Design, analysis, and control of a hybrid field-controlled axial-flux permanent-magnet motor. *IEEE Trans. Ind. Electron.* **2010**, *57*, 78–87. [CrossRef]

16. Lubin, T.; Rezzoug, A. 3-D analytical model for axial-flux adjust speed eddy-durrent couplings and brakes under steady-state conditions. *IEEE Trans. Magn.* **2015**, *51*, 1–12. [CrossRef]

17. Mohammadi, S.; Mirsalim, M.; Vaez-Zadeh, S. Nonlinear modeling of eddy-current couplers. *IEEE Trans. Energy Convers.* **2014**, *29*, 224–231. [CrossRef]

18. Mohammadi, S.; Mirsalim, M. Double-sided permanent-magnet radial-flux adjust speed eddy-current couplings: Three-dimensional analytical modelling, static and transient study, and sensitivity analysis. *IET Electr. Power Appl.* **2013**, *7*, 665–679. [CrossRef]

19. Li, Z.; Wang, D.; Zheng, D.; Yu, L. Analytical modeling and analysis of magnetic field and torque for novel axial flux eddy current couplers with PM excitation. *AIP Adv.* **2017**. [CrossRef]

20. Li, Y.; Lin, H.; Yang, H.; Fang, S.; Wang, H. Analytical analysis of a novel flux adjustable permanent magnet eddy-current coupling with a movable stator ring. *IEEE Trans. Magn.* **2018**, *3*, 1–4. [CrossRef]

21. Wang, J.; Zhu, J. A simple method for performance prediction of permanent magnet eddy current couplings using a new magnetic equivalent circuit model. *IEEE Trans. Ind. Electron.* **2018**, *65*, 2487–2495. [CrossRef]

22. Russell, R.L.; Norsworthy, K.H. Eddy currents and wall losses in screened-rotor induction motors. *Proc. IEE Part A Power Eng.* **1958**, *105*, 163–175. [CrossRef]

Article

Online MTPA Trajectory Tracking of IPMSM Based on a Novel Torque Control Strategy

Jianxia Sun, Cheng Lin *, Jilei Xing and Xiongwei Jiang

School of Mechanical Engineering, Beijing Institute of Technology, Beijing 100081, China
* Correspondence: lincheng@bit.edu.cn; Tel.: +86-137-0135-3142

Received: 1 August 2019; Accepted: 22 August 2019; Published: 24 August 2019

Abstract: The maximum-torque-per-ampere (MTPA) scheme is widely used in the interior permanent magnet synchronous machine (IPMSM) drive system to reduce copper losses. However, MTPA trajectory is complicated to solve analytically. In order to realize online MTPA trajectory tracking, this paper proposes a novel torque control strategy. The torque control is designed to be closed form. Considering the machine reluctance torque as the torque feedback, when this is compared with the torque reference, then the excitation torque reference can be obtained. Since the excitation torque is proportional to the q-axis current, the q-axis current reference can be fed by the excitation torque reference through a proportional regulator. Once the q-axis current reference is given, the d-axis current reference can be calculated based on the per-unit model, which aims to simplify the calculation and make the control strategy independent of machine parameters. In this paper, the stability of the control system is demonstrated. Meanwhile, simulation and experiment results show this torque control strategy can realize MTPA trajectory tracking online and have success in transients.

Keywords: maximum-torque-per-ampere (MTPA); torque control; per unit; IPMSM

---

1. Introduction

Due to outstanding characteristics, such as high-power density, wide speed range, and a high torque-to-inertia ratio, interior permanent magnet synchronous machines (IPMSMs) are suitable for many industrial applications [1–3]. Differing from the surface permanent magnet synchronous machine (SPMSM), the IPMSM contains not only an excitation torque but also a reluctance torque. Moreover, the excitation torque is proportional to the q-axis current and the reluctance torque is relative to both d- and q-axis stator current. In general, the d-axis current reference is set to zero, and the reluctance torque is consequently null. Then, the torque of IPMSM is proportional to the q-axis current. As a result, this method is easy to implement, but inefficient. In order to improve the efficiency, the maximum-torque-per-ampere (MTPA) scheme, which aims to minimize copper losses of IPMSM during operating, is proposed [4,5]. Under the MTPA condition, the reluctance torque is fully employed and the amplitude of stator current is minimum.

One way for MTPA implementation is finding MTPA operating points directly. For a fixed torque, the optimal current can be solved by a quartic equation, resulting from making the differentiation of the torque equation with respect to stator current equal to zero [6]. In [7], the quartic equation, yielded by the optimal problem, was solved by utilizing Ferrari's method. However, the analytic solution is so complicated that it is hard to utilize online. Therefore, some approximate methods are proposed to avoid complex calculation. For instance, look-up-table (LUT) is a widely used method in IPMSMs. In the LUT method, the optimal current is calculated in advance and sorted in a table. During operating, the optimal current is taken from the premade table. In order to improve accuracy, the LUT method requires accurate machine parameters. In [8], a 3-D lookup table combined with a PI compensator was proposed, to eliminate the undesired deviation which results from the machine parameter uncertainties. However, the lookup table, especially the 3-D lookup table, is huge and consumes system memory.

In order to avoid complex calculation procedures and get rid of machine parameters, some approaches have been proposed—for instance, the searching algorithm and signal injection method. Machine parameters are not required in these methods, which ensures robustness against parameter variations. In the searching algorithm [9–11], the MTPA operation is achieved by continuously disturbing the current reference, until the minimum of the stator current at the given load torque is found. The advantage of the searching algorithm is to get lower parameter dependence. However, if the convergence rate of the searching algorithm is slow, the algorithm may fail in transients. The signal injection method achieves the MTPA operation by injecting a real or a virtual signal into the reference to obtain the optimum operating points [12–15]. The signal injection method can rapidly respond to transients, whereas the injection of the perturbation signal results in disturbance and harmonics, thus, the control accuracy is reduced.

Instead of finding the optimal current directly, some approaches propose using an outer loop controller for MTPA operation [16–18]. In these methods, the relationship between the d- and q-axis stator currents is used for MTPA operation. The d-axis current reference is generated from the q-axis current, which is provided by an outer loop controller. However, the equation derived from the relationship is complex. Besides this, the outer loop controller is usually a proportional–integral (PI) controller, which results in a high-order mathematical model of the control system, and it is difficult to design a proper PI controller which guarantees global stability.

In this paper, a novel torque control strategy is proposed to track the MTPA trajectory online. The strategy utilizes the per unit model to simplify the calculation and make the model independent of machine parameters. Combined with the MTPA condition expressed in the per unit model, the torque controller can provide the reference stator currents for the current regulators. Besides this, the torque controller is a simple proportional controller. This strategy can respond to transients rapidly. This paper is organized as follows. The per unit modeling procedure and the MTPA condition are described in Section 2. The torque control strategy, system stability analysis, and convergence rate estimation are reported in Section 3. The simulation and experiment results are discussed in Section 4. Finally, the conclusive considerations are drawn in Section 5.

2. System Model and MTPA Condition

2.1. Per Unit Model of IPMSM

The torque equation of the IPMSM in a rotor reference frame is given as follows:

$$T_e = 1.5P\varphi_f i_q + 1.5P\left(L_d - L_q\right)i_d i_q, \tag{1}$$

where T_e is the electromagnetic torque; P is the number of pole pairs; φ_f is the magnetic flux linkage; i_d, i_q are the d- and q-axis stator currents; and L_d, L_q are the d- and q-axis inductances. The electromagnetic torque T_e consists of reluctance torque T_1 and excitation torque T_2. Moreover, the equation of T_1 and T_2 can be expressed as:

$$\begin{cases} T_1 = 1.5P\left(L_d - L_q\right)i_d i_q \\ T_2 = 1.5P\varphi_f i_q \end{cases}. \tag{2}$$

Equations (1) and (2) reveal that different machine parameters lead to varied torques. Therefore, the universality of the model is poor. In order to solve this problem, a per unit model of IPMSM is presented in this paper.

To build the per unit model, the base values must be defined. Moreover, the relationship between the base current and the base torque should correspond to (1). In this paper, the base current I_b is defined as $\frac{\varphi_f}{2(L_q - L_d)}$. Consequently, the base torque T_b is equal to:

$$T_b = 1.5P\varphi_f I_b + 1.5P(L_d - L_q)I_b^2 = 0.75P\varphi_f I_b. \tag{3}$$

According to (1), the torque of IPMSM in per unit model can be expressed as:

$$\begin{cases} T_{en} = 2i_{qn} - i_{qn}i_{dn} \\ T_{1n} = -i_{qn}i_{dn} \\ T_{2n} = 2i_{qn} \end{cases}, \tag{4}$$

where, $T_{en} = \frac{T_e}{T_b}$, $T_{1n} = \frac{T_1}{T_b}$, $T_{2n} = \frac{T_2}{T_b}$, $i_{dn} = \frac{i_d}{I_b}$, and $i_{qn} = \frac{i_q}{I_b}$. T_{en} is the per unit value of T_e, T_{1n} is the per unit value of T_1, T_{2n} is the per unit value of T_2, i_{dn} is the per unit value of i_d, and i_{qn} is the per unit value of i_q.

The per unit model can be applied to all IPMSMs, because it is independent of machine parameters. The machine parameters can be measured offline or estimated online [19].

2.2. MTPA Condition

The magnitude of stator current I_s is equal to $\sqrt{i_d^2 + i_q^2}$, which can be represented in per unit values as:

$$I_{sn}^2 = i_{dn}^2 + i_{qn}^2, \tag{5}$$

where $I_{sn} = \frac{I_s}{I_b}$ is the per unit value of I_s. According to (4), the relationship between i_{qn} and i_{dn} can be expressed as:

$$i_{qn} = \frac{T_{en}}{2 - i_{dn}}. \tag{6}$$

Equation (6) reveals that the relationship between i_{dn} and i_{qn} is fixed for a given torque T_{en}. Inserting (6) into (5), the relationship between I_{sn} and i_{dn} can be obtained as:

$$I_{sn}^2 = i_{dn}^2 + \frac{T_{en}^2}{(2 - i_{dn})^2}. \tag{7}$$

The minimum value of I_{sn} can be obtained by differentiating (7) with respect to i_{dn}, and setting the resulting expression to zero yields,

$$\frac{dI_{sn}^2}{di_{dn}} = 2i_{dn} + \frac{2T_{en}^2}{(2 - i_{dn})^3} = 0. \tag{8}$$

Equation (8) is a quartic equation and hard to solve, i.e., it is difficult to get the optimal solution of i_{dn} and i_{qn} for a given torque T_{en} in MTPA condition. In order to solve this problem, a novel torque control strategy is proposed in this paper, which will be described in the following sections.

3. Torque Control and Stability Analysis

3.1. Torque Control with MTPA Trajectory Tracking

For the sake of MTPA trajectory tracking, a novel torque control strategy is introduced (Figure 1).

The reference torque T_{e_ref} is given by an outer controller, and only positive reference torque will be taken into account in this paper. Considering the machine reluctance torque T_1 as the torque feedback, comparing with the torque reference T_{e_ref}, then the excitation torque reference T_2 can be obtained. Since the excitation torque is proportional to the q-axis current, the q-axis current reference i_q^* can be fed by the excitation torque reference T_2 through a proportional regulator.

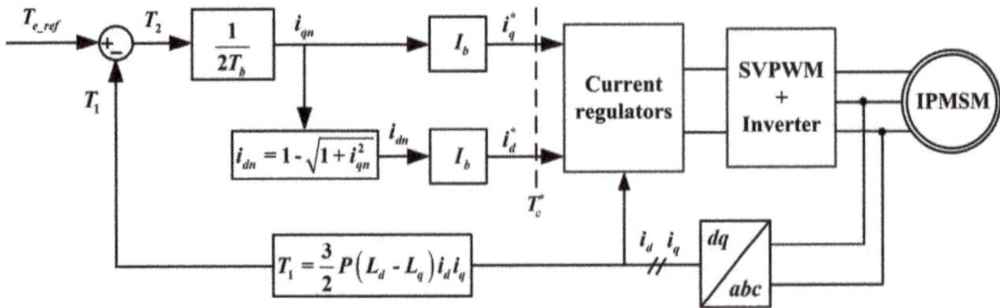

Figure 1. The torque control strategy structure with maximum-torque-per-ampere (MTPA) conditions.

For the sake of simplicity, the per unit model is used in this strategy. Combing (2) and (4), i_{qn} is equal to $T_2/2T_b$. Once i_{qn} is given, the per unit value i_{dn} in MTPA condition can be solved as follows. By combining (6), Equation (8) can be simplified to:

$$i_{qn}^2 + (2 - i_{dn})i_{dn} = 0. \tag{9}$$

Solving i_{dn} gives:

$$i_{dn} = 1 - \sqrt{1 + i_{qn}^2}. \tag{10}$$

Since i_{dn} and i_{qn} have been solved, the currents reference vector (i_d^*, i_q^*) can be derived easily. Within the effects of torque control, (i_d^*, i_q^*) will converge asymptotically to the MTPA operating point $(i_{d,mtpa}^*, i_{q,mtpa}^*)$.

The dynamic analysis findings are as follows: if the q-axis current reference $i_q^* > i_{q,mtpa}^*$, then the d-axis current reference $i_d^* < i_{d,mtpa}^*$, the machine reluctance torque T_1 increases, then excitation torque reference T_2 will decrease, and i_q^* will decrease consequently; in the case of $i_q^* < i_{q,mtpa}^*$, it is exactly the same way. This torque controller keeps adjusting i_q^* until it is equal to $i_{q,mtpa}^*$ and then i_d^* is equal to $i_{d,mtpa}^*$, i.e., $(i_{d,mtpa}^*, i_{q,mtpa}^*)$ is the equilibrium point of (i_d^*, i_q^*). The relationship between the given torque and MTPA operating point is shown in Figure 2.

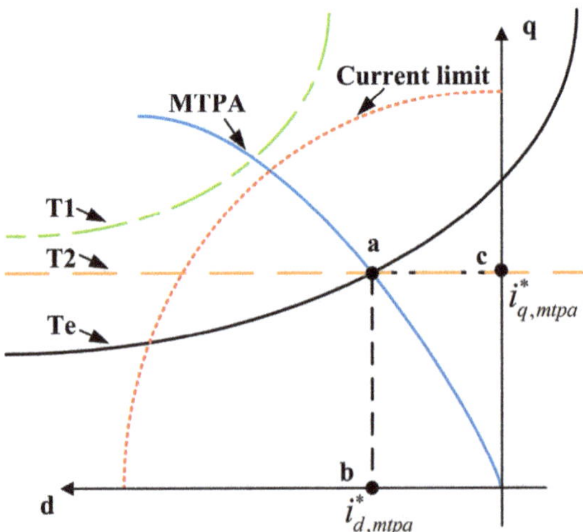

Figure 2. The relationship between the given torque and MTPA operating point.

The reference currents can be used properly in the field-oriented control (FOC) scheme to achieve MTPA trajectory tracking.

3.2. Stability Analysis and Convergence Rate Estimation

The dynamic analysis describes roughly the process of tracking MTPA operating points. In this section, the stable performance of torque control strategy will be proven as follows.

Within the proper design of vector control, the d- and q-axis current closed loop systems can be approximated by a low pass filter [20], i.e.,:

$$\begin{cases} \dot{i}_d = \frac{1}{\tau}\left(i_d^* - i_d\right) \\ \dot{i}_q = \frac{1}{\tau}\left(i_q^* - i_q\right) \end{cases},$$ (11)

where, τ is the time constant of current closed loop. Denoting the torque T_e^*, T_1^* and T_2^* are given by i_d^* and i_q^*. Meanwhile, their relationships should correspond to (1) and (2), i.e., $T_1^* = 1.5P\left(L_d - L_q\right)i_d^* i_q^*$, $T_2^* = 1.5P\varphi_f i_q^*$, and $T_e^* = T_1^* + T_2^*$.

Making the time derivative of T_1 and combining the result with (11), the following equation can be obtained:

$$\tau \dot{T}_1 = -2T_1 + 1.5P\left(L_d - L_q\right)i_d i_q^* + 1.5P\left(L_d - L_q\right)i_q i_d^*.$$ (12)

Differentiating (12) with respect to time again and then inserting the result into (12), it follows that:

$$\tau^2 \ddot{T}_1 + 3\tau \dot{T}_1 + 2T_1 = 2T_1^*.$$ (13)

As expressed in Figure 1, T_2^* equals to T_2, and T_2 equals to $T_{e\_ref} - T_1$. Thus, $\dot{T}_2^* = -\dot{T}_1$, $\ddot{T}_2^* = -\ddot{T}_1$, and $T_1 = T_{e\_ref} - T_2^*$. Inserting the results into (13), it holds that:

$$\tau^2 \ddot{T}_2^* + 3\tau \dot{T}_2^* = 2(T_{e\_ref} - T_e^*).$$ (14)

According to the singular perturbation approximation principle [9], the second order term $\tau^2 \ddot{T}_2^*$ can be neglected because τ^2 is far less than 3τ. Combing with $T_2^* = 1.5P\varphi_f i_q^*$, the derivative of i_q^* is given by:

$$\dot{i}_q^* = \frac{4(T_{e\_ref} - T_e^*)}{9\tau P\varphi_f}.$$ (15)

Based on (10), the derivative of i_d^* can be expressed as:

$$\dot{i}_d^* = -\frac{8i_q^*(T_{e\_ref} - T_e^*)}{9\tau P\varphi_f \sqrt{1 + \left(i_q^*/I_b\right)^2}}.$$ (16)

Apparently, $T_{e\_ref} = T_e^*$ when $\dot{i}_q^* = 0$ and $\dot{i}_d^* = 0$, i.e., $T_{e\_ref} = T_e^*$ when (i_d^*, i_q^*) converge to the MTPA operating point. Considering the Lyapunov function $V = \frac{1}{2}\left(T_{e\_ref} - T_e^*\right)^2$, the derivative of V is given by (17):

$$\begin{aligned} \dot{V} &= \dot{T}_e^*\left(T_e^* - T_{e\_ref}\right) = \left(\frac{\partial T_e^*}{\partial i_d^*}\frac{di_d^*}{dt} + \frac{\partial T_e^*}{\partial i_q^*}\frac{di_q^*}{dt}\right)\left(T_e^* - T_{e\_ref}\right) \\ &= -\left[\frac{3}{2}P\left(L_d - L_q\right)i_d^* + \frac{3}{2}P\varphi_f - \frac{3P\left(L_d - L_q\right)i_q^{*2}}{\sqrt{1 + \left(i_q^*/I_b\right)^2}}\right]\frac{4(T_{e\_ref} - T_e^*)^2}{9\tau P\varphi_f} \end{aligned}$$ (17)

Since only positive reference torque T_{c_ref} is taken into account, $i_q^* > 0$ and $i_d^* < 0$. Besides, $L_d < L_q$. Therefore, $V \geq 0$ and $\dot{V} \leq 0$, i.e., this system is asymptotically stable.

According to (17), $\dot{V} \leq -\frac{4V}{3\tau}$, solving V gives:

$$V \leq V_0 e^{-\frac{4(t-t_0)}{3\tau}},\tag{18}$$

where, t_0 is the initial time, V_0 is the initial value of V. This shows that the convergence rate of V is higher than $\frac{4}{3\tau}$.

According to (11), solving (i_d, i_q) gives:

$$\begin{cases} i_d = i_d^* - \left(i_d^* - i_{d0}\right)e^{-\frac{1}{\tau}(t-t_0)} \\ i_q = i_q^* - \left(i_q^* - i_{q0}\right)e^{-\frac{1}{\tau}(t-t_0)} \end{cases},\tag{19}$$

where (i_{d0}, i_{q0}) are the initial values of (i_d, i_q). Equation (19) shows the convergence rate of (i_d, i_q) equals $\frac{1}{\tau}$. Obviously, the convergence rate of V is higher than (i_d, i_q). That means the convergence rate of the torque loop is higher than the current loop, i.e., T_e^* coverages to T_{e_ref} and (i_d^*, i_q^*) converge to $(i_{d,mtpa}^*, i_{q,mtpa}^*)$ before (i_d, i_q) converge to (i_d^*, i_q^*). Therefore, this torque control strategy is suitable for dynamic MTPA trajectory tracking.

4. Results

4.1. Simulation Results

To validate the proposed torque control strategy, the whole system was simulated in Matlab/Simulink. The parameters of interior permanent magnet synchronous machine (IPMSM) are tabulated in Table 1.

Table 1. Parameters of IPMSM.

Symbol	Parameter	Value
P	pole pairs	5
L_d	d-axis inductance	0.017961 H
L_q	q-axis inductance	0.023747 H
R_s	Stator resistance	0.768 Ω
φ_f	Magnet flux linkage	0.2364 Wb
T_b	base torque	18.11
I_b	base current	20.4286

The rated speed was set to 1000 rpm. The machine speed was controlled by an outer speed controller, which provided the reference torque T_{e_ref}. In order to maintain a constant speed, T_{e_ref} was varied with the load torque T_L. Based on the different state of T_L, the simulation was divided into three stages. Stage 1: 0 s~0.2 s, T_L increased from 0 Nm to 50 Nm equably. Stage 2: 0.2 s~0.4 s, T_L maintained at 50 Nm. Stage 3: 0.4 s~1 s, T_L dropped to 40 Nm at 0.4 s, then remained constant until 1 s. Figure 3 shows the torque and current response of MTPA trajectory tracking under the torque control strategy. Figure 3a shows the drop to 40 Nm at 0.4 s, then remaining constant until 1 s. Figure 3 shows the torque and current response of MTPA trajectory tracking under the torque control strategy. Figure 3a shows the response of the machine speed. Figure 3b shows the response of reference torque T_{e_ref} and machine torque T_e. Figure 3c shows the response of the d-axis reference current i_d^* and stator current i_d. Figure 3d shows the response of the q-axis reference current i_q^* and stator current i_q. It can be seen that T_e followed T_{e_ref} closely in these load conditions, even if T_L changed continuously. Besides, i_d and i_q varied with T_{e_ref}.

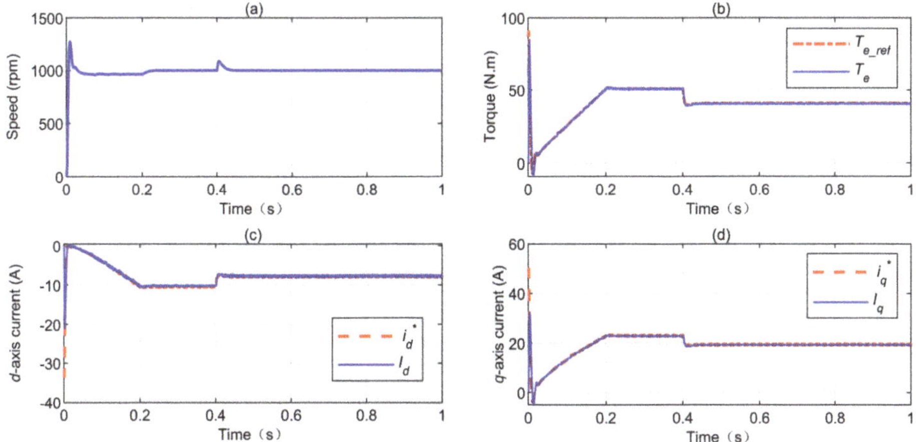

Figure 3. The simulation results: (**a**) machine speed; (**b**) reference and machine torque; (**c**) d-axis reference current and stator current; (**d**) q-axis reference current and stator current.

Figure 4 reports the transient current locus and the desired MTPA trajectory. It can be seen that (i_d, i_q) converged to the MTPA operating point asymptotically. The simulation results reveal that this online MTPA trajectory tracking strategy is excellent in steady and transient conditions.

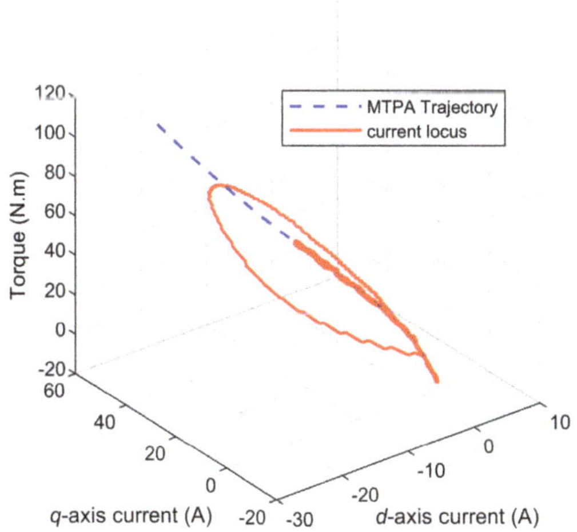

Figure 4. The transient current locus and the desired MTPA trajectory.

4.2. Experimental Results

The proposed control approach was validated through experimental investigations, which were conducted on a laboratory IPMSM drive system, shown in Figure 5.

The experiment platform contained a 3 kW IPMSM, which was used for applying the proposed strategy and an asynchronous motor operated as a programmable load, which was controlled by the PC. The tested motor was driven by the three-phase voltage source PWM (Pulse Width Modulation) inverter and the controller was implemented in the TMS320F28335 DSP (Digital Signal Processor). The position of the tested motor was estimated by a sliding-mode observer. The machine reluctance torque was calculated based on the machine parameters. The parameters of the IPMSM were tabulated in Table 1.

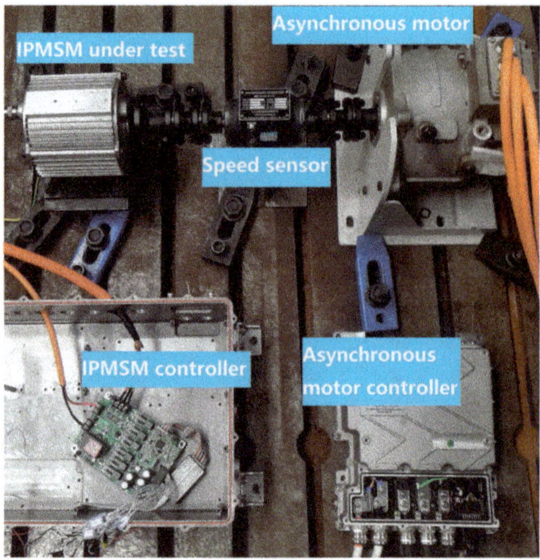

Figure 5. Experimental setup.

The rated speed was set to 1000 rpm. The machine speed response is shown in Figure 6a. The response of the reference torque T_{e_ref} is shown in Figure 6b. The response of the d-axis reference current i_d^* and stator current i_d are shown in Figure 6c. The response of the q-axis reference current i_q^* and stator current i_q are shown in Figure 6d. Figure 7 shows the transient response of the reference torque and q-axis reference current. It can be seen that the stator current can follow the current reference closely. The torque reference was adjusted with speed and the current reference was adjusted with torque reference. The experiment results reveal that this strategy is active.

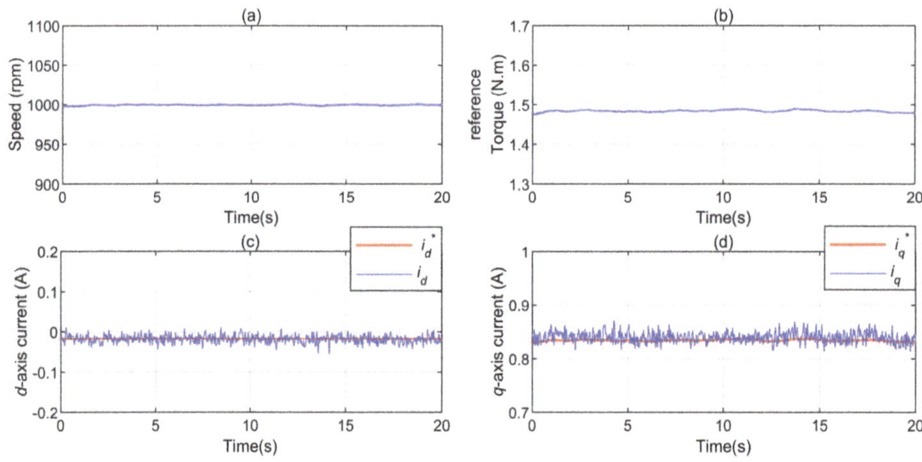

Figure 6. The experiment result: (**a**) machine speed; (**b**) reference torque; (**c**) d-axis reference current and d-axis stator current; (**d**) q-axis reference current and q-axis stator current.

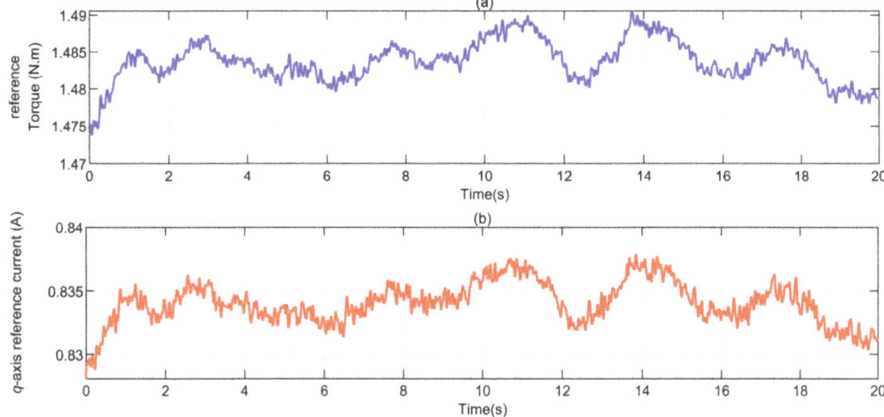

Figure 7. The experiment result: (**a**) reference torque; (**b**) q-axis reference current.

5. Conclusions

A novel torque controller was presented in this paper. Combined with the MTPA condition, this torque control strategy can track the MTPA trajectory online. The reference torque consisted of two parts, excitation torque and reluctance torque. The excitation torque was fed to the proportional regulator, which produced the q-axis current reference. Once the q-axis current reference was given, the d-axis current reference could be calculated online based on the MTPA condition in the per unit model. Moreover, the implementation of the per unit model can simplify the MTPA condition, reduce the calculation burden, and make the machine parameters independent. In addition, parameter estimation algorithms can be conveniently combined with the torque controller to improve accuracy. This strategy combined the MTPA condition with the torque controller to realize MTPA trajectory tracking online. Moreover, the stability and dynamic performance were proven analytically. The simulation and experiment results show the excellent performance of the strategy in steady and transient conditions.

Author Contributions: Data curation, X.J.; Formal analysis, J.S.; Investigation, J.S.; Methodology, J.S.; Project administration, C.L.; Software, J.X.; Supervision, C.L.

Funding: This work was supported by Ministry of Science and Technology of the People's Republic of China under Grant 2017YFB0103801.

Conflicts of Interest: The authors declare no conflict of interest.

References

1. Lin, F.J.; Hung, Y.C.; Chen, J.M.; Yeh, C.M. Sensorless IPMSM Drive System Using Saliency Back-EMF-Based Intelligent Torque Observer With MTPA Control. *IEEE Trans. Ind. Inform.* **2014**, *10*, 1226–1241.
2. Lu, X.M.; Iyer, K.L.V.; Mukherjee, K.; Ramkumar, K.; Kar, N.C. Investigation of Permanent-Magnet Motor Drives Incorporating Damper Bars for Electrified Vehicles. *IEEE. Trans. Ind. Electron.* **2015**, *62*, 3234–3244. [CrossRef]
3. Yang, Z.; Shang, F.; Brown, I.P.; Krishnamurthy, M. Comparative Study of Interior Permanent Magnet, Induction, and Switched Reluctance Motor Drives for EV and HEV Applications. *IEEE Trans. Transp. Electron.* **2015**, *1*, 245–254. [CrossRef]
4. Ni, R.G.; Xu, D.G.; Wang, G.L.; Ding, L.; Zhang, G.Q.; Qu, L.Z. Maximum Efficiency Per Ampere Control of Permanent-Magnet Synchronous Machines. *IEEE Trans. Ind. Electron.* **2015**, *62*, 2135–2143. [CrossRef]
5. Ge, H.; Miao, Y.; Bilgin, B.; Nahid-Mobarakeh, B.; Emadi, A. Speed Range Extended Maximum Torque Per Ampere Control for PM Drives Considering Inverter and Motor Nonlinearities. *IEEE T. Power Electron.* **2017**, *32*, 7151–7159. [CrossRef]

6. Jahns, T.M.; Kliman, G.B.; Neumann, T.W. Interior Permanent-Magnet Synchronous Motors for Adjustable-Speed Drives. *IEEE Trans. Ind. Appl.* **1986**, *22*, 738–747. [CrossRef]

7. Jung, S.-Y.; Hong, J.; Nam, K. Current Minimizing Torque Control of the IPMSM Using Ferrari's Method. *IEEE Trans. Power Electron.* **2013**, *28*, 5603–5617. [CrossRef]

8. Nalepa, R.; Orlowska-Kowalska, T. Optimum Trajectory Control of the Current Vector of a Nonsalient-Pole PMSM in the Field-Weakening Region. *IEEE Trans. Ind. Electron.* **2012**, *59*, 2867–2876. [CrossRef]

9. Dianov, A.; Kim, Y.-K.; Lee, S.-J.; Lee, S.-T. Robust Self-Tuning MTPA Algorithm for IPMSM Drives. In Proceedings of the 34th Annual Conference of IEEE Industrial Electronics 2008, Orlando, FL, USA, 10–13 November 2008.

10. Morimoto, S.; Sanada, M.; Takeda, Y. Wide-speed operation of interior permanent magnet synchronous motors with high-performance current regulator. *IEEE Trans. Ind. Appl.* **1994**, *30*, 920–926. [CrossRef]

11. Ahmed, A.; Sozer, Y.; Hamdan, M. Maximum Torque per Ampere Control for Buried Magnet PMSM Based on DC-Link Power Measurement. *IEEE Trans. Power Electron.* **2017**, *32*, 1311. [CrossRef]

12. Lin, F.J.; Liu, Y.T.; Yu, W.A. Power Perturbation Based MTPA With an Online Tuning Speed Controller for an IPMSM Drive System. *IEEE Trans. Ind. Electron.* **2018**, *65*, 3677–3687. [CrossRef]

13. Liu, G.H.; Wang, J.; Zhao, W.X.; Chen, Q. A Novel MTPA Control Strategy for IPMSM Drives by Space Vector Signal Injection. *IEEE Trans. Ind. Electron.* **2017**, *64*, 9243–9252. [CrossRef]

14. Antonello, R.; Carraro, M.; Zigliotto, M. Maximum-Torque-Per-Ampere Operation of Anisotropic Synchronous Permanent-Magnet Motors Based on Extremum Seeking Control. *IEEE Trans. Ind. Electron.* **2014**, *61*, 5086–5093. [CrossRef]

15. Lin, F.J.; Chen, S.G.; Liu, Y.T.; Chen, S.Y. A power perturbation-based MTPA control with disturbance torque observer for IPMSM drive system. *Trans. Inst. Meas. Control* **2018**, *40*, 3179–3188. [CrossRef]

16. Mohamed, A.R.I.; Lee, T.K. Adaptive self-tuning MTPA vector controller for IPMSM drive system. *IEEE Trans. Energy Convers.* **2006**, *21*, 636–644. [CrossRef]

17. Morimoto, S.; Ueno, T.; Sanada, M.; Yamagiwa, A.; Takeda, Y.; Hirasa, T. Effects and Compensation of Magnetic Saturation in Permanent Magnet Synchronous Motor Drives. In Proceedings of the 1993 IEEE Industry Applications Conference Twenty-Eighth IAS Annual Meeting, Toronto, ON, Canada, 2–8 October 1993.

18. Consoli, A.; Scarcella, G.; Scelba, G.; Testa, A. Steady-State and Transient Operation of IPMSMs Under Maximum-Torque-per-Ampere Control. *IEEE Trans. Ind. Appl.* **2010**, *46*, 121–129. [CrossRef]

19. Underwood, S.J.; Husain, I. Online Parameter Estimation and Adaptive Control of Permanent-Magnet Synchronous Machines. *IEEE Trans. Ind. Electron.* **2010**, *57*, 2435–2443. [CrossRef]

20. Liu, Q.; Hameyer, K. High-Performance Adaptive Torque Control for an IPMSM With Real-Time MTPA Operation. *IEEE Trans. Energy Convers.* **2017**, *32*, 571–581. [CrossRef]

Article

Analysis and Design of Innovative Magnetic Wedges for High Efficiency Permanent Magnet Synchronous Machines [†]

Lucia Frosini [1,*] and Marco Pastura [2]

[1] Department of Electrical, Computer and Biomedical Engineering, University of Pavia, Via Ferrata 5, 27100 Pavia, Italy

[2] Department of Engineering Enzo Ferrari, University of Modena and Reggio Emilia, Via P. Vivarelli 10, 41125 Modena, Italy; marco.pastura@unimore.it

* Correspondence: lucia.frosini@unipv.it

† This paper is an extended version of our paper published in the 2019 21st European Conference on Power Electronics and Applications (EPE '19 ECCE Europe), Genova, Italy, 3–5 September 2019.

Received: 16 November 2019; Accepted: 2 January 2020; Published: 4 January 2020

Abstract: The global decarbonization targets require increasingly higher levels of efficiency from the designers of electrical machines. In this context, the opportunity to employ magnetic or semi-magnetic wedges in surface-mounted permanent magnet machines with fractional-slot concentrated winding has been evaluated in this paper, with the aim to reduce the power losses, especially in the magnets. Since an analytical calculation is not sufficient for this evaluation, finite element methods with two different software have been employed, by using a model experimentally validated on a real motor. The effects of wedges with different values of permeability and different magnetization characteristics have been evaluated on flux density, back electromotive force, and inductances, in order to choose the more suitable wedge for the considered motor. Furthermore, a new wedge consisting of different portions of materials with different magnetic permeability values is proposed. The effects of both conventional and unconventional magnetic wedges were assessed to optimize the motor performance in all working conditions.

Keywords: permanent magnet motor; synchronous motor; efficiency; brushless drive; industrial application

---

1. Introduction

Permanent magnet synchronous machines (PMSMs) with fractional-slot concentrated winding (FSCW) provide numerous benefits with respect to machines with traditional distributed winding [1]. The main advantages are related to the possibility of attaining many poles with a relatively small number of slots, and hence with a compact radial dimension of the machine; additionally, the slot fill factor is higher (up to 60%) and the end windings are shortened, with consequent savings in conductive material and reduced axial dimension.

From a performance point of view, compared to similar motors with distributed winding, in most cases, PMSMs with FSCW present:

1. Greater efficiency, essentially due to the reduction of Joule losses (at the expense of an increase in iron and magnet losses);
2. Greater power density, thanks to the reduced size of the machine;
3. Higher field weakening capability;
4. Fault tolerance.

Considering the overall system, consisting of the PMSM and its load or prime mover, the FSCW permits a direct coupling to be achieved, when the required speed at a rated frequency is low, so a gearbox would be unnecessary and therefore the efficiency of the entire system would be increased, reducing also its size, weight, noise, maintenance costs, and failure rate.

From the manufacturing point of view, this design allows for easier and faster insertion of the coils into the slots, which, generally, have to be open, with the following drawbacks:

1. Increase in the Carter factor and the equivalent length of the airgap, with a consequently greater reluctance of the magnetic path;
2. Worsening of the harmonic content of the magnetic field, with particular negative consequences in surface-mounted permanent magnet (SPM) machines;
3. Possible increasing in the cogging torque, due to the considerable discontinuity of the magnetic permeability of the stator core.

In order to mitigate the problems mentioned above deriving from the design of the stator slot opening, it is possible to use magnetic or semi-magnetic wedges, with the effect of simulating semi-closed slots. In the literature, different solutions have been proposed for this purpose for different types of electrical machines, but very few papers have focused on radial PMSMs and in particular on high torque and low speed SPM machines, as in this research.

1.1. State of the Art of the FSCW Machines

FCSW has been considered in the last 15 years as a valid alternative for PMSMs with respect to the traditional distributed winding. Many researches are present in the literature, mainly related to three-phase radial-flux machines [1–5], but also on axial-flux [6], tubular and flux-switching machines, and to machines with more than three phases [7]. Nevertheless, in this paper, the analysis of FCSW has been limited to the most widespread application, i.e., three-phase radial-flux machines. The first most significant researches on this topic are reported in [1–5], but more recently, other interesting papers have been published and the conclusions of some of them are reported in the following.

One of the drawbacks arising from FCSW is the increase in the rotor losses, due to the large amount of magnetomotive force harmonics produced by this type of winding. In [8], the effects of multilayer windings on rotor losses, especially the eddy-current losses in the permanent magnets of IPM machines with FSCW, were studied by means of the 2-D finite element analysis.

Another winding method is presented in [9] to eliminate both the sub and super space harmonics of the FCSWs in which the stator slot numbers are integer multiples of six. The proposed method is based on three concentrated single-layer windings in which the number of turns in one layer is optimized while the other layers are moved to cancel the subharmonics. The application of the proposed winding concept to a PMSM demonstrates the cancellation of dominant sub and super harmonics, reduction of the total harmonic distortion (THD), reduction of torque ripple, and minimization of core loss.

In [10], a multi-layer winding was designed to achieve an innovative symmetrical FSCW with unconventional slot–pole combinations. A general design methodology for this purpose is presented and validated by finite element analysis. The application of an unconventional FSCW to a shipboard surface permanent-magnet machine prototype is presented to illustrate the possible practicality of this type of winding; tests on the prototype for experimental validation are reported.

In [11], a method is proposed to reduce the torque pulsation of the three-phase FSCW interior permanent magnet (IPM) motors by lowering the subharmonics. The key principle of this method is the selection of an optimized six-layer winding with a different number of conductors. The optimized six-layer winding is used to suppress the first subharmonic and ensure that the second subharmonic is zero in a novel 9-slot/8-pole IPM motor.

Based on the consideration that the increase in the number of pole pairs leads to a lower electromagnetic yoke, and therefore to lower vibration and magnetic noise, in [12], the influence of different numbers of pole pairs on the vibroacoustic design aspects of the machine was studied, considering the natural frequencies under a variable speed analysis. The paper aimed to determine the optimal number of pole pairs for a low-speed and high-torque PMSM with FSCWs for wind turbine applications. Both 3-D finite element analysis and experimental investigations were used.

From this analysis, it is clear that in the last years that the research in this field has been focused in particular on the design of special windings to solve the drawbacks arising from FCSWs; however, this choice could impose many constraints for the design of the machine and involve manufacturing complications and higher costs for its realization.

1.2. State of the Art of the Magnetic Wedges Employed in Rotating Electrical Machines

The first researches on the use of magnetic wedges date back to at least 50 years ago and mainly concern the application to induction motors. The problem starts from the consideration that high-power rotating electrical machines, with the stator winding connected to the voltage in the order of a few kilovolts, require the manufacture of form-wound coils, which are made and insulated separately, and then inserted into the stator slots, which must be open. However, the open slots have some drawbacks, which are derived from the variations in the airgap permeance produced by the relatively wide opening of the slots. These drawbacks include the high frequency no-load losses, higher magnetizing current due to the increase in the effective airgap, consequent reduction in efficiency, and the power factor. The influence of the stator slot opening is stronger in induction motors, which have a higher "slot-opening/airgap" ratio than synchronous machines [13]. This is related to the fact that induction machines require a smaller airgap with respect to synchronous ones, due to the need to contain the value of the magnetizing current. Even if this requirement is valid, especially for large machines, the first studies have been applied to small ones.

A first investigation on a 3.7 kW motor showed that the use of magnetic wedges in induction motors can improve their performance, especially at full load [13]. However, the paper does not answer questions related to the mechanical strength and durability of the wedges under service conditions in large machines. A second study proved that soft ferrite magnetic wedges in a 0.75 kW induction motor decrease the 17th and 19th asynchronous torques (this effect is equivalent to the rotor skewing), but they are not able to reduce the 5th asynchronous and synchronous torques [14].

A more recent paper, focused on both induction and synchronous motors, starts from the assumption that it is possible to obtain a high power density in rotating electrical machines by increasing the flux density at the airgap, and in particular by shortening the length of the airgap [15]. Since this solution increases the ripple flux density of the tooth and, consequently, the losses on the surface of the poles, magnetic wedges have been proposed to reduce these losses. Magnetic wedges with a relative permeability, μ_r, from 10 to 30 were evaluated and the calculated and measured results on large motors (500 and 745 kW) indicated the effectiveness of the magnetic wedges, selecting $\mu_r = 10$ as the most suitable.

In [16], the effects of magnetic wedges with $\mu_r = 8$ and different dimensions were evaluated in the stator slot ripple components of the airgap flux density distribution and in the rotor current of a 55 kW squirrel cage induction motor. The simulated and experimental results showed a reduction of these components but also revealed that a thicker magnetic wedge can induce an increase of the slot leakage flux and a decrease of the power factor.

The impact of magnetic wedges in high power induction motors has also been studied in recent years. An interesting study is reported in [17], where the effects of two commercial magnetic wedges were simulated by finite element (FE) analysis and experimentally tested on a squirrel cage induction motor of 1660 kW, 6.6 kV. Both wedges have a relative permeability of few units, based on the value of the magnetic flux density. The authors prove that magnetic wedges offer the following advantages: (i) Reduction of the magnetizing current (less no-load current); (ii) consequent reduction of the Joule losses in the stator winding and increase of the efficiency of the machine, and (iii) in the case of totally enclosed fan cooled (TEFC) motors, a further benefit is the reduction in temperature rise.

However, a qualification process is needed to ensure a reliable magnetic wedge system, as the flux density increases with greater permeability and the forces acting on the magnetic wedges also increase.

In [18], the influence of semi-magnetic wedges with $\mu_r = 10$ on the harmonic content of the magnetic field at the airgap and on the electromagnetic characteristics of a 4 kW induction motor was studied through FE analysis. The results indicate that these wedges lead to a decrease in the airgap magnetic field space harmonic content. In addition, the electromagnetic torque is characterized by significantly lower oscillations, resulting in a longer bearing lifecycle. This behavior leads to an improvement of the motor efficiency but also to a lower power factor at the rated speed. The performances of a 2.2 kW induction motor with magnetic wedges of different permeability were evaluated in [19] by FE analysis. The authors concluded that this solution improves the motor performance, even if it tends to decrease drastically when $\mu_r > 20$, as most of the flux lines is shorted in the wedges. Another recent paper reports a study on a 120 kW, 350 V induction motor with a squirrel cage rotor [20]. The FE analysis and the experimental tests showed that the magnetic wedges with $\mu_r = 4$ significantly reduced the load current (thanks to a reduction of the magnetizing current) and increased both the power factor and the efficiency. However, when the relative permeability of the magnetic wedges was greater than 30, the torque of the motor decreased and, consequently, both the motor efficiency and the power factor tended to decrease due to the leakage fluxes. The optimal value of the magnetic wedge relative permeability, corresponding to the maximum efficiency of the motor, can be found for each case by using FE analysis. The use of magnetic wedges has also been evaluated in a special type of induction machine, i.e., a 250 kW brushless doubly fed machine, which can be employed for offshore wind power generation. The results show that magnetic wedges increase the series inductance in the equivalent circuit, which significantly affects the reactive power control and the rating of the converter [21]. A very recent paper explored the impact of adopting semi-magnetic wedges in wound field synchronous machines and concludes that they can be useful for reducing the pole face losses, the stator air-gap Carter coefficient, and, as a consequence, the rotor DC current required for the rated flux together with the associated loss and temperature rise [22].

Over the past 20 years, some papers have addressed axial flux permanent magnet machines (AFPMs) with slotted winding, which are suitable for electric traction applications, such as wheel motors. The slotted winding allows a slight flux weakening capability, but it generates a significant cogging torque and a large harmonic content in the back electromotive force (e.m.f.), due to the magnetic anisotropy of the stator. The use of magnetic wedges, with a relative permeability in the range between 10 and 100, causes an increase in the phase inductance, as well as a good reduction of the cogging torque. In [23], a 2D FE procedure, validated with experimental tests, was applied to a three-phase machine, with a single central stator and two rotors, one slot per pole and per phase, with a nominal power of 8 kW. Magnetic wedges significantly increase the phase inductance while the mutual inductance is nearly not affected. As a drawback, this solution increases the axial force between the stator and rotor by about 20% in comparison to standard non-magnetic wedges. A similar machine was simulated in [24,25], by means of 2D and 3D FE models, considering magnetic wedges of different heights and with a variable gap in the center of the wedge. The performances of this machine were analyzed both at no-load and at load, highlighting the advantages and disadvantages offered by the various solutions.

A new magnetic wedge design was rated for an AFPM in [26–28]. It was obtained from a normal magnetic wedge by milling a straight hollow groove along the length of the wedge. The groove was filled with a non-magnetic bar, which had a purely mechanical function of increasing the strength of the wedge. The main influences of this magnetic wedge design can be observed in terms of the cogging torque and stator phase slot leakage inductance. These two effects determine the most convenient cross-section design of the wedge, depending on the specific application. The large inductance allows to advantageously reduce the flux-weakening current and helps to reduce the short-circuit current, which is very important, especially in permanent magnet machines, to avoid demagnetization in the case of failure.

The influence of magnetic wedges on a 110 kW radial flux interior permanent magnet (IPM) motor with concentrated winding was investigated in [29], using flux 2D analysis and experimental tests. The machine was designed for traction applications, with a torque demand in the low speed area up to 2.5 times the nominal torque, and in the high speed area, a good field weakening capability. Several wedge materials were evaluated, with a relative permeability of 1, 1.5, 3, 30, 300, and 500. The authors showed that magnetic wedges reduce iron losses and, above all, permanent magnet losses, around the rated speed. However, magnetic wedges with a relative permeability greater than 3 do not provide an advantage but rather a disadvantage, since they significantly reduce the maximum torque.

Recently, magnetic wedges were considered for applications to switched reluctance machines, which has two important drawbacks: Torque ripple and stator vibrations. Using FE simulations, the authors demonstrated that, although the introduction of magnetic wedges degrades the average torque, it does reduce the radial force and torque ripples [30].

For a better understanding, two illustrations of wedges for FSCW machines are reported in Figure 1: Note that these wedges are non-magnetic, but the shapes of magnetic wedges are the same and also the method to assemble them in the stator slots.

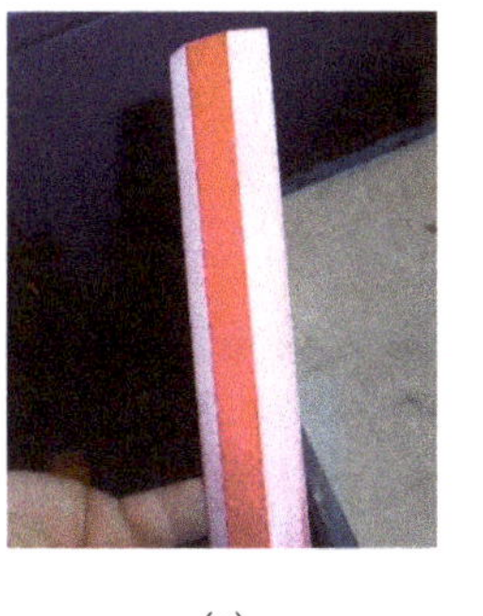

(a)

(b)

Figure 1. Typical wedges for FSCW machines: (**a**) a single wedge; (**b**) some wedges inserted in the stator slots.

1.3. Novelties of the Paper

In this paper, first of all, the main issues in the design of FSCW-SPM machines will be investigated, with particular attention to the effect of this type of winding on the losses of the machine. After that, a preliminary study on the effects of magnetic wedges in SPM machines will be carried out, in order to choose the best compromise for the value of magnetic permeability of these wedges. Finally, a case study will be introduced and the most suitable characteristics of the wedges for this case will be studied, in terms of material, geometry, and their combinations, through finite element (FE) simulations, with the aim of optimizing motor performance. Particular attention will be paid to the reduction of power losses in the magnets. Different configurations of the magnetic wedge will be evaluated, not only for the specific application but also to evaluate the effects derived from this type of wedge on machines with other characteristics, in order to obtain a broader estimate of the benefits and critical problems deriving from this change in the stator design.

The main differences and contributions of this paper compared to the literature are given by: (i) The proposal to use magnetic wedges to solve the problems provided by FSCW for SPM machines with radial flux for applications with high torque and low speed; and (ii) the evaluation of an innovative unconventional magnetic wedge that allows optimization of the overall performance of this type of machine.

2. Design of a FSCW-SPM Synchronous Machine

As explained in the introduction, the FSCW can provide advantages to PMSMs, but its design has to take into account several requirements to be optimized for the specific application. Multiple combinations are possible to obtain a FSCW, but only some are convenient in order to have:

1. High winding factor for the fundamental component of the magnetomotive force (m.m.f.), which maximizes the torque;
2. Low winding factor for the other harmonics of the m.m.f., which reduces the torque ripple;
3. Reduced cogging torque;
4. Null unbalance in the electromagnetic force.

The first requirement is related to the maximization of the torque, but it may be in contrast with the second one, which, on the other hand, is related to the reduction of the losses in the rotor iron and magnets, due to the presence of harmonics in the m.m.f. It has been demonstrated in [2,3] that a high winding factor for the fundamental component ($k_{w1} \geq 0.866$) needs a q number of slots per pole and per phase comprises between 0.25 and 0.5:

$$0.25 \leq q = Q/(3 \cdot p) \leq 0.5, \tag{1}$$

where Q is the total number of stator slots, p is the number of poles, and 3 is the number of phases. A good choice may be, for example, $p = 10$, $Q = 12$, and $q = 2/5$, which determines $k_{w1} = 0.933$. The same value of k_{w1} is obtained when $Q = 36$ and $p = 30$, because the q number is equal [2]. These two choices allow a cogging torque to also be obtained with a short period, which is, as a consequence, negligible.

In general, the period, τ, of the cogging torque is given by the following formula:

$$\tau = 360°/\text{LCM } (Q, p), \tag{2}$$

where LCM is the least common multiple. This formula provides a low value when the p number of poles and the Q number of slots are similar, as in the case of a FSCW, contrary to a distributed winding. In Figure 2, a comparison between the cogging torque for two small machines (1.4 kW) with the same number of slots and different number of poles is reported. The simulations are realized by Motor-CAD software, by maintaining the same dimensions of the machine. It is possible to observe that the peak value of the cogging torque is reduced by 75% in the case of a higher number of poles, even with a relatively small number of slots.

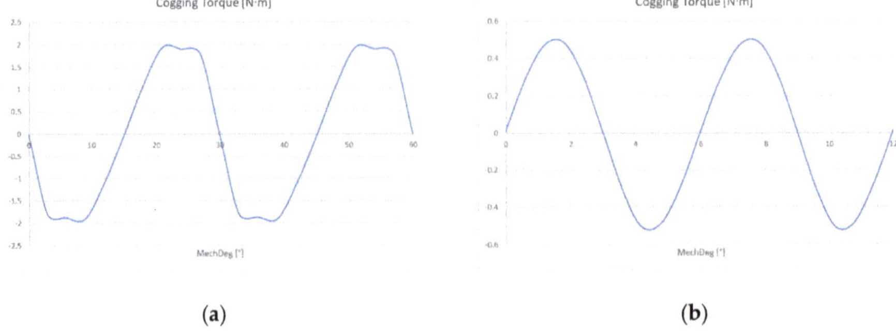

(a)   (b)

Figure 2. Cogging torque for a 1.4 kW machine with: (a) $Q = 12$, $p = 4$, $\tau = 30°$; (b) $Q = 12$, $p = 10$, $\tau = 6°$.

Considering again the winding factor, an important difference is given by the single- or double-layer winding. In the same conditions of the number of slots and poles, a single-layer winding presents a winding factor of the fundamental higher than a double-layer winding, which implies a higher value of constant torque but also a higher value of the total harmonic distortion (THD) of the m.m.f.; moreover, the harmonic content of the flux density and the back electromotive force (BEMF) is also higher. As a consequence, a single-layer winding produces a higher torque ripple and higher losses in the magnets and iron. For this reason, normally, the double layer is preferred (see an example in Table 1).

Table 1. Comparison between the single and double layer for a 1.4 kW machine with $Q = 12$, $p = 10$.

	THD BEMF [%]	T_{ripple} [%]	k_{w1}
Single layer	9.16	10.54	0.966
Double layer	6.45	8.87	0.933

It is important to remember that a FSCW determines not only the harmonics multiple of the fundamental, as in the case of distributed winding, but also sub-harmonics, which can have a noticeable amplitude, sometimes even higher than the fundamental, especially in the case of single layer. Note also that more than two layers can be realized, even if with a more complex structure.

The fourth requirement, i.e., a null unbalance in the electromagnetic force, can be achieved by the presence of a "symmetry" in the winding. For example, by considering two double-layer windings, the first one with $Q = 9$ and $p = 10$, and the second one with $Q = 36$ and $p = 40$, we can observe that the latter presents a lower torque ripple and a null unbalance in the electromagnetic force, contrarily to the first one. In Figure 3, the flux density distribution at no-load for the same machine with two different windings is shown. Note that the range of the flux density is comprised between almost zero (blue color) and 1.7 T (red color).

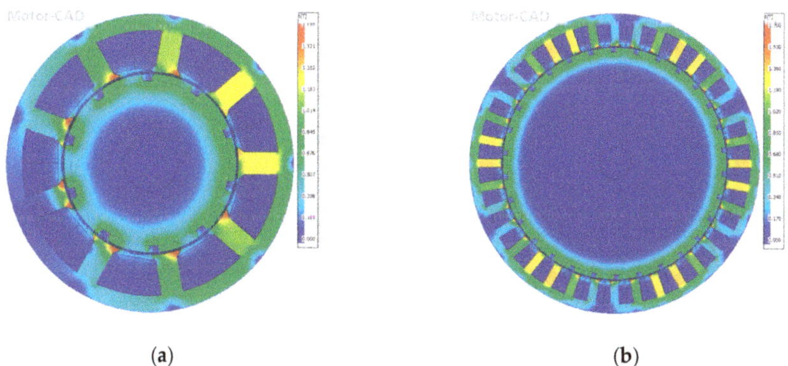

(a)   (b)

Figure 3. Flux density distribution for a machine with: (a) $Q = 9$, $p = 10$; (b) $Q = 36$, $p = 40$.

Since the machine with 9 slots and 10 poles does not present any symmetry in the flux density distribution, a resultant unbalance force is generated, despite no eccentricity being present (see Figure 4). In general, this unbalance appears when $p = Q \pm 1$ [3].

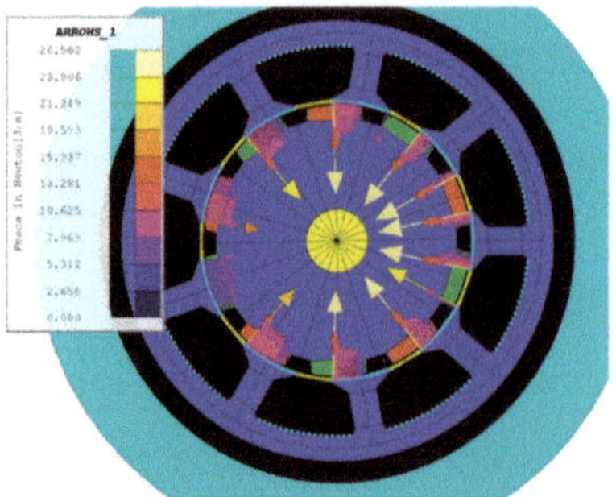

Figure 4. Magnetic force distribution of the for a machine with $Q = 9$, $p = 10$.

3. Power Losses in FSCW SPM Synchronous Machines

3.1. Joule Losses

Stator Joule losses cover most parts of the power losses in a PMSM supplied at low frequency (less or equal to the industrial frequency). At the same conditions of current density, J, these losses increase with the total volume of the conductor material. From this point of view, FSCW gives two advantages: On the one hand, the slot fill factor can reach values until 60%, much higher than a traditional distributed winding. In this way, the section of the conductor can be larger and therefore the current can be higher, by maintaining the same value of J, thus obtaining higher power at parity of dimensions or a more compact machine for the same power. On the other hand, an FSCW presents shorter end-windings with respect to a distributed winding, with consequent lower Joule losses. This effect becomes particularly relevant in machines with an active length that is relatively short with respect to the diameter at the airgap, as in the case of machines with a large number of poles. The best solutions is given by a two-layer winding, with a vertical division in the slot, which allows the minimum end-winding volume to be obtained with respect to a single or a double layer with horizontal separation [4].

3.2. Iron Losses

Stator iron losses can be reduced with standard methods employed for all types of AC electrical machines, e.g., by using steel sheets with low specific iron loss, especially in the case of machines supplied at high frequency.

Rotor iron losses are only due to the harmonics of a rotating magnetic field, and for this reason, they are much lower than stator iron losses. In particular, sub-harmonics due to the FSCW, not present in the case of distributed winding, cause the main part of these losses. In [5], it was demonstrated that these losses are higher in the case of single layer with respect to double or quadruple layer. These two latter windings present very similar rotor iron losses, which, in general, increase with the rotor speed. Moreover, [5] puts in evidence that the flux generated by the permanent magnets (PMs) mitigate the effect of the harmonics produced by the FSCW.

3.3. Losses in the Magnets

Losses in the magnets, which can be named also as "magnet losses", are mainly due to the eddy currents arousing from the rotating magnetic field and therefore they will be higher in the case of FSCW, especially with a single layer, with respect to distributed winding. Beside the issue due to this additional component in the losses, there is the problem given by the overheating of the magnets, which increases the possibility of demagnetization. This phenomenon has to be absolutely avoided, because it can greatly compromise the operation and the performance of the machine. The methods that can be applied to reduce these losses comprise the segmentation of the magnets, axially or along the circumference; the higher the segmentation, the lower the eddy currents. This benefit is more relevant in the case of segmentation parallel to the axis of the rotor, as in Figure 5a, even if it leads to greater manufacturing difficulties. On the contrary, a segmentation parallel to the circumference of the rotor, as in Figure 5b, becomes necessary in the case of a large power machine with a long rotor, due to the problem of assembling long magnets. In Figure 6, the results of a simulation with Motor-CAD are shown, highlighting the increase of magnet losses with the square of the rotor speed, for three different configurations of magnets.

(a)   (b)

Figure 5. Magnet segmentation: (**a**) Segmentation A, parallel to the axis (two magnets per pole); (**b**) Segmentation B, parallel to the circumference (three magnets per pole).

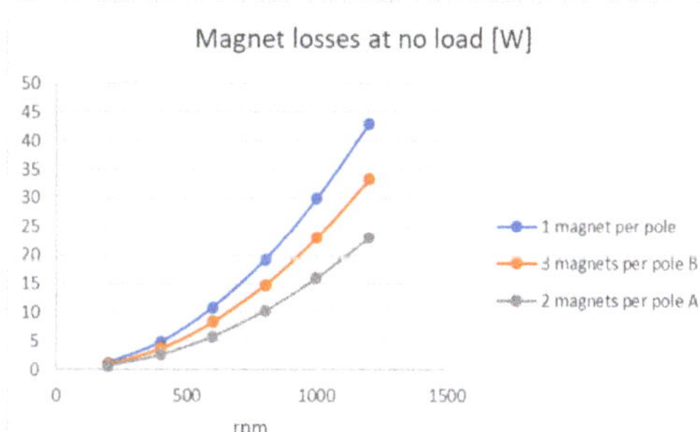

Figure 6. Magnet losses under no load for a 1.4 kW machine, $Q = 12$, $p = 10$ (simulation results).

Nevertheless, in general, segmentation can cause problems of mechanical strength, due to the repulsive forces among the magnets composing the same pole, because of their same polarity and short gap between them; the repulsive forces depend on the materials magnetic properties and are proportional to H^2, where H is the magnetic field strength (A/m).

Another solution to reduce the magnet losses consists in the use of magnetic wedges, instead of the traditional non-magnetic ones, to close the stator slots, which, in the case of FSCW, are generally open and therefore worsen the harmonic content of the flux density at the airgap. In the following paragraph, the state of the art related to the use of this type of wedge is presented.

4. Application of Magnetic Wedges in SPM Synchronous Machines

Based on the literature review reported in the introduction, the authors decided to study the possible benefits of using magnetic wedges in a different type of machine, such as an SPM motor with FSCW, with particular attention to low speed and high torque applications. The contents of this section do not aim to optimize the design of the magnetic wedges but to examine the effects of these wedges, with different values of magnetic permeability, on the behavior and performances of this kind of machine, with particular reference to the flux distribution, BEMF, and inductances.

4.1. Characteristics of Magnetic Wedges

Typically, a magnetic wedge is composed by 70% iron powder, 20% epoxy resin, and 10% glass material, and its relative permeability is higher than one [17]. For each machine, an optimal value of relative permeability may be found to maximize its performance. Usually, this value is in the order of the magnitude of some units. This kind of wedge is defined as semi-magnetic ($\mu_r < 10$). Wedges with $\mu_r > 10$, even until some hundreds, are known as soft magnetic composite [29]. The exact composition of the magnetic wedges is often covered by trade secrecy; however, it is possible to assume that a different concentration of the ferromagnetic powder corresponds to a different relative permeability.

In the design phase of an electrical machine, an analytical calculation is not sufficient to evaluate the effects of precise values of the relative permeability and thickness of a magnetic wedge, but FE analysis is necessary. As a first approximation, the relative permeability can be assumed as constant, in order to identify its optimal value for the specific application; then, the wedge with the most suitable characteristics may be chosen, taking into account that its relative permeability tends to decrease beyond a certain value of flux density, due to the saturation.

4.2. Effects of the Magnetic Wedges on the Flux Density and BEMF

In the following, some results from FE simulations are reported. They were carried out with Flux 2D and Motor-CAD (v11, Motor Design Ltd., UK) software related to an SPM motor with 12 slots, 10 poles, 130 mm external stator diameter, 80 mm internal stator diameter, 100 mm stator pack length, and rated power of 1.4 kW at 100 Hz. Figure 7a shows the average absolute value of the flux density under no load in a stator tooth with a thickness of 7.5 mm, as function of its relative position with respect to the rotor. As for the flux density at the airgap, its rms value is higher when a wedge with a larger permeability is used. However, after further increasing the relative permeability above 100, there are no longer any improvements in the trend of the flux density, which tends to saturate. In this case, the average flux density does not exceed 1.39 T for $\mu_r \geq 100$, so higher permeability wedges are useless. For this reason, only curves with μ_r until 100 are reported in Figure 7.

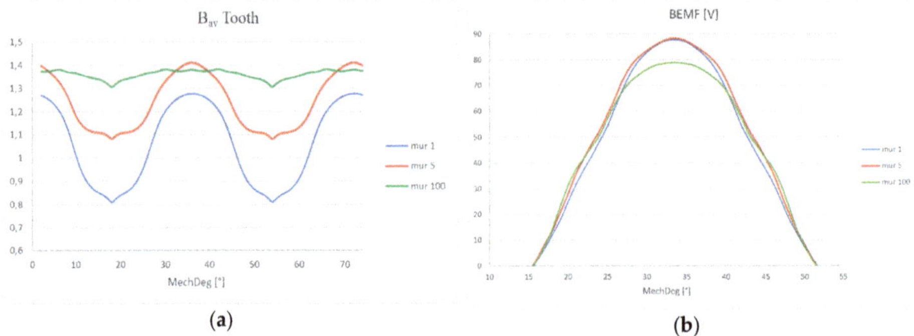

Figure 7. FE simulation results under no load: (**a**) mean values of flux density in the tooth; (**b**) BEMF in one phase.

Apparently, wedges with $\mu_r = 100$ seem to guarantee higher flux linkage and BEMF, due to the higher average flux density in the tooth. However, this higher average flux density in the tooth is due to the large values of the flux density reached in the polar shoe. From here, a part of the flux lines is deviated by the wedge with higher permeability, increasing the leakage flux. If the polar shoe is neglected, the average flux density in the top of the tooth is slightly higher with the wedge with $\mu_r = 5$. This result confirms the higher BEMF obtained with $\mu_r = 5$, as shown in Figure 7b.

In Figure 8, the flux density within the wedges is shown, in case of $\mu_r = 5$ (a) and $\mu_r = 100$ (b). As it can be observed, by increasing the relative permeability, the flux density increases, as the tendency to saturation. Note that the color scale in the two cases does not reflect exactly the same values of the flux density. For $\mu_r = 5$ (a), the maximum value is 2.282 T, and for $\mu_r = 100$ (b), it is 2.486 T. In Table 2, a short summary of the simulation findings obtained with FLUX 2D is reported. As a conclusion of this first investigation, we may state that it is possible to identify specific values of the relative permeability, according to the characteristics of the machine, which slightly increase the effective value of both the flux linkage, ψ, and BEMF. Moreover, these values allow a reduction of the harmonic content of the waveforms of these electromagnetic quantities. In general, it is difficult to optimize all the parameters of the machine, but it is possible to determine a valid compromise: In the considered case, $\mu_r = 5$ or $\mu_r = 20$.

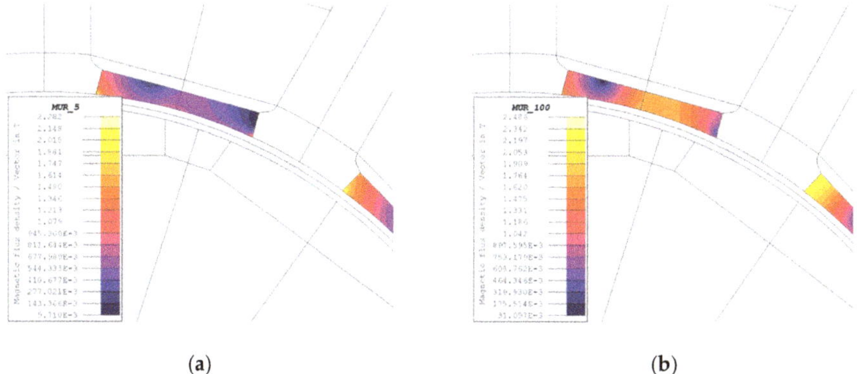

(a)   (b)

Figure 8. Flux density in the magnetic wedges under no load with: (**a**) $\mu_r = 5$; (**b**) $\mu_r = 100$.

Table 2. BEMF (rms value, peak value, and THD) and leakage flux (rms value) of one phase, as a function of the magnetic permeability of the wedges.

	$\mu_r = 1$	$\mu_r = 5$	$\mu_r = 20$	$\mu_r = 100$	Unit
BEMF_rms	57.89	60	58.93	55.87	V
BEMF_peak	87.71	88.35	84	78.85	V
ψ_rms	91	95	94	89	mWb
THD_BEMF	9.66	6.64	4.13	3.78	%

4.3. Effects of the Magnetic Wedges on the Inductances

The auto-inductance of a single phase, L_{ph}, can be defined as:

$$L_{ph} = L_a + L_\sigma, \tag{3}$$

where L_a is the magnetizing inductance, related to the flux linkage with the rotor, which crosses the airgap and produces the electromechanical conversion, while L_σ is the leakage inductance, related to the leakage flux, which has a path that does not pass through the airgap, but, for example, from a tooth to another one through a magnetic wedge. Providing the machine with paths with higher relative permeability can increase both L_a and L_σ, and so L_{ph}.

In the conventional reference system used in this paper, the d axis is parallel to the flux coming from one pole of the machine and the i_d current can be used to regulate the effective flux linkage, defined as:

$$\varphi = \varphi_m - L_d\, i_d, \tag{4}$$

where φ_m is the flux linkage due to the magnets (which would be equal to ψ in the rated condition) and L_d is the d axis inductance. In order to reduce φ, i_d has to be increased. With a higher L_d, the same flux linkage reduction can be obtained with a lower i_d, so a higher L_d allows a better flux weakening capability. The SPM machine is ideally isotropic, so the saliency ($L_q - L_d$) is very small and nearly zero when saturation is low. The magnetic wedges also change the q axis inductance, L_q; however, only L_d is involved in the flux weakening capability. So, a little change in the saliency will not significantly affect the flux weakening capability, which, for this type of machine, is highly linked only to L_d. The torque formula adopted here is:

$$T = 3/4 \, p \, (\varphi - (L_d - L_q) \cdot I_d) \cdot I_q. \tag{5}$$

The simulations show an overall improvement in the auto-inductance, due to both greater values of leakage inductance, L_σ, and magnetizing inductance, L_a, the latter caused by a reduced reluctance of the magnetic circuit. The result is an increase in the inductance along the direct axis, L_d, which allows a better flux weakening capability to be obtained. Moreover, higher values of L_d give rise to lower short circuit currents.

By increasing the permeability of the wedges, the inductance increases and therefore even the voltage drops. Consequently, the maximum current that can flow in the stator windings decreases, taking into account the voltage limit imposed by the converter. The torque is directly proportional to i_q. If a significant decrease in the maximum torque is not desired, wedges with a relative permeability of a few units should be used.

4.4. Effects of the Saturation on the Magnetic Wedges

The simulations carried out in this section have taken into account the saturation phenomenon on the magnetic wedge, even if it may tend to be established earlier, with lower values of flux density. A wedge with more pronounced saturation allows a lower improvement in the efficiency compared to another with less tendency to saturation. Moreover, the values of the inductance along the direct axis tend to be slightly lower, with consequently lower benefits in the weakening operation.

Regarding the power losses in the magnets, it can be proven that any phenomenon that tends to increase the saturation of the wedges has a negative impact on these losses. Wedges with a higher tendency to saturation are therefore more sensitive to load variations. Wedges with a high relative permeability reach saturation faster and consequently they are not very suitable for applications with rather high currents. In such cases, it is therefore preferable to use semi-magnetic wedges.

Therefore, on the basis of the results presented in this paragraph, for the case study, we chose to model magnetic wedges with a more marked saturation.

5. Case Study

The main motor data are shown in Table 3 and three images of it in Figure 9; other details are specified in [31]. The magnets are made of NdFeB and the stator is water cooled. The curves μ_r–B and B–H of the magnetic wedge were extrapolated from the technical sheet provided by a supplier and are reported in Figure 10. These magnetic wedges can be produced by sintering resin and iron powder.

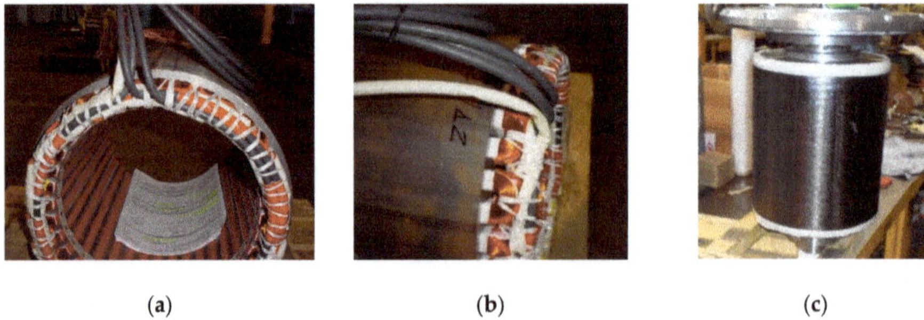

(a)   (b)   (c)

Figure 9. SPM motor: (**a**) wound stator; (**b**) stator end-windings; (**c**) carbon fiber covered rotor.

Table 3. Main data of the analyzed motor.

Parameter	Value	Unit of Measurement
Rated power	23.1	kW
Rated torque	1700	N·m
Rated speed	130	rpm
Rated frequency	32.5	Hz
Rated current	46.5	A
Rated voltage	381	V
Rated efficiency	87	%
Number of poles	30	-
Number of slots	36	-
Stator core length	300	mm
External stator diameter	359	mm
Internal stator diameter	270	mm
Shaft diameter	150	mm
Magnet length	25	mm
Magnet width	6	mm

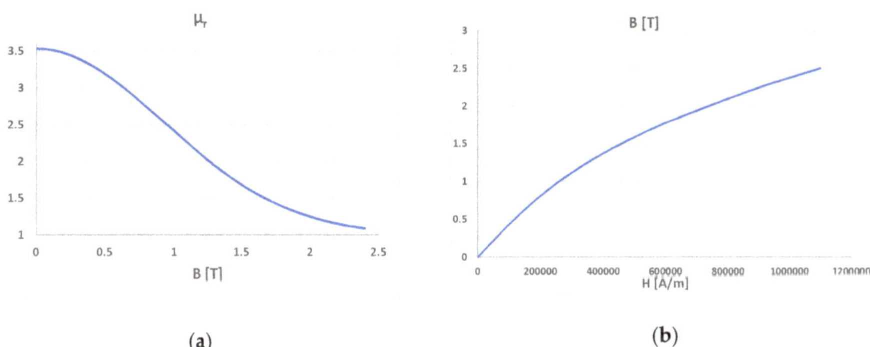

(a) (b)

Figure 10. Magnetic wedge characteristics: (**a**) relative permeability; (**b**) magnetization characteristics.

The behavior of the motor with different types of wedges was simulated using two FE software, Motor-CAD and FLUX 2D (2018 version, Altair Engineering Inc., USA). Initially, Motor-CAD was used because the original motor, with standard non-magnetic wedges, was modelled with this software and validated by the company with experimental measurements. Therefore, it is reasonable that the simulations implemented with this model, varying the characteristics of the wedges, are reliable. In addition, this software takes little time to get results from simulations. However, some calculations require FLUX 2D, in particular for the *spectrum analysis* function, which provides the harmonics of any quantity, e.g., the airgap flux density. The distribution of the fractional-slot concentrated winding of the motor was calculated automatically with Motor-CAD, as shown in Figure 11a. The magnets are not segmented radially, but axially, as displayed in Figure 11b: Each pole comprises a series of 12 magnets, in order to reduce losses in the magnets due to eddy currents and to facilitate the assembly.

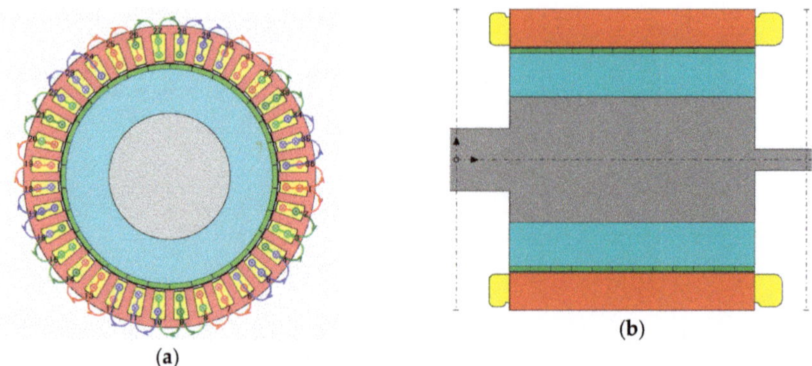

(a)   (b)

Figure 11. SPM motor: (**a**) fractional-slot concentrated winding; (**b**) axial section.

6. Behavior of SPM Synchronous Machines with Uniform Magnetic Wedges

The behavior of the machine with magnetic wedges, with the characteristics shown in Figure 10, was evaluated for different load percentages and compared with the same motor with conventional non-magnetic wedges.

6.1. Effect of Uniform Magnetic Wedges under No Load

Figure 12a shows the radial component of the airgap flux density under no load, both in the case of the traditional wedge and the magnetic wedge. It can be noted that the flux density is slightly less irregular in the case of a magnetic wedge and its rms value is greater than in the traditional wedge ($B_r = 0.762$ T instead of $B_r = 0.738$ T). The flux linkage also increases, as in Figure 12b, causing a slight increase of the rms value of BEMF (+1.34%) and the maximum value of the flux density in the tooth (from 1.726 to 1.759 T, increase of 1.88%.

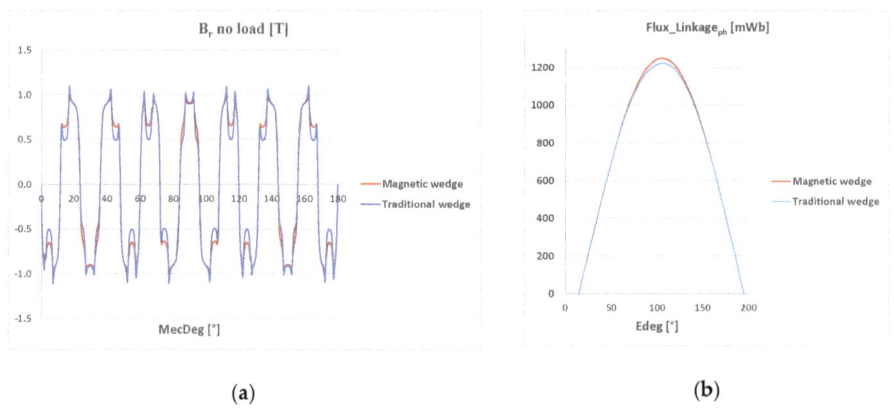

(a)   (b)

Figure 12. (**a**) Radial component of airgap flux density under no load; (**b**) Flux linkage with one phase.

The cogging torque period is 2° mechanical degrees and has not particularly high values, but the magnetic wedge permits to achieve a significant decrease, greater than 90%, even if its relative permeability is not too high, as in Figure 13a.

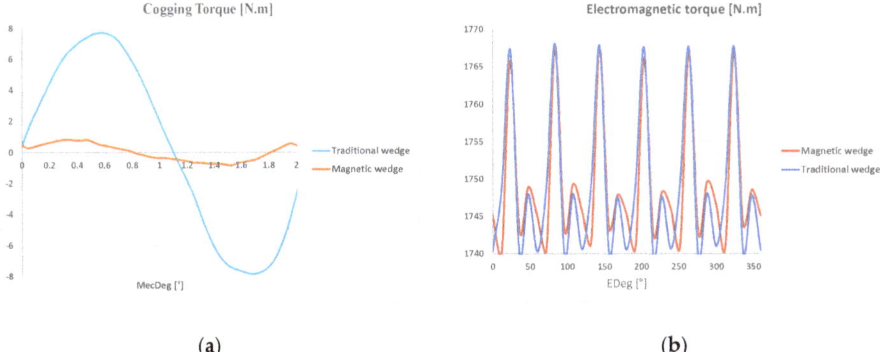

Figure 13. (**a**) Cogging torque under no load; (**b**) Torque at full load.

Without load, there is a net drop in magnet losses (−63%) and a growth in total iron losses (+7.4%); the latter is due to the fact that the rotor iron losses decrease, but their contribution is negligible compared to the stator iron losses, which increase, especially in the teeth, due to the increase in the flux density in this part of the machine. Note that the stator iron losses could further increase in the case of wedges with greater permeability. Overall, without load, total losses decrease slightly (−1.9%) and the performance of the machine is enhanced by the use of magnetic wedges, as the magnets are the most sensitive part of the machine and they are better protected with this solution. Nevertheless, to decide the convenience given by the use of magnetic wedges, it is also necessary to consider the behavior of the motor when loaded.

6.2. *Effect of Uniform Magnetic Wedges at Full Load*

In order to evaluate the effect of magnetic wedges on the behavior of the machine at full load, a comparison was made with traditional wedges on an equal value of the electromagnetic torque (1749 Nm), calculated by means of FE analysis with Motor-CAD, to correctly compare the losses. The average torque over an electrical period remains constant; however, in the two cases, there is a different ripple, as in Figure 13b. The magnetic wedge allows a slight reduction of 0.18%, from 1.89% to 1.71%. This result is obtained with a small increase of the line current (from 46.50 to 46.95 A), also due to a slight reduction of the power factor (from to 0.88 to 0.87).

At full load, the magnetic wedges provide an increase in self-inductance, due to the reduction in the reluctance of the magnetic circuit. Its average value rises by only 10.7%, both for the not very high permeability of the wedge and for the saturation present in some parts of the machine, as in Figure 14a. The mutual inductance also increases, by about 20%, as in Figure 14b, however, its absolute values remain low and do not cause any problem. Due to the saturation phenomena, the fully loaded machine behaves as not perfectly isotropic from the magnetic point of view. The inductance along the d axis increases as well and the flux linkage along the d axis decreases (Table 4).

Table 4. Mean values of inductances and flux linkage at full load.

Parameter	Traditional Wedge	Magnetic Wedge	Δ (%)
L_d	10.38 mH	11.82 mH	13.87
L	20.65 mH	22.86 mH	10.7
|M|	1.42 mH	1.70 mH	19.7
ψ_d	1131 mWb	1094 mWb	−3.3
ψ_q	705 mWb	876 mWb	11.6

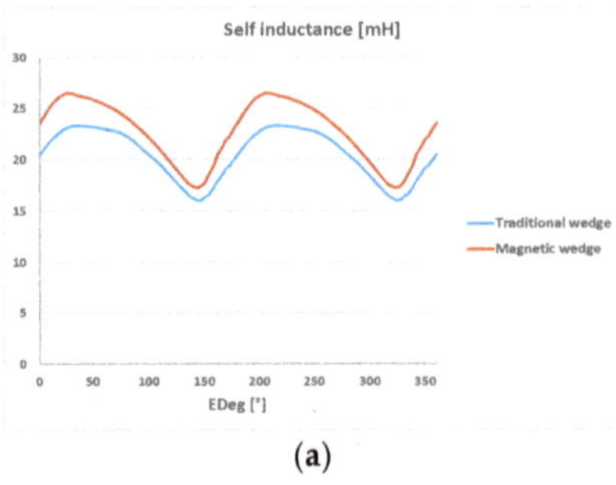
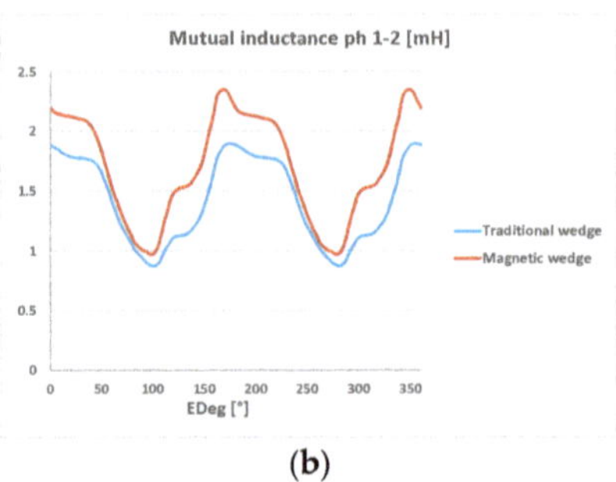

Figure 14. (**a**) Self-inductance of phase 1 at full load; (**b**) Absolute value of mutual inductance between phase 1 and 2 at full load.

As for the losses, since the current necessary to have the same torque is greater with the magnetic wedges, the Joule losses increase slightly (+1.9%). The rotor iron losses decrease, but the stator iron losses increase, and the total iron losses also increase. On the contrary, the losses in the magnets decrease (−29.4%), but the total losses slightly increase (+1.6%). Again, the magnets are better protected, but, overall, the use of magnetic wedges does not seem convenient at full load, especially considering that these wedges are more expensive and more prone to breaking than traditional ones. The reduced advantage at full load is given by the higher saturation induced by the higher currents, which decreases the positive effect of the magnetic wedges, which, in general, tend to accentuate the saturation locally.

6.3. Effect of Uniform Magnetic Wedges at Intermediate Load and in Field Weakening Operation

With intermediate load percentages, magnetic wedges offer greater advantages, since the losses are proportionally similar to those in the absence of load. The behavior of the machine was also assessed in field weakening area, for currents equal to the nominal one. As displayed in Figure 15a, at 500 rpm, the torque is zero with traditional wedges while with magnetic wedges, the machine is still able to provide a torque equal to 0.26 p.u. (456 N·m). This difference not negligible. Figure 15b underlines that, although at the rated speed the performance of the machine is better with the standard wedges, its behavior changes for a speed greater than 230 rpm. The magnetic wedges provide an efficiency greater than 80% even at 600 rpm while with traditional wedges, the efficiency drops below this threshold, starting from 440 rpm. It is clear that magnetic wedges offer greater flexibility in these working conditions.

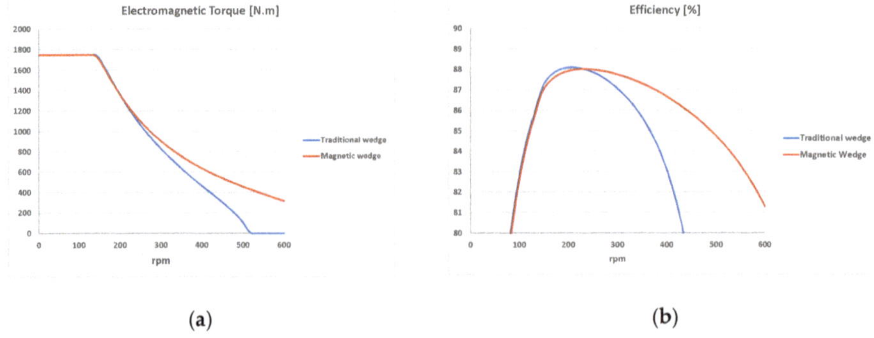

Figure 15. (**a**) Electromagnetic torque at variable speed; (**b**) Efficiency at variable speed.

6.4. Effect of Uniform Magnetic Wedges on the Radial Forces

Finally, the radial forces acting on the internal stator surface were evaluated. The idea of this evaluation comes from the evidence that magnetic wedges are more brittle and hence more susceptible to failure compared to non-magnetic ones, as highlighted in [32]. An interesting analysis referring to a large synchronous wound field machine of 5 MW was reported in [33], where the results showed that the shape of the magnetic wedge also has significant effect on the stator interior surface force distribution.

From our analysis, with magnetic wedges, these forces are distributed more evenly and their maximum value decreases (−15.5%), while their average value is higher than the standard wedges (+26%). Therefore, the stress on the wedges increases with their permeability; nevertheless, the values achieved do not compromise their integrity, as shown in Figure 16a.

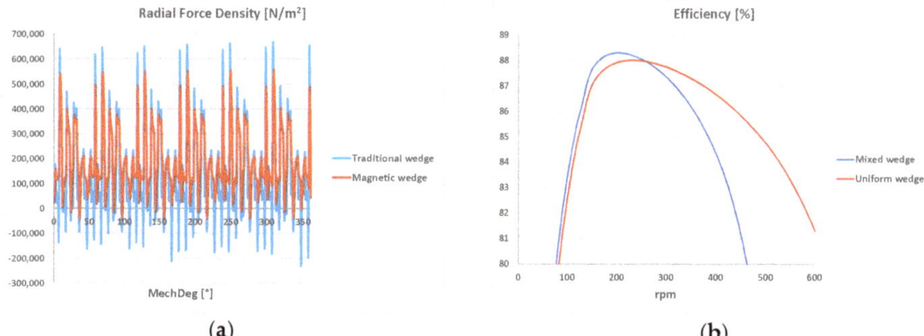

Figure 16. (a) Radial component of the radial force density acting on the internal stator surface; (b) Electromagnetic torque at a variable speed.

7. Optimization of a Non Conventional Magnetic Wedge

The simulations reported in the previous paragraph permitted an estimation of the effects of commercial magnetic wedges with uniform permeability on the performance of an FSCW-SPM motor. It is clear that an increase in μ_r reduces the magnet losses but increases the leakage flux lines throughout the wedges. The consequence of these wedges is similar to that produced by a more or less accentuated closing of the slot. For this reason and on the basis of previous works in the literature [26–28], a different distribution of the relative permeability within the wedge was investigated, in order to imitate the behavior of a semi-closed slot and optimize the full load operation of the motor. The proposed wedge is divided in five portions of materials with three different values of μ_r: The central part with $\mu_r = 1$, the ends with high permeability, and two intermediate portions with μ_r in the order of some units. In general, simulations have shown that an improvement in efficiency can be achieved with a permeability that increases from the center to the ends of the wedge. Many combinations of geometry and permeability were assessed, in particular by varying the length and permeability of the intermediate portions. The slot opening of the original motor is 10.8 mm wide and the best configuration found after many simulations has the following parameters: $\mu_r = 1$ (central part), $\mu_r = 4$ (intermediate parts), $\mu_r = 100$ (ends). It was confirmed that values of $\mu_r > 100$ in the ends of the wedge do not arouse any significant improvement in efficiency. It has to be noted that the configuration of this unconventional wedge differs from that proposed in [26–28], which has two zones with different permeability instead of three.

7.1. Effect of Non-Conventional Magnetic Wedges at Full Load

The comparison of the performance of the motor at full load, with a uniform and non-uniform (mixed) magnetic wedge, is reported in Table 5, together with the maximum values of the flux density in different parts of the machine. With the mixed wedge, the induction in the teeth is slightly higher (excluding the polar shoe). This also leads to a slight increase in iron losses, mainly located in the stator teeth.

Table 5. Power losses, efficiency, and flux densities of the motor at full load.

Parameter	Uniform Wedge	Mixed Wedge	Δ (%)
P_{Cu}	3512 W	3286 W	−6.4
P_{mag}	56.62 W	55.7 W	−1.6
P_{Festat}	354.2 W	361.1 W	1.9
P_{Ferot}	1.86 W	1.87 W	0.5
P_{tot}	3924.68 W	3704.67 W	−5.6
Efficiency	85.63%	86.32%	0.8%
Airgap flux density	1.18 T	1.13 T	4.4%
Flux density in the tooth (no polar shoe)	1.75 T	1.77 T	−1.1%
Flux density in the polar shoe of the tooth	2.45 T	2.43 T	0.8%
Flux density in the stator yoke	1.25 T	1.23 T	1.6%

The non-uniform magnetic wedge has a lower L_d than the uniform magnetic wedge, in particular it has a lower leakage inductance, L_σ, so lower voltage drops and higher flux linkage, while it produces similar magnet losses at full load. Therefore, it can provide the same torque with a lower current. By considering Figure 15b and Figure 16b, it is possible to note that the magnetic wedge with uniform magnetic permeability does not increase the efficiency for this machine at full load with respect to the traditional wedge ($\mu_r = 1$), due to the high saturation. Conversely, the non-uniform magnetic wedge is able to do so. As illustrated in Table 5, the main advantage is given by the lower losses in the magnets.

7.2. Effect of Non-Conventional Magnetic Wedges in Field Weakening Operation

As for the field weakening operation, the trend with the mixed wedge is very similar to that achieved with a traditional wedge. The improvement over the latter is in fact minimal. Hence, if speeds above 2 p.u are desired, it is more convenient to use the uniform magnetic wedge, which consequently cannot be completely replaced by that with mixed permeability. The main reason is given by the different inductances in the two cases: The greater inductance due to the uniform wedge slightly worsens the power factor, but it permits a more flexible operation in terms of speed.

The efficiency reachable with mixed wedge, displayed in Figure 16b, is higher up to 260 rpm, i.e., up to a rotational speed of 2 p.u. Conversely, starting from 470 rpm, it drops below 80%. For applications up to 260 rpm, this solution is certainly more suitable, but it can also allow higher speeds to be reached, within the limits just mentioned.

7.3. Effect of Non-Conventional Magnetic Wedges at Low Load

The impact of this type of wedge is different for lower currents. In Table 6, it is possible to observe how the efficiency improvement tends to decrease and to become even slightly lower than the application with a traditional wedge, or a uniform magnetic one, with a load equal to one-quarter of the rated one.

Table 6. Efficiencies, Joule losses, iron losses, and magnets losses at low loads.

Parameter	Uniform Wedge	Mixed Wedge	Δ (%)
Efficiency @ $P_n/2$	91.18%	91.04%	−0.2%
Efficiency @ $P_n/4$	91.77%	92.02%	0.3%
Stator Joule losses @ $P_n/2$	855.2 W	817.3 W	−4.4
Stator Joule losses @ $P_n/4$	212.5 W	203.4 W	−4.3
Total iron losses @ $P_n/2$	296.55 W	314.18 W	5.9
Total iron losses @ $P_n/4$	281.07 W	304.52 W	8.3
Magnets losses @ $P_n/2$	24.88 W	25.15 W	1.1
Magnets losses @ $P_n/4$	16.82 W	17.44 W	3.7

The half load efficiency is still slightly higher, mainly thanks to the lower Joule losses. Hence, the main benefit of the mixed wedge lies in the reduction of the current necessary to supply the same electromagnetic torque; moreover, it guarantees greater protection of the magnets. Its use would therefore be particularly appropriate for nominal loads but also for half loads. On the contrary, it is counterproductive for low currents. Table 6 shows a summary of the losses. Joule losses are always lower with the mixed wedge, but when currents decrease, their impact on performance decreases, while that due to the losses in iron and magnets increases. The losses in iron are the highest with this type of wedge. For this reason, its use for low loads is not opportune.

By reducing the currents, their influence on the losses due to eddy currents induced in the magnets also decreases, then the wedge with a uniform relative permeability becomes better performing, although slightly.

8. Discussion

The aim of this research was to investigate the possibility to mitigate the main drawbacks given by the FSCW for SPM machines, with a more general manufacturing solution with respect to the methods based on the design of special windings, which impose many constraints in the project and realization of the machine. This solution is represented by the magnetic wedges, which have been considered in the literature for various types of rotating electrical machines, but not for SPM machines with radial flux, high torque, and low speed.

The investigation mainly considered the impact of wedges with various values of magnetic permeability and also a novel wedge composed of multiple materials with different magnetic permeability. Even the thickness and the shape of the wedge could be taken into account to find the best solution, but they are constrained by the shape and dimensions of the stator slots, which, on the other hand, are bound by the electromagnetic design of the machine. So, a study on the impact of the thickness and shape of the wedge would be limited to a specific machine, whereas the study on the impact of the relative permeability is more general. For these reasons, in this paper, the optimization of the wedge was mainly based on the distribution of the relative permeability.

9. Conclusions

The application of two different types of magnetic wedge in SPM synchronous motors with fractional-slot concentrated winding was studied in this paper, with the aim of overcoming the drawbacks deriving from this type of winding and improving the performance of this machine. Magnetic wedges were shown to reduce the harmonic content of the airgap flux density, with positive effects on magnet losses, and can achieve the same torque with lower currents, as well as a reduction in its ripple. Nevertheless, the degree of these improvements depends quite a lot on the type of load. The high currents also cause a greater saturation of the magnetic wedges with a resulting lower reduction of the losses in the magnets.

The uniform magnetic wedge is more appropriate in the case of field weakening operation and at low loads, where its saturation is limited, while the unconventional wedge lets a considerable decrease in the stator Joule losses, since the same torque is attained with a lower current, as well as a reduction in magnet losses. Conversely, this wedge does not seem to guarantee efficiency improvements at low loads.

In conclusion, the innovative magnetic wedge presented here combines the positive aspects of a traditional wedge with those of a uniform magnetic wedge, even if the absolute best magnetic wedge configuration cannot be defined, since the benefits depend on the specific application.

Author Contributions: Methodology, L.F. and M.P.; software, validation, formal analysis, investigation, M.P.; writing—original draft preparation, review and editing, supervision, L.F. All authors have read and agreed to the published version of the manuscript.

Funding: This research received no external funding.

Acknowledgments: The authors acknowledge: Gianpietro Pacinotti of COMER S.r.l., for his fundamental contribution related to experimental data and measurements for the validation of the model and for his proposal to investigate the effects due to the magnetic wedge on the performance of the machine; Alessandro Tassi of SPIN S.r.l. for the software technical support.

Conflicts of Interest: The authors declare no conflict of interest.

References

1. EL-Refaie, A.M. Fractional-slot concentrated-windings synchronous permanent magnet machines: Opportunities and challenges. *IEEE Trans. Ind. Electron.* **2010**, *57*, 107–121. [CrossRef]
2. Meyer, F. Permanent Magnet Synchronous Machines with Non-Overlapping Concentrated Windings for Low-Speed Direct-Drive Applications. Ph.D. Thesis, Royal Institute of Technology, Stockholm, Sweden, 2008.
3. Martinez, D. Design of a Permanent-Magnet Synchronous Machine with Non-Overlapping Concentrated Windings for the Shell Eco Marathon Urban Prototype. Master's Thesis, Royal Institute of Technology, Stockholm, Sweden, 2012.
4. Salminen, P. Fractional Slot Permanent Magnet Synchronous Motors for Low Speed Applications. Ph.D. Thesis, Lappeenranta University of Technology, Lappeenranta, Finland, 2004.
5. Alberti, L.; Bianchi, N. Theory and design of fractional-slot multilayer windings. *IEEE Trans. Ind. Appl.* **2013**, *49*, 841–849. [CrossRef]
6. Chen, Q.; Liang, D.; Jia, S.; Ze, Q.; Liu, Y. Loss analysis and experiment of fractional-slot concentrated-winding axial flux PMSM for EV applications. In Proceedings of the 2018 IEEE ECCE, Portland, OR, USA, 23–27 September 2018.
7. Abdel-Khalik, A.S.; Massoud, A.; Ahmed, S. Nine-phase six-terminal pole-amplitude modulated induction motor for electric vehicle applications. *IET Electric Power Appl.* **2019**, *13*, 1696–1707. [CrossRef]
8. Sun, A.; Li, J.; Qu, R.; Li, D. Effect of multilayer windings on rotor losses of interior permanent magnet generator with fractional-slot concentrated-windings. *IEEE Trans. Magn.* **2014**, *50*. [CrossRef]
9. Islam, M.S.; Kabir, M.A.; Mikail, R.; Husain, I. Method to minimize space harmonics of fractional slot concentrated windings in AC machines. In Proceedings of the 2018 IEEE ECCE, Portland, OR, USA, 23–27 September 2018.
10. Tessarolo, A.; Ciriani, C.; Bortolozzi, M.; Mezzarobba, M.; Barbini, N. Investigation into multi-layer fractional-slot concentrated windings with unconventional slot-pole combinations. *IEEE Trans. Energy Convers.* **2019**, *34*, 1985–1996. [CrossRef]
11. Liu, G.; Zhai, F.; Chen, Q.; Xu, G. Torque pulsation reduction in fractional-slot concentrated-windings IPM motors by lowering sub-harmonics. *IEEE Trans. Energy Convers.* **2019**, *34*, 2084–2095. [CrossRef]
12. Asef, P.; Bargalló Perpiñà, R.; Lapthorn, A.C. Optimal pole number for magnetic noise reduction in variable-speed permanent magnet synchronous machines with fractional-slot concentrated windings. *IEEE Trans Transport. Electrific.* **2019**, *5*, 126–134. [CrossRef]
13. Chalmers, B.J.; Richardson, J. Performance of some magnetic slot wedges in an open-slot induction motor. *Proc. IEEE* **1967**, *114*, 258–260. [CrossRef]
14. Anazawa, Y.; Kaga, A.; Akagami, H.; Watabe, S.; Makino, M. Prevention of harmonic torques in squirrel cage induction motors by means of soft ferrite magnetic wedges. *IEEE Trans. Magn.* **1982**, *18*, 1550–1552. [CrossRef]
15. Takeda, Y.; Yagisawa, T.; Suyama, A.; Yamamoto, M. Application of magnetic wedges to large motors. *IEEE Trans. Magn.* **1984**, *20*, 1780–1782. [CrossRef]
16. Mikami, H.; Ide, K.; Arai, K.; Takahashi, M.; Kajiwara, K. Dynamic harmonic field analysis of a cage type induction motor when magnetic slot wedges are applied. *IEEE Trans. Energy Convers.* **1997**, *12*, 337–343. [CrossRef]
17. Gaerke, T.R.; Hernandez, D.C. The temperature impact of magnetic wedges on TEFC induction motors. In Proceedings of the 2012 Annual IEEE Pulp and Paper Industry Technical Conference, Portland, OR, USA, 17–21 June 2012.

18. Gyftakis, K.N.; Panagiotou, P.A.; Kappatou, J. The influence of semi-magnetic wedges on the electromagnetic variables and the harmonic content in induction motors. In Proceedings of the 2012 XXth ICEM, Marseille, France, 2–5 September 2012.
19. Lavanya, M.; Selvakumar, P.; Vijayshankar, S.; Easwarlal, C. Performance analysis of three phase induction motor using different magnetic slot wedges. In Proceedings of the 2014 IEEE 2nd ICEES, Chennai, India, 7–9 January 2014.
20. Madescu, G.; Moţ, M.; Greconici, M.; Biriescu, M.; Vesa, D. Performances analysis of an induction motor with stator slot magnetic wedges. In Proceedings of the 2016 International Conference on Applied and Theoretical Electricity, Craiova, Romania, 6–8 October 2016.
21. Abdi, S.; Abdi, E.; McMahon, R. Numerical analysis of stator magnetic wedge effects on equivalent circuit parameters of brushless doubly fed machines. In Proceedings of the 2018 XIII ICEM, Alexandroupoli, Greece, 3–6 September 2018.
22. Donaghy-Spargo, C.; Spargo, A. Use of fractional-conductor windings and semi-magnetic slot wedges in synchronous machines. *J. Eng.* **2019**, *2019*, 4396–4400. [CrossRef]
23. Di Napoli, A.; Honorati, O.; Santini, E.; Solero, L. The use of soft magnetic materials for improving flux weakening capabilities of axial flux PM machines. In Proceedings of the 2000 IEEE Industry Applications Conference, Rome, Italy, 8–12 October 2000.
24. De Donato, G.; Giulii Capponi, F.; Caricchi, F. Influence of magnetic wedges on the no-load performance of axial flux permanent magnet machines. In Proceedings of the 2010 IEEE International Symposium on Industrial Electronics, Bari, Italy, 4–7 July 2010.
25. De Donato, G.; Giulii Capponi, F.; Caricchi, F. Influence of magnetic wedges on the load performance of axial flux permanent magnet machines. In Proceedings of the IECON 2010, Glendale, AZ, USA, 7–10 November 2010.
26. Tessarolo, A.; Luise, F.; Bortolozzi, M.; Mezzarobba, M. A new magnetic wedge design for enhancing the performance of open-slot electric machines. In Proceedings of the 2012 Electrical Systems for Aircraft, Railway and Ship Propulsion, Bologna, Italy, 16–18 October 2012.
27. Tessarolo, A.; Luise, F.; Mezzarobba, M.; Bortolozzi, M.; Branz, L. Special magnetic wedge design optimization with genetic algorithms for cogging torque reduction in permanent-magnet synchronous machines. In Proceedings of the 2012 Electrical Systems for Aircraft, Railway and Ship Propulsion, Bologna, Italy, 16–18 October 2012.
28. Tessarolo, A.; Branz, L.; Mezzarobba, M. Optimization of a SPM machine using a non-isotropic magnetic wedge with an analytical method for cogging torque estimation. In Proceedings of the 2016 XXII ICEM, Lausanne, Switzerland, 4–7 September 2016.
29. Lindh, P.M.; Pyrhönen, J.J.; Ponomarev, P.; Vinnikov, D. Influence of wedge material on losses of a traction motor with tooth-coil windings. In Proceedings of the IECON 2013, Vienna, Austria, 10–13 November 2013.
30. Belhadi, M.; Krebs, G.; Marchand, C.; Hannoun, H.; Mininger, X. Evaluation of a switched reluctance motor with magnetic slot wedges. In Proceedings of the 2014 ICEM, Berlin, Germany, 2–5 September 2014.
31. Frosini, L.; Pastura, M.; Pacinotti, G. Performance improvement of SPM synchronous machines with non-conventional stator slot magnetic wedges. In Proceedings of the EPE'19 ECCE Europe, Genova, Italy, 2–6 September 2019.
32. Wang Lee, K.; Hong, J.; Hyun, D.; Bin Lee, S.; Wiedenbrug, E.J.; Teska, M.; Lim, C. Detection of stator-slot magnetic wedge failures for induction motors without disassembly. *IEEE Trans. Ind. Appl.* **2014**, *50*, 2410–2419.
33. Chu, K.H.; Anand Prabhu, M.; Pou, J.; Ramakrishna, S.; Gupta, A.K. Analysis of local forces acting on stator teeth and magnetic wedges in large synchronous machines. In Proceedings of the 2018 ACEPT, Singapore, Singapore, 30 October–2 November 2018.

Supercapacitor Storage Sizing Analysis for a Series Hybrid Vehicle

**Massimiliano Passalacqua [1], Mauro Carpita [2], Serge Gavin [2], Mario Marchesoni [1,*],
Matteo Repetto [3], Luis Vaccaro [1] and Sébastien Wasterlain [2]**

[1] Department of Electrical, Electronic, Tlc Engineering and Naval Architecture (DITEN), University of Genoa, via all'Opera Pia 11a, 16145 Genova, Italy; massimiliano.passalacqua@edu.unige.it (M.P.); luis.vaccaro@unige.it (L.V.)

[2] Institute of Energy and Electrical Systems (IESE), University of Applied Sciences of Western Switzerland, route de Cheseaux 1, CH 1400 Yverdon-les-Bains, Switzerland; mauro.carpita@heig-vd.ch (M.C.); serge.gavin@heig-vd.ch (S.G.); sebastien.wasterlain@heig-vd.ch (S.W.)

[3] Department of Mechanical, Energy, Management and Transportation Engineering (DIME), University of Genova, via all'Opera Pia 15, 16145 Genova, Italy; salabi@unige.it

* Correspondence: marchesoni@unige.it

Received: 9 April 2019; Accepted: 5 May 2019; Published: 9 May 2019

Abstract: The increasing interest in Hybrid Electric Vehicles led to the study of new powertrain structures. In particular, it was demonstrated in the technical literature how series architecture can be more efficient, compared to parallel one, if supercapacitors are used as storage system. Since supercapacitors are characterized by high efficiency and high power density, but have low specific energy, storage sizing is a critical point with this technology. In this study, a detailed analysis on the effect of supercapacitor storage sizing on series architecture was carried out. In particular, in series architecture, supercapacitor storage sizing influences both engine number of starts and the energy that can be stored during regenerative braking. The first aspect affects the comfort, whereas the second aspect directly influences powertrain efficiency. Vehicle model and Energy Management System were studied and simulations were carried out for different storage energy, in order to define the optimal sizing.

Keywords: Hybrid Electric Vehicle (HEV); series architecture; supercapacitor; Energy Management System (EMS); storage sizing; energy efficiency

---

1. Introduction

Hybrid Electric Vehicles (HEVs) have experienced a great interest in the last decades thanks to increasing attention both in carbon emission and local pollutant reduction. On the one hand, HEVs, by increasing overall powertrain efficiency, can reduce CO_2 emissions; on the other hand, allowing electric traction in urban missions and smoothing Internal Combustion Engine (ICE) accelerations, they can reduce local pollutant emissions. In the technical literature, different hybrid topologies and many Energy Management Systems (EMS) were proposed [1,2] nevertheless hybrid architectures can be mainly divided in two categories: parallel and series hybrid vehicles. Regarding parallel architecture, this structure is mainly used on medium size cars and can be realized either with one electrical machine [3–7], two electrical machines and a planetary gearbox (also known as power split or series/parallel hybrid) [8–14] can be realized with a compound structured permanent-magnet motor [15]. As far as series architecture is concerned, this topology was mainly designed, until now, for urban bus applications [16–18]. However, innovations and developments in storage systems and power electronics may change this paradigm. Analyzing storage system innovations, the supercapacitor significantly increased their performance in terms of energy density [19,20], so that their use become possible with a proper sizing; indeed many studies focused on the combined use of supercapacitors and batteries as storage system in hybrid vehicle applications [21–25]. In addition to that, the availability of low-loss power electronics devices, such as Silicon Carbide (SiC) metal-oxide-semiconductor field-effect transistor (MOSFET) [26,27], allows to design high efficiency power converter. Storage systems and power converters efficiency highly influence series architecture fuel consumption. As a matter of fact, in [28], it was shown how parallel architecture fuel consumption is slightly affected by storage and converters efficiency; indeed, these components are mainly used to recover (and reuse) backward energy deriving from regenerative braking. On the contrary, in the series architecture, all the energy provided by the ICE incurs in a double conversion (i.e., from mechanical to electrical and from electrical to mechanical) and the instantaneous difference between the power provided by the ICE and the power required by the electric motor is balanced by the storage system. As a consequence, energy flows, both in power converters and storage system, are significantly higher compared to parallel architecture and the efficiency of these components highly influences powertrain fuel consumption.

Authors in [29] proposed a series architecture for medium size cars based on supercapacitor storage and they showed the benefit of this solution in terms of fuel consumption in comparison to parallel architecture. Although all the energy provided by the ICE is affected by the generator, generator inverter, motor inverter and motor losses, series architecture allows the ICE to work in optimal working condition (i.e., high efficiency) and, as a result, the overall powertrain efficiency is higher. In particular, they proposed an EMS to properly use the supercapacitor storage, which aims both at increasing the system efficiency and the comfort (i.e., reducing ICE number of starts and ICE provided power at low vehicle speed). That study was carried out hypothesizing a 465 Wh storage system.

However, a detailed analysis on storage optimal sizing has not yet been carried out and remains a topic of great interest. As a matter of fact, storage sizing is a current issue and was investigated in the technical literature for electric vehicle using a combined battery-supercapacitor storage [30] and for fuel cell hybrid vehicle, again using a combined battery-supercapacitor storage [31]. Nevertheless, since series hybrid architecture on a medium size car using only supercapacitor storage was proposed for the first time in [29], a study on storage sizing in this application is still missing. In this paper, the effect of storage sizing on hybrid powertrain will be shown. In particular, the impact on both powertrain efficiency and ICE management will be investigated. The study will be performed on a significant number of road missions, in order to consider a wide range of possible working conditions. Standard cycles will be taken into account: Highway Fuel Economy Test (HWFET), Urban Dynamometer Driving Schedule (UDDS) and the Supplemental Federal Test Procedures (SFTP or US06) [32]; nevertheless, being these type approvals in plane missions, backward energy is limited and therefore they are not critical for storage sizing. For this reason, a large number of experimentally measured missions will be considered, including also mountain roads with a great change in altitude.

On balance, the main innovative contributions of this research can be pinpointed as follows: (1) analysis of supercapacitor storage sizing influences on ICE number of starts in a series hybrid vehicle, (2) supercapacitor storage sizing influences on fuel consumption, evaluated both with a spark-ignition engine and a diesel engine, and (3) feasibility of a series hybrid vehicle using only supercapacitor as a storage system.

The paper is structured as follows: in Section 2 series architecture is presented and EMS is described, focusing, in particular, on storage system parameters. Storage zones analysis and sizing optimization are shown in Section 3, whereas road missions are shown in Section 4. Simulation results are reported in Section 5. Finally, conclusions are pointed out in Section 6.

2. Series Architecture and Energy Management System (EMS)

Series architecture structure is depicted in Figure 1, where contour maps show powertrain components efficiency. The aim of the study was the analysis of storage sizing on powertrain efficiency, a quasi-stationary model was created in MATLAB/Simulink environment [29,33]. As a matter of fact, dynamic behavior of the different components is not relevant to energy evaluation. Each component was modeled, taking into account its efficiency contour map, as reported in Figure 1. Machine efficiency is given as a function of per-unit torque and per-unit speed, according to what was presented in [33]; in this way, the same contour map can be used also for different machine sizing. The same approach was used to model inverter losses, which are given in per-unit (i.e., referred to inverter nominal power). DC/DC converter efficiency is given as a function of storage voltage and current, supposing a 650 V DC-link. Vehicle longitudinal dynamic parameters (i.e., vehicle mass, rolling coefficient, aerodynamic drag coefficient and front section) are also reported in Figure 1. Differential gear efficiency is considered as a constant parameter. Powertrain fuel consumption is evaluated both using a spark ignition engine and a diesel engine. The aim of the study is to properly size the supercapacitor storage (red circle in Figure 1).

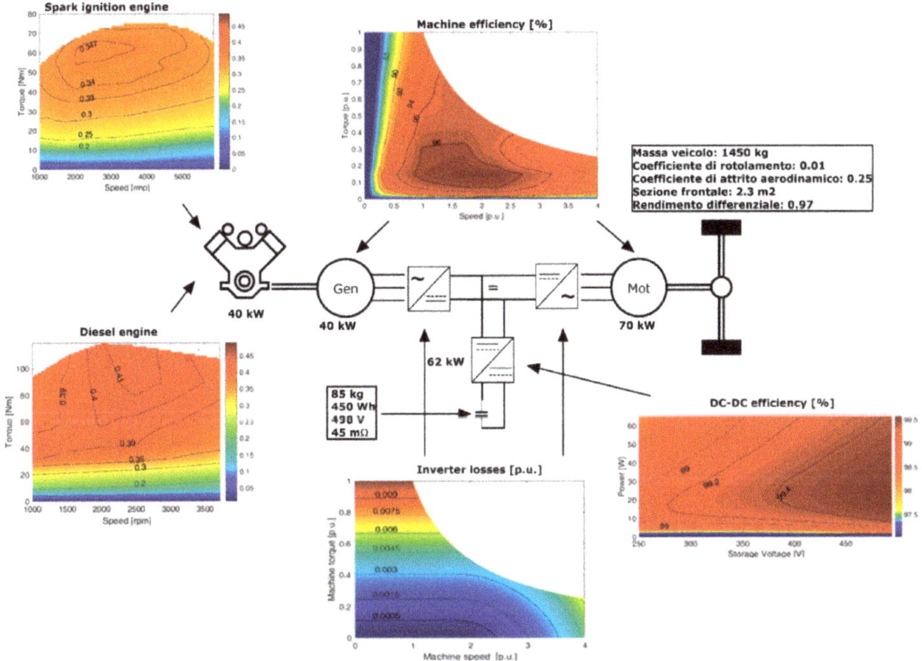

Figure 1. Series architecture structure. Contour maps show components efficiency.

Supercapacitor main features are reported in Table 1 both for Electrostatic Double-Layer capacitors (EDLCs) [19] and Lithium supercapacitors [20]. Fuel consumption evaluation was carried out considering EDLC supercapacitor; nevertheless, since lithium supercapacitor efficiency is quite close to EDLC, the resulting trend can be considered correct.

Table 1. Supercapacitor (SC) features.

Feature	EDLC SC	Lithium SC
Cell energy density	7–8 Wh/kg	10–13 Wh/kg
Package energy density [1]	3.5–4 Wh/kg	5–7 Wh/kg
Cell maximum voltage	2.8–3 V	3.8 V
Round trip efficiency at 120 s [2]	96–98%	91–97%

[1] Energy density considering cells, package, connections between cells and cell management system weight. [2] Charge—discharge efficiency with a DC current that can charge (or discharge) the storage in 60 s. EDLC: Electrostatic Double-Layer capacitors.

According to the EMS proposed in [29], one can define four storage voltage working zones (three allowed zones and a forbidden one):

- Braking zone (UVL $< V < V_{max}$):
- Normal zone (LVL $< V <$ UVL)
- Low speed zone ($V_{limit} < V <$ LVL)
- Forbidden zone ($V < V_{limit}$)

where UVL (Upper Voltage Limit) and LVL (Lower Voltage Limit) are two settable parameters, V_{max} is the storage rated voltage and V_{limit} is the storage lower allowed voltage (which can be either a storage or a converter constraint). In this study, V_{limit} is set to 0.5 V_{max}, whereas LVL and UVL definition is one of the objective of the research. V_{ref} is another settable parameter, the aim of which is explained in detail in page 5.

The braking zone is devoted to regenerative brakings, whereas the low speed zone is devoted to electrical traction at low vehicle speed. In addition to storage zone definition, and consequently LVL and UVL definition, it is necessary to identify also two vehicle speed thresholds: a Lower Speed Limit (LSL) and an Upper Speed Limit (USL).

Once these parameters are defined, one can illustrate the Engine Ignition Management System (EIMS), which regulates ICE starts and stops. The EIMS decision depends on storage state of charge, vehicle speed and ICE condition (i.e., if engine is already on or not), as shown in Figure 2.

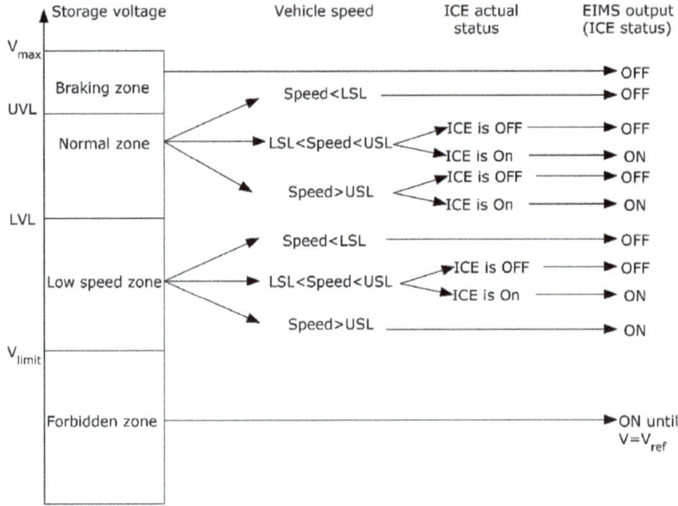

Figure 2. Engine Ignition Management System (EIMS).

The EIMS decides whether the ICE has to be turned on, turned off or it has to remain in the current condition; nevertheless, once the EIMS has decide the ICE should be turned on, the proposed Engine Power Management System (EPMS) [29] controls the ICE output power, as shown in Figure 3.

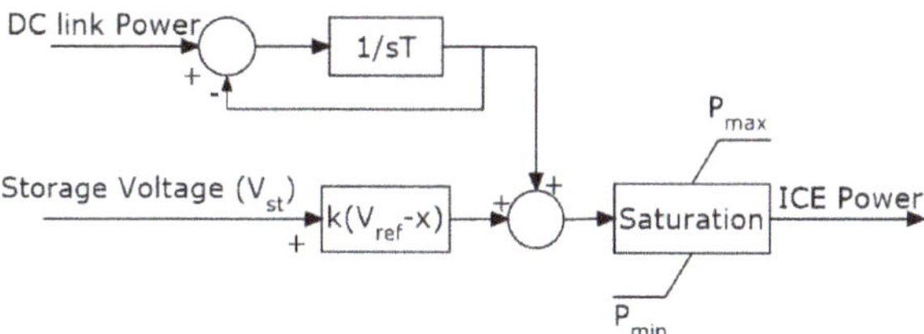

Figure 3. Engine Power Management System (EPMS).

The output power of the EPMS is given by the filtered DC-link required power and a contribution which depends on storage voltage (i.e., storage state of charge). This structure avoids high engine speed (i.e., high engine power) when the required power is low. As a matter of fact, high engine speed when the required power is low (e.g., when vehicle speed is low) is an uncomfortable condition. The power required at the DC-link is filtered in order to smooth acceleration and therefore reduce pollutant emissions [34,35].

The EPMS output is saturated between P_{max} (i.e., ICE rated power), and P_{min}, which is the ICE output power until which engine efficiency is still high; from ICE contour maps in Figure 1 indeed, it is possible to notice that efficiency worsens significantly in low power working conditions. When the DC-link required power is higher than P_{min} (e.g., in highway missions) the EPMS keeps storage voltage around V_{ref}. On the contrary, when the required power is lower than P_{min}, the ICE charges the storage system until UVL and then turns off until storage voltage reaches LVL. For this reason, if speed is higher than USL and the vehicle is not involved in long braking, storage voltage is always in the normal zone. The aim of the low speed zone definition is to keep the ICE off when vehicle starts and until it reaches USL. It is necessary to define both USL and LSL in order to reduce ICE number of starts when vehicle speed keeps around USL (or equally around LSL) for a long time.

The parameter T was set to 10 s, its optimization depends on ICE dynamic behavior and it is not considered in this study. Analogously, V_{ref} optimization depends on road mission profile (high values for highway missions, low values for long downhill road missions) and it can be set either manually by the driver or in a predictive way if the mission is known; in this study, it was set in the middle of UVL and LVL for all the missions. One has to note that V_{ref} has to be inside the normal zone. Parameter k was set to 0.06 kW/V (been V_{max} 500 V in all storage configurations). LSL and USL were set to 15 km/h and 30 km/h, respectively, whereas LVL and UVL optimization is one of the aims of this study, as shown in the next section. P_{min} was set to 6.6 kW for spark-ignition engine (power which corresponds to 34% efficiency, with maximum efficiency being 34.7%) and to 4.7 kW for diesel engine (power which corresponds to 39.5% efficiency, maximum efficiency being 41.5%). P_{max} was 40 kW for both spark-ignition and diesel engine.

3. Storage Zones Analysis and Sizing Optimization

The aim of the study is to analyze the influence of storage sizing on powertrain efficiency and on ICE number of starts (which affects comfort). Even if the sizing optimization of each of the three zones (the forbidden zone is not optimizable) is not independent from each other, the sizing of each zone has a precise consequence on powertrain performance:

- Braking Zone (BZ): if the braking zone is too small, there is a small quantity of energy available for regenerative braking. This influences powertrain efficiency, especially in missions characterized by long downhill roads. As a matter of fact, when the storage system reaches V_{max} the extra backward energy has to be dissipated on mechanical brakes.
- Normal Zone (NZ): if this zone is too small, the storage system charges and discharges quickly; as a consequence, ICE number of starts increases significantly.

- Low Speed Zone (LSZ): the aim of this zone is to avoid engine use at low speed. Indeed, one of hybrid vehicle aims is to guarantee electric traction at low speed. If the low speed zone is limited, ICE starts when vehicle speed is low (i.e., below USL) may occur frequently.

Although it is possible to identify a precise effect on powertrain management for each zone sizing, they still depend on each other. As an example, if the braking zone is limited but the energy associated to normal zone is high, there will be, on average, enough to store backward energy in the normal zone. Indeed, if no predictive control on the road mission is implemented, the braking phase can occur with the same probability in each point of the normal zone. Analogously, if the energy associated to the low speed zone is limited, but, again, the normal zone is large, ICE starts at low speed are infrequent. Nevertheless, low speed zone and braking zone sizing do not influence each other. Moreover, even if normal zone sizing influences the average storable energy during regenerative braking, braking zone sizing does not significantly influence the ICE number of starts; indeed, if storage voltage is in the braking zone, ICE is turned off until LVL is reached. In the same way, normal zone sizing influence low speed zone sizing, but low speed zone sizing do not significantly influences ICE number of starts when vehicle speed is above USL, which are only related to normal zone sizing.

The above considerations on zone sizing influences are summarized in the flowchart shown in Figure 4.

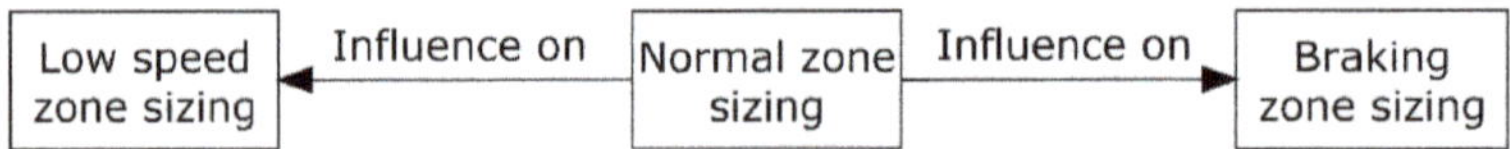

Figure 4. Zone sizing influences.

Taking into account the influences of Figure 4, the storage sizing analysis will be carried out as follows. Starting with an a priori defined sizing on low speed and braking zones, normal zone sizing influence on ICE number of starts will be analyzed. The analysis will be carried out on all the road missions considered in this study. Once a proper sizing has been defined, which will be a compromise between ICE number of starts and zone sizing, low speed zone and braking zone sizing will be investigated, the definition of which is almost independent from other zones once normal zone sizing has been established.

4. Road Missions

At first, standard cycles were considered: US06, UDDS and HWFET [32]. As can be noticed from Table 2, where road mission parameters are reported, these type approvals do not consider altitude variations; therefore, they are not critical for the supercapacitor storage, since there is no potential energy to recover during downhill roads.

Table 2. Road mission features.

Missions	Average Speed [km/h]	Maximum Speed [km/h]	Length [km]	Time [Minutes]	Change of Altitude [m]
US06	78	130	13	10	-
UDDS	31	90	12	23	-
HWFET	78	90	16.5	13	-
Urban	24	57	11.4	25	-
Fast-urban	27	68	22	52	62
Extra-urban 1	45	80	36	50	300
Mountain mission 1	48	85	24	30	500
Extra-urban 2	62	96	57	55	190
Mountain mission 2	51	90	60	70	710
Highway + mountain	87	125	480	330	1700

For this reason, experimentally measured missions were considered. Mission were measured with high precision GPS and barometric altimeter, with a 1 s sampling period. These road profiles were acquired in the north-west area of Italy (Liguria, Piemonte, Lombardia and Valle d'Aosta regions). The measured missions are:

- Urban: 11 km inside Aosta (Italy) town center.
- Fast-urban: from University of Genova, Albaro district (Genoa, Italy) to Bogliasco (Genoa, Italy) and return.
- Extra-urban 1: from University of Genova, Albaro district (Genoa, Italy) to Lavagna (Genoa, Italy) via Bargagli (Genoa, Italy).
- Mountain mission 1: Champorcher valley (Aosta, Italy) and return.
- Extra-urban 2: from Boves (Cuneo, Italy) to Mondovì (Cuneo, Italy).
- Mountain mission 2: from Boves (Cuneo, Italy) to Limonetto (Cuneo, Italy) and return.
- Highway + Mountain: from Milan (Italy) to Pont (Valsavaranche, Aosta, Italy) and return.

Road mission main features are summarized in Table 2, whereas speed and altitude profiles are shown in Figure 5 (Urban), Figures 6 and 7 (Fast-urban), Figures 8 and 9 (Extra-urban 1), Figures 10 and 11 (Mountain mission 1), Figures 12 and 13 (Extra-urban 2), Figures 14 and 15 (Mountain mission 2), Figures 16 and 17 (Highway + mountain mission). Since altitude change is negligible in urban mission, only speed profile is shown. Please note that, conventionally, initial altitude is set to zero in all missions; therefore, altitude has to be intended as the positive altitude difference during the mission and not as the absolute altitude value.

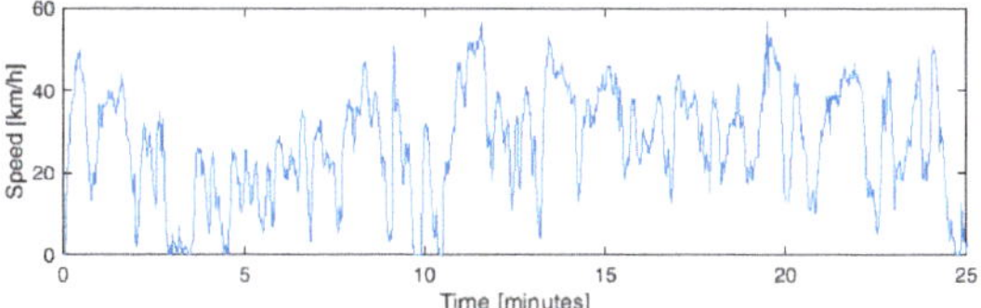

Figure 5. Urban.

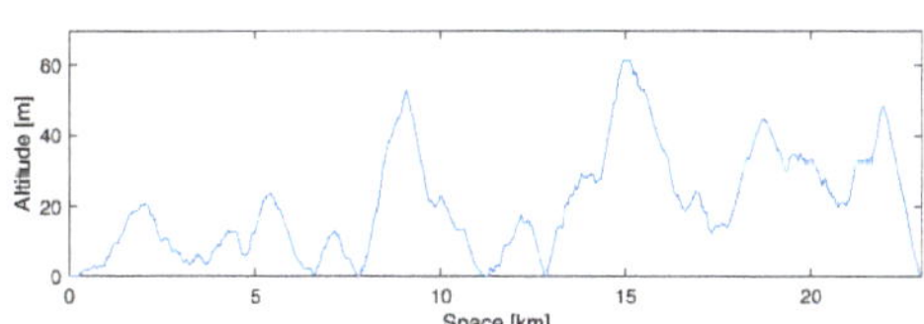

Figure 6. Fast-urban altitude profile.

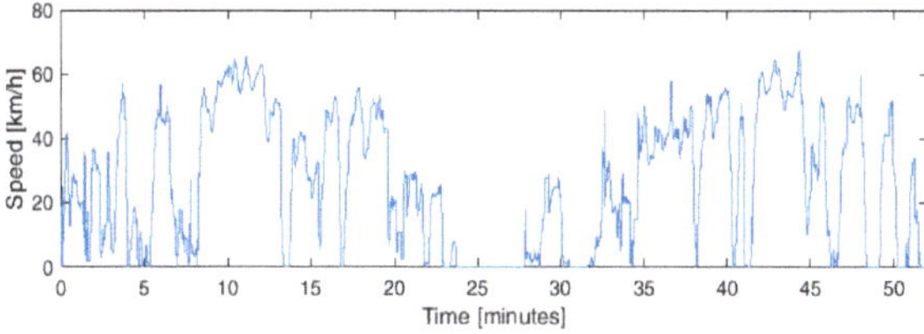

Figure 7. Fast-urban speed profile.

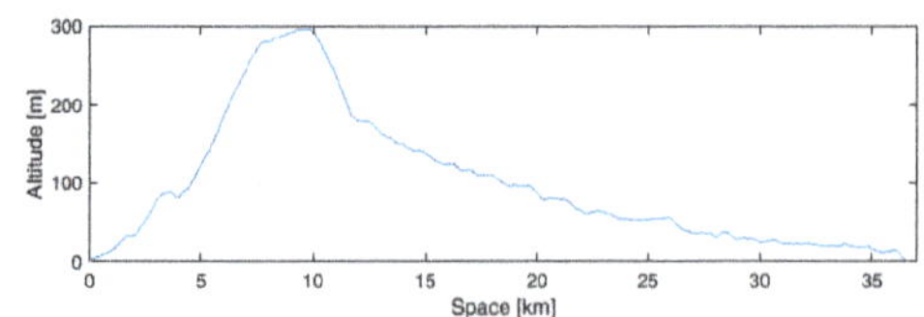

Figure 8. Extra-urban 1 altitude profile.

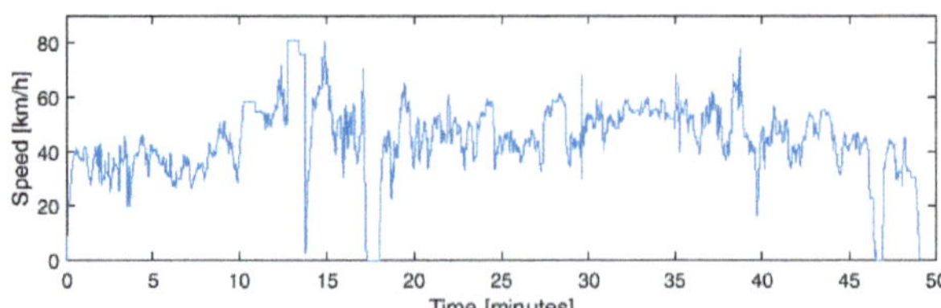

Figure 9. Extra-urban 1 speed profile.

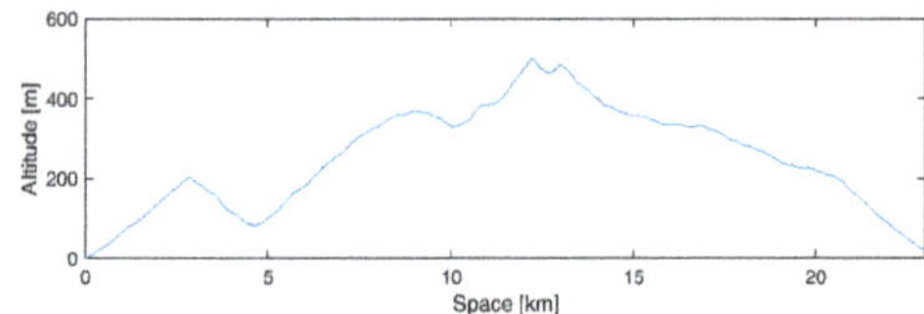

Figure 10. Mountain mission 1 altitude profile.

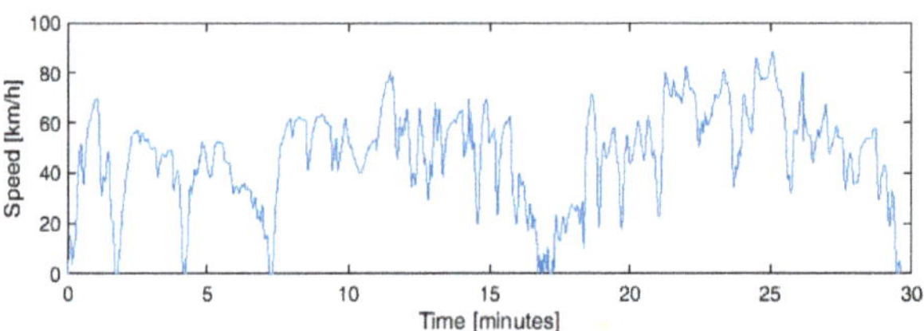

Figure 11. Mountain mission 1 speed profile.

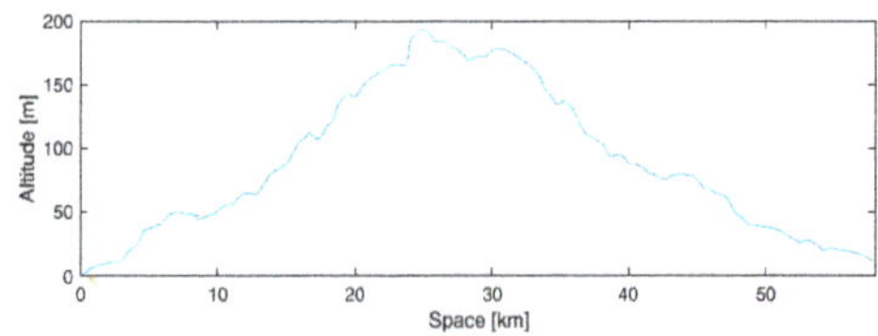

Figure 12. Extra-Urban 2 altitude profile.

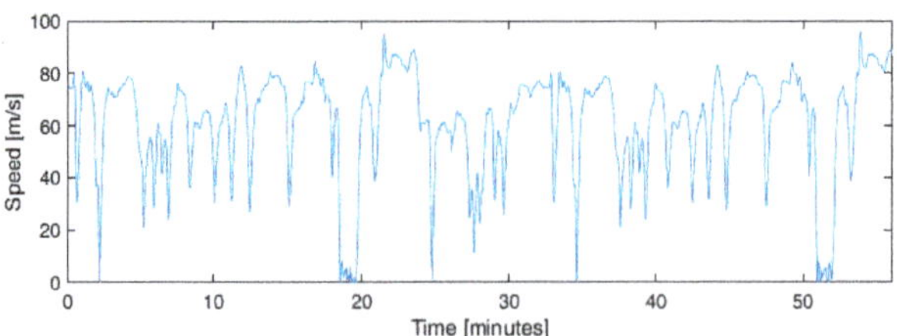

Figure 13. Extra-Urban 2 speed profile.

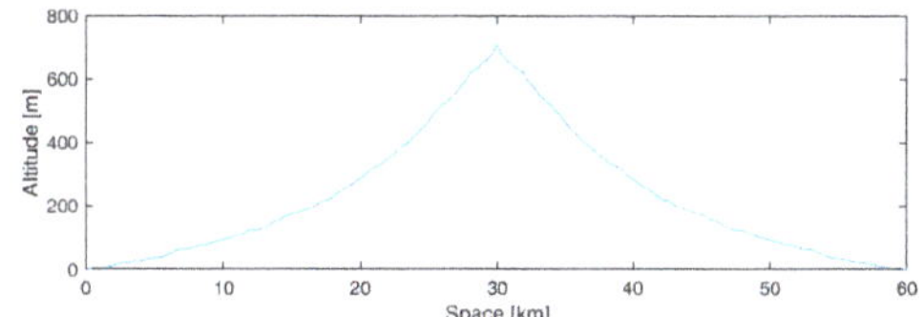

Figure 14. Mountain mission 2 altitude profile.

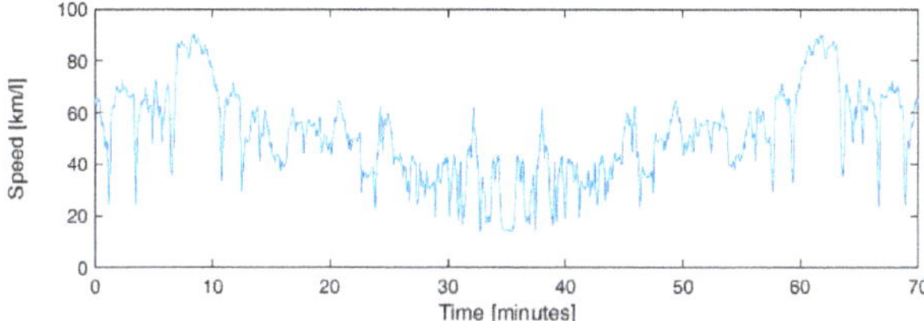

Figure 15. Mountain mission 2 speed profile.

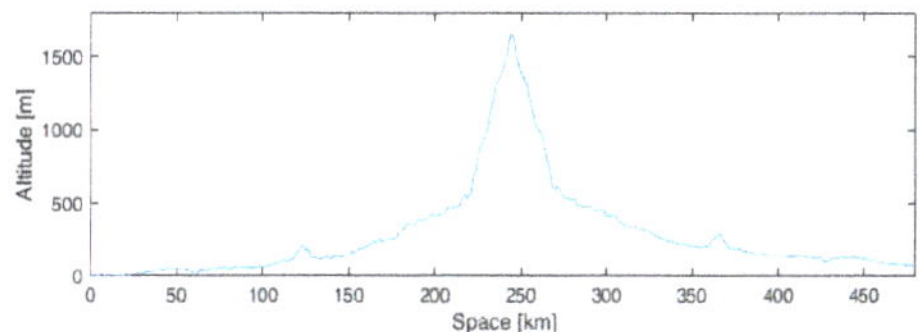

Figure 16. Highway + mountain mission altitude profile.

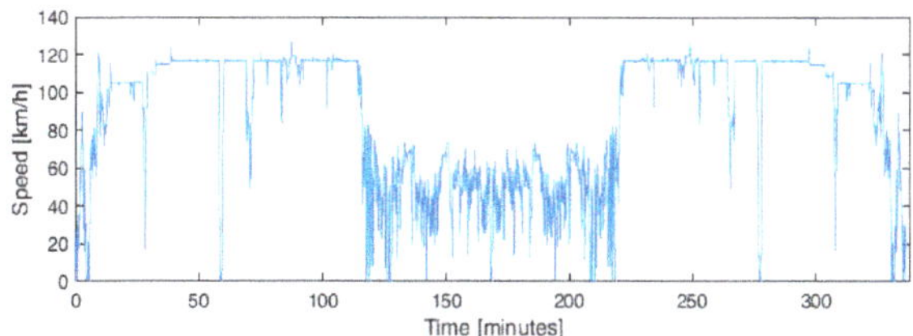

Figure 17. Highway + mountain mission speed profile.

5. Simulation Results

According to what previously shown in Section 3, NZ sizing is the first zone to be optimized. LSZ is a priori fixed to 30 Wh and braking zone to 100 Wh, indeed they do not have great influence on NZ sizing. Once NZ has been optimized, LSZ and BZ will be sized with the established NZ sizing. The monitored parameters, which are simulation outputs, vary stochastically on the same mission; as a matter of fact, ICE starts can occur at a certain moment depending on storage State Of Charge (SOC) and, analogously, storable energy available during braking depends on storage SOC. For this reason, each road mission is repeated various times with different initial SOC. The output for that specific mission is considered as the averaged value over the different simulations.

5.1. Normal Zone Sizing

Normal zone aims at reducing ICE number of starts; according to EIMS shown in Figure 2, if speed is above USL, ICE turns on when SOC is at LVL, and turns off when UVL is reached. To establish if NZ is properly sized, one has to observe ICE starts when speed is above USL. As a matter of fact, ICE starts when speed is below USL depends on LSZ, and will be investigated in Section 5.2. Simulation results are shown in Table 3, where spark-ignition engine number of starts, while speed is above USL, is reported as a function of NZ sizing. Same results are reported in Table 4 for diesel engine. Indeed, been P_{min} different for spark-ignition engine and diesel engine, simulation results are different.

Table 3. Spark-ignition engine number of starts while speed is above Upper Speed Limit (USL) as a function of Normal Zone (NZ) sizing (i.e., NZ Wh). Values approximated to the nearest integer.

Missions	18 Wh	30 Wh	42 Wh	54 Wh	70 Wh	160 Wh
US06	6	5	4	4	4	4
UDDS	19	14	13	12	10	8
HWFET	5	4	3	3	2	1
Urban	26	17	12	11	11	10
Fast-urban	21	16	16	15	12	10
Extra-urban 1	35	26	18	14	11	5
Mountain mission 1	12	7	7	6	6	5
Extra-urban 2	39	35	30	28	21	11
Mountain mission 2	29	23	16	14	11	5
Highway + mountain	109	86	74	58	47	27

Table 4. Diesel engine number of starts while speed is above USL as a function of NZ sizing (i.e., NZ Wh). Values approximated to the nearest integer.

Missions	18 Wh	30 Wh	42 Wh	54 Wh	70 Wh	160 Wh
US06	7	7	6	5	4	3
UDDS	18	14	13	12	11	9
HWFET	5	4	3	2	2	1
Urban	21	20	18	14	14	10
Fast-urban	20	19	17	16	15	11
Extra-urban 1	29	20	16	12	10	4
Mountain mission 1	11	9	8	7	6	6
Extra-urban 2	34	31	27	27	21	9
Mountain mission 2	21	15	13	11	10	5
Highway + mountain	102	87	66	53	43	27

Since P_{min} is lower for diesel engine, the number of starts are lower in Table 4 for same NZ sizing. Indeed, diesel efficiency decrease slower for decreasing power demand compare to spark-ignition engine, as can be notice from Figure 1; as a consequence, storage system is charged slower in low-power missions and number of starts is reduced. Number of starts for NZ size of 18 Wh is very high and it decreases if zone sizing is increased. Nevertheless, moving from 70 Wh to 160 Wh, the number of starts slightly decrease; therefore, 70 Wh is considered the optimal compromise. In urban mission number of starts is anyway lower than a traditional vehicle with start and stop system. Moreover, Extra-urban 2 and Highway + Mountain missions are characterized by high number of ICE starts, but they are 55 min–57 km and 5.5 h–480 km, respectively; therefore, values in Table 4 are absolutely tolerable.

5.2. Low Speed Zone Sizing

Been NZ sizing fixed at 70 Wh, low speed zone sizing analysis was performed considering 70 Wh and 100 Wh for NZ and BZ, respectively. LSZ aim is to reduce number of ICE starts when speed is below USL; as a matter of fact one of hybrid vehicle objective is to perform electric traction at low speed. In Table 5, the ICE number of starts when speed is below USL are shown. While ICE ignitions when vehicle speed is above USL occurs normally and the NZ aim is just to limit them, ICE ignition when vehicle speed is below USL is an unwanted situation and therefore just very few of them are tolerable. One has to note that ICE starts when speed is below USL do not depend on P_{min}; therefore, Table 5 values concern both spark-ignition and diesel engines.

Table 5. Internal Combustion Engine (ICE) number of starts when speed is below USL, as a function of Low Speed Zone (LSZ) sizing (i.e., LSZ Wh).

Missions	10 Wh	15 Wh	20 Wh	30 Wh	50 Wh
US06	0	0	0	0	0
UDDS	0	0	0	0	0
HWFET	0	0	0	0	0
Urban	2	1.4	1.2	1	0.2
Fast-urban	4.2	3.2	3.2	2	1
Extra-urban 1	0	0	0	0	0
Mountain mission 1	0	0	0	0	0
Extra-urban 2	0	0	0	0	0
Mountain mission 2	3	3	2	2	0
Highway + mountain	1	0	0	0	0

As aforementioned, each mission was repeated several times in order to average simulation outputs; for this reason, decimal values are reported in Table 5.

One can observe that in standard type approvals, even a very small LSZ is sufficient not to have engine starts when speed is below USL. This aspect shows the need of introducing experimentally measured missions in order to evaluate powertrain behavior in real working conditions. As a matter of fact, values reported in Table 5 in 10 Wh sizing for urban missions (Urban and Fast-urban) and Mountain mission 2 are quite high. Low speed zone sizing at 30 Wh is considered sufficient to reduce ICE starts at a tolerable value.

5.3. Braking Zone Sizing

During the braking zone sizing study, NZ and LSZ sizing were fixed at 70 Wh and 30 Wh, respectively. BZ sizing directly influences powertrain efficiency, since backward energy has to be wasted on mechanical brakes once storage is fully charged. In Table 6 wasted energy on mechanical brakes as a percentage of energy provided by the generator is reported. Additionally, in this case, type approvals would not give enough information for a complete evaluation of storage sizing; indeed, energy waste is negligible already with 50 Wh BZ sizing. Moreover, in mission with long downhill roads (mountain mission 1, mountain mission 2 and highway + mountain), impact on fuel economy moving from 100 Wh to 270 Wh is not enough significant to justify such a great increase in BZ sizing. On balance, braking zone optimal sizing can be considered between 50 Wh and 100 Wh. As a matter of fact, this sizing is enough to recover all the backward energy in more than half of the proposed missions; in addition, in mountain missions, the benefit in terms of fuel consumption reduction, moving to a significantly higher BZ (e.g., 270 Wh), would be negligible. Fuel economy for spark-ignition engine is reported in Figure 18 and for diesel engine in Figure 19. One has to note that Table 6 data are the same for both type of ICE; indeed energy provided by the generator is the same in both cases, as well as wasted energy, while fuel economy is different been ICE efficiency different.

Table 6. Wasted energy on mechanical braking as a percentage of energy provided by generator, for various Braking Zone (BZ) sizing.

Missions	50 Wh	100 Wh	270 Wh
US06	3.7%	0	0
UDDS	0	0	0
HWFET	1%	0	0
Urban	0	0	0
Fast-urban	1.4%	0	0
Extra-urban 1	6%	4.7%	0
Mountain mission 1	28.6%	26.9%	19%
Extra-urban 2	0	0	0
Mountain mission 2	16.7%	16.5%	13%
Highway + mountain	3.7%	3.3%	3.0%

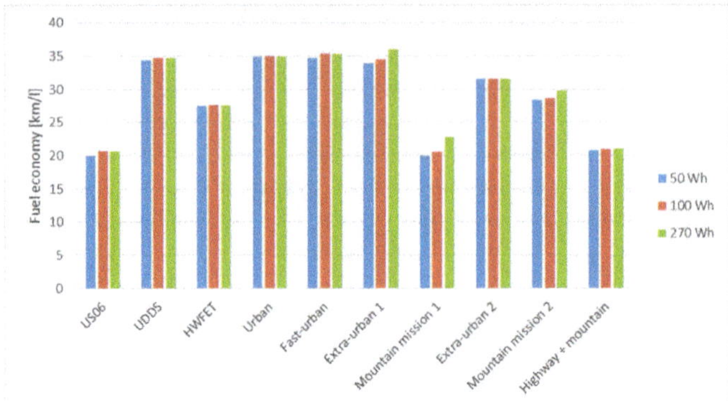

Figure 18. Fuel economy as a function of BZ sizing, spark-ignition engine.

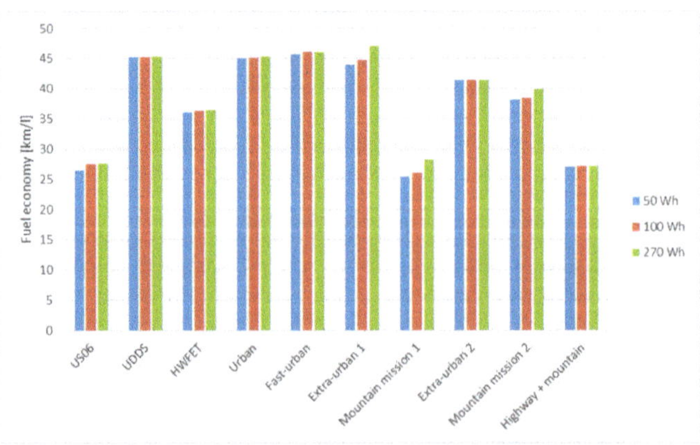

Figure 19. Fuel economy as a function of BZ sizing, diesel engine.

6. Conclusions

An analysis of supercapacitor storage sizing effect on hybrid series architecture was carried out in this study; a medium size car (i.e., 1450 kg) was considered. The impact both on ICE number of starts (which affects comfort) and on wasted energy on mechanical brakes (which influences powertrain efficiency) was taken into account. Simulation results demonstrated that standard type approvals do not properly model vehicle behavior; indeed, a storage sizing of about 100 Wh (e.g., 10 Wh for LSZ, 42

Wh for NZ and 50 Wh for BZ) would be enough to properly achieve all storage requirements. For this reason, seven experimentally measured missions were considered. Analyzing real working conditions, 150–200 Wh storage sizing is required (e.g., 30 Wh for LSZ, 70 Wh for NZ and 50–100 Wh for BZ). To recover all backward energy in mountain missions, storage sizing should be highly increased, becoming incompatible with vehicle applications. The proposed storage sizing (i.e., 150–200 Wh) is, instead, highly compatible with hybrid vehicle application, since it involves a storage weight of about 40–50 kg using EDLC supercapacitors, which can be further decreased to about 25–35 kg using lithium-ion supercapacitors.

Concluding, the increasing performance of supercapacitors has led to the possibility of realizing a series hybrid vehicle using only supercapacitors as storage system. The benefit of such types of architecture was shown in the technical literature and the feasibility, with regard to supercapacitor storage sizing, has been analyzed in detail in this paper. Moreover, the constantly increasing performance of such type of storage, together with the decreasing price, could further increase the interest for the proposed hybrid structure.

Author Contributions: M.P. conceived the article, developed the simulation model and wrote the article. S.G., S.W. and L.V. performed simulations and handled result diagrams. M.R. provided useful information on internal combustion engines and developed, together with M.P., the EMS. M.M. and M.C. supervised the all work and revised the article.

Funding: This research received no external funding.

Conflicts of Interest: The authors declare no conflict of interest.

References

1. Capata, R. Urban and extra-urban hybrid vehicles: A technological review. *Energies* **2018**, *11*, 2924. [CrossRef]
2. Salmasi, F.R. Control strategies for hybrid electric vehicles: Evolution, classification, comparison, and future trends. *IEEE Trans. Veh. Technol.* **2007**, *56*, 2393–2404. [CrossRef]
3. Uebel, S.; Murgovski, N.; Tempelhahn, C.; Bäker, B. Optimal energy management and velocity control of hybrid electric vehicles. *IEEE Trans. Veh. Technol.* **2017**, *67*, 327–337. [CrossRef]
4. Millo, F.; Cubito, C.; Rolando, L.; Pautasso, E.; Servetto, E. Design and development of an hybrid light commercial vehicle. *Energy* **2017**, *136*, 90–99. [CrossRef]
5. Chen, S.-Y.; Wu, C.-H.; Hung, Y.-H.; Chung, C.-T. Optimal strategies of energy management integrated with transmission control for a hybrid electric vehicle using dynamic particle swarm optimization. *Energy* **2018**, *160*, 154–170. [CrossRef]
6. Zhang, F.; Liu, H.; Hu, Y.; Xi, J. A supervisory control algorithm of hybrid electric vehicle based on adaptive equivalent consumption minimization strategy with fuzzy pi. *Energies* **2016**, *9*, 919. [CrossRef]
7. Hu, Y.; Li, W.; Xu, H.; Xu, G. An online learning control strategy for hybrid electric vehicle based on fuzzy q-learning. *Energies* **2015**, *8*, 11167–11168. [CrossRef]
8. Finesso, R.; Misul, D.; Spessa, E.; Venditti, M. Optimal design of power-split hevs based on total cost of ownership and co2 emission minimization. *Energies* **2018**, *11*, 1705. [CrossRef]
9. Burress, T.A.; Campbell, S.L.; Coomer, C.L.; Ayers, C.W.; Wereszczak, A.A.; Cunningham, J.P.; Marlino, L.D.; Seiber, L.E.; Lin, H.T. *Evaluation of the 2010 Toyota Prius Hybrid Synergy Drive System*; Power Electronics and Electric Machinery Research Facility, Technical Report ORNL/TM2010/253; Oak Ridge National Laboratory (ORNL): Oak Ridge, TN, USA, 2011.
10. Kim, N.; Cha, S.; Peng, H. Optimal control of hybrid electric vehicles based on pontryagin's minimum principle. *IEEE Trans. Control Syst. Technol.* **2011**, *19*, 1279–1287.
11. Bonfiglio, A.; Lanzarotto, D.; Marchesoni, M.; Passalacqua, M.; Procopio, R.; Repetto, M. Electrical-loss analysis of power-split hybrid electric vehicles. *Energies* **2017**, *10*, 2142. [CrossRef]
12. Cipek, M.; Pavković, D.; Petrić, J. A control-oriented simulation model of a power-split hybrid electric vehicle. *Appl. Energy* **2013**, *101*, 121–133. [CrossRef]
13. Pei, H.; Hu, X.; Yang, Y.; Tang, X.; Hou, C.; Cao, D. Configuration optimization for improving fuel efficiency of power split hybrid powertrains with a single planetary gear. *Appl. Energy* **2018**, *214*, 103–116. [CrossRef]

14. Xiang, C.; Ding, F.; Wang, W.; He, W. Energy management of a dual-mode power-split hybrid electric vehicle based on velocity prediction and nonlinear model predictive control. *Appl. Energy* **2017**, *189*, 640–653. [CrossRef]

15. Xu, Q.; Mao, Y.; Zhao, M.; Cui, S. A hybrid electric vehicle dynamic optimization energy management strategy based on a compound-structured permanent-magnet motor. *Energies* **2018**, *11*, 2212. [CrossRef]

16. Chen, J.; Du, J.; Wu, X. Fuel economy analysis of series hybrid electric bus with idling stop strategy. In Proceedings of the 2014 9th International Forum on Strategic Technology (IFOST), Cox's Bazar, Bangladesh, 21–23 October 2014; pp. 359–362.

17. Kim, M.; Jung, D.; Min, K. Hybrid thermostat strategy for enhancing fuel economy of series hybrid intracity bus. *IEEE Trans. Veh. Technol.* **2014**, *63*, 3569–3579. [CrossRef]

18. Zhao, Y.; Yao, J.; Zhong, Z.M.; Sun, Z.C. The research of powertrain for supercapacitor-based series hybrid bus. In Proceedings of the 2008 IEEE Vehicle Power and Propulsion Conference, Harbin, China, 3–5 September 2008; pp. 1–4.

19. Maxwell Supercapacitor. Available online: www.maxwell.com (accessed on 8 May 2019).

20. Ultimo Supercapacitor. Available online: https://www.jsrmicro.be/emerging-technologies/lithium-ion-capacitor/products/ultimo-lithium-ion-capacitor-laminate-cells-lic (accessed on 8 May 2019).

21. Cao, J.; Emadi, A. A new battery/ultracapacitor hybrid energy storage system for electric, hybrid, and plug-in hybrid electric vehicles. *IEEE Trans. Power Electron.* **2012**, *27*, 122–132.

22. Marchesoni, M.; Vacca, C. New DC–DC converter for energy storage system interfacing in fuel cell hybrid electric vehicles. *IEEE Trans. Power Electron.* **2007**, *22*, 301–308. [CrossRef]

23. Laldin, O.; Moshirvaziri, M.; Trescases, O. Predictive algorithm for optimizing power flow in hybrid ultracapacitor/battery storage systems for light electric vehicles. *IEEE Trans. Power Electron.* **2013**, *28*, 3882–3895. [CrossRef]

24. Marchesoni, M.; Savio, S. Reliability analysis of a fuel cell electric city car. In Proceedings of the 2005 European Conference on Power Electronics and Applications, Dresden, Germany, 11–14 September 2005; p. 10.

25. Carpaneto, M.; Ferrando, G.; Marchesoni, M.; Savio, S. A new conversion system for the interface of generating and storage devices in hybrid fuel-cell vehicles. In Proceedings of the IEEE International Symposium on Industrial Electronics, ISIE 2005, Dubrovnik, Croatia, 20–23 June 2005; pp. 1477–1482.

26. Kim, H.; Chen, H.; Zhu, J.; Maksimović, D.; Erickson, R. Impact of 1.2 kV SIC-MOSFET EV traction inverter on urban driving. In Proceedings of the 2016 IEEE 4th Workshop on Wide Bandgap Power Devices and Applications (WiPDA), Fayetteville, AR, USA, 7–9 November 2016; pp. 78–83.

27. Ding, X.; Cheng, J.; Chen, F. Impact of silicon carbide devices on the powertrain systems in electric vehicles. *Energies* **2017**, *10*, 533. [CrossRef]

28. Lanzarotto, D.; Marchesoni, M.; Passalacqua, M.; Prato, A.P.; Repetto, M. Overview of different hybrid vehicle architectures. *IFAC-PapersOnLine* **2018**, *51*, 218–222. [CrossRef]

29. Passalacqua, M.; Lanzarotto, D.; Repetto, M.; Vaccaro, L.; Bonfiglio, A.; Marchesoni, M. Fuel economy and energy management system for a series hybrid vehicle based on supercapacitor storage. *IEEE Trans. Power Electron.* **2019**. [CrossRef]

30. Zhang, L.; Hu, X.; Wang, Z.; Sun, F.; Deng, J.; Dorrell, D.G. Multiobjective optimal sizing of hybrid energy storage system for electric vehicles. *IEEE Trans. Veh. Technol.* **2018**, *67*, 1027–1035. [CrossRef]

31. Snoussi, J.; Elghali, S.B.; Benbouzid, M.; Mimouni, M.F. Optimal sizing of energy storage systems using frequency-separation-based energy management for fuel cell hybrid electric vehicles. *IEEE Trans. Veh. Technol.* **2018**, *67*, 9337–9346. [CrossRef]

32. Epa. Available online: https://www.epa.gov/vehicle-and-fuel-emissions-testing/dynamometer-drive-schedules (accessed on 8 May 2019).

33. Passalacqua, M.; Lanzarotto, D.; Repetto, M.; Marchesoni, M. Advantages of using supercapacitors and silicon carbide on hybrid vehicle series architecture. *Energies* **2017**, *10*, 920. [CrossRef]

34. Di Cairano, S.; Liang, W.; Kolmanovsky, I.V.; Kuang, M.L.; Phillips, A.M. Power smoothing energy management and its application to a series hybrid powertrain. *IEEE Trans. Control Syst. Technol.* **2013**, *21*, 2091–2103. [CrossRef]
35. Ortner, P.; Del Re, L. Predictive control of a diesel engine air path. *IEEE Trans. Control Syst. Technol.* **2007**, *15*, 449–456. [CrossRef]

 *energies*

Article

Turbocompound Power Unit Modelling for a Supercapacitor-Based Series Hybrid Vehicle Application

Matteo Repetto [1], **Massimiliano Passalacqua** [2], **Luis Vaccaro** [2], **Mario Marchesoni** [2,*] **and Alessandro Pini Prato** [1]

[1] Department of Mechanical, Energy, Management and Transportation Engineering (DIME), University of Genova, via all'Opera Pia 15, 16145 Genova, Italy; matteo.repet@gmail.com (M.R.); salabi@unige.it (A.P.P.)

[2] Department of Electrical, Electronic, Tlc Engineering and Naval Architecture (DITEN), University of Genoa, via all'Opera Pia 11a, 16145 Genova, Italy; massimiliano.passalacqua@edu.unige.it (M.P.); luis.vaccaro@unige.it (L.V.)

* Correspondence: marchesoni@unige.it

Received: 11 December 2019; Accepted: 13 January 2020; Published: 16 January 2020

Abstract: In this paper, starting from the measurements available for a 2000 cm^3 turbocharged diesel engine, an analytical model of the turbocharger is proposed and validated. The model is then used to extrapolate the efficiency of a power unit with a diesel engine combined with a turbocompound system. The obtained efficiency map is used to evaluate the fuel economy of a supercapacitor-based series hybrid vehicle equipped with the turbocompound power unit. The turbocompound model, in accordance with the studies available in the technical literature, shows that the advantages (in terms of efficiency increase) are significant at high loads. For this reason, turbocompound introduction allows a significant efficiency improvement in a series hybrid vehicle, where the engine always works at high-load. The fuel economy of the proposed vehicle is compared with other hybrid and conventional vehicle configurations.

Keywords: turbocompound; turbocharger; supercapacitor; hybrid electric vehicle (HEV); series architecture; fuel economy; efficiency

---

1. Introduction

The increasing interest in sustainable mobility has led to the study and development of new hybrid architectures, which can be mainly divided in three main categories [1]: parallel [2–6], series/parallel (also known as power-split) [6–14], and series hybrid [15–21]. Passalacqua et al. [22] proposed a new series architecture configuration for medium size cars, with a storage system based only on supercapacitors and they highlighted the benefits in terms of internal combustion engine (ICE) optimal working points. This topology may further benefits from innovation in engine design, such as the introduction of turbocompound (TC) systems. As observed by Jain et al. [23], turbocompound conditions are quite occasional in real road missions, thus, they do not render beneficial the introduction of the TC system; as a matter of fact, turbocompounding benefits are substantial at high load [23–28]. This aspect explains why today this technology is not used on medium size cars and is developed only for sportive cars or trucks and buses, characterized by frequent high load demand. For this reason, many studies [26–30] focused on turbocompound application to diesel engines above 12 L. In particular, Katsanos et al. [27] show a fuel consumption reduction of 4% in a 330-kW diesel engine at full load and 4% fuel consumption reduction was also found by Kant et al. [28]. Zhao et al. [29] found a brake-specific fuel consumption (BSFC) reduction from 3.1% to 7.8%, whereas Teo Sheng Jye et al. [26] found a 2% average BSFC reduction for a 13,000 cm^3 diesel engine. In [30], an analytical model is developed to show BSFC of a turbocompound diesel engine with different power turbine expansion ratio.

Various studies also concern turbocompounds on spark-ignition engines. In [25] a theoretical analysis of turbocharging benefits on a 2000 cm^3 spark-ignition engine is shown whereas in [24] turbocharging benefits on a 1000 cm^3 spark-ignition engine are supported by experimental results.

In this paper, starting from measured quantities on a four cylinder 2000 cm^3 direct ignition turbocharged (variable geometry turbine—VGT) diesel engine, ICE efficiency and turbocharger efficiency are calculated. Indeed, ICE and turbocharger efficiency cannot be measured directly but calculated from experimental measures. An analytical analysis is then carried out in order to quantify efficiency improvements in various working conditions (i.e., mean effective pressure and engine speed) with turbocompound introduction. Once the efficiency improvement is determined, BSFC (or equally engine efficiency) map is available both for turbocharged and turbocompound engines. After turbocompound modelling, its impact on the hybrid series architecture proposed in [22] is analyzed. Indeed, despite in traditional vehicles and parallel hybrid vehicles the engine works frequently at low load, in the abovementioned architecture the engine works constantly near full load, thus turbocompound can lead to significant advantages.

In contrast, the main novel contribution of this paper can be pinpointed as follow: (1) efficiency map for a 2000 cm^3 turbocompound diesel engine is evaluated starting from measured values on a traditional turbocharged diesel engine; (2) turbocompound contribution to a hybrid series architecture medium size car based on supercapacitor storage is analyzed.

In Section 2 turbocompound modelling is carried out and efficiency maps, both for turbocharged and turbocompound diesel engines are reported; in Section 3 turbocompound influence on series hybrid architecture is discussed. Conclusions are finally pointed out in Section 4.

2. Turbocompound Efficiency Evaluation

2.1. Turbocompound Overview

Turbocompound technology consists of a power unit where the ICE is combined with a turbine which recovers exhaust gas enthalpy. Differently from traditional turbochargers, where the energy produced by the turbine is used only to move the compressor (C), in the turbocompound system part of the power produced by the turbine is delivered and used outside the turbocharger. One of the solutions to exploit this available energy, which is considered in this study, is to connect an electric generator (Gen) to the turbine (T), as shown in Figure 1.

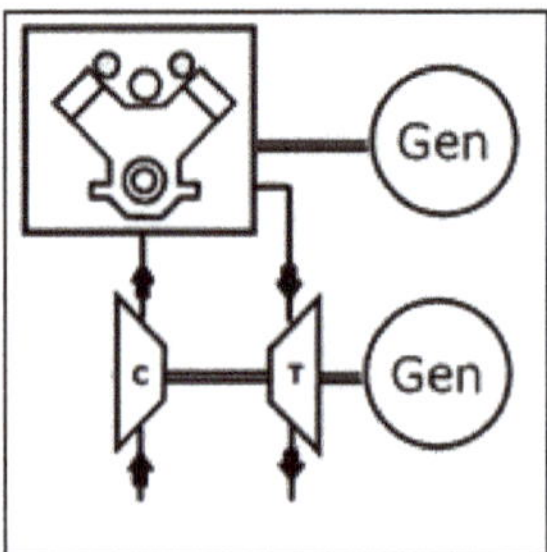

Figure 1. Turbocompound power unit scheme.

The interaction between the ICE and the turbine is quite complex since the power provided by the turbine is generated with a reduction in ICE generation; therefore, the optimization of the power unit cannot consider the optimization of the two subsystems separately, but has to be a combined optimization of all of the power unit. Moreover, since in road traction application the ICE rarely works in steady state condition, all load and speed working conditions have to be taken into account in the power unit optimization.

The first step in power unit optimization starts from evaluating the behavior of turbochargers usually used on current vehicles. Traditional turbochargers work by self-sustaining, i.e., the power provided by the turbine is used to move the compressor and to overcome friction; therefore, the turbocharger do not provide power outside.

Various works focus on thermal and mechanical losses of turbochargers [31,32].

Other works focus on turbine coupling with the ICE which has an outflow characterized by cyclic variation [31,33]. These studies show that the energy provided to the turbine is higher than the amount of energy evaluated taking into account the average pressure values before and after the turbine. The ratio between the energy provided to the turbine and the estimated energy using pressure average values is known in the technical literature as "pulsating factor" [34,35].

In order to evaluate turbine design criteria and turbocharger control logic while power is provided to the electric generator, it is necessary to divide as much as possible the various phenomena:

- Thermal losses;
- Mechanical efficiency;
- Effects of outflow cyclic variation (pulsating factor).

In this study, an analytical model of the turbocharger will be shown. After its validation (i.e., the comparison between the model output and the experimental measurements) the analytical model will be used to predict turbocompound behavior (i.e., the power unit efficiency).

2.2. Analytical Model of a Turbocharger via Experimental Measurements

In order to estimate efficiency improvement with a turbocompound system combined with a diesel engine, measurements on a four cylinders, turbocharged (variable geometry turbine—VGT), common rail, Euro 5, 2000 cm^3 engine were performed for various operating points (i.e., engine speed and mean effective pressure) [36]. It should be clarified that in the following approach, behavior during transient was neglected; as a matter of fact, even if this aspect is evaluated in various studies [37,38] transient behavior is negligible in series hybrid application, where ICE dynamics are smoothed by the storage system [22]. Moreover, operating conditions and turbocharger control have a great influence on turbine efficiency; however, since experimental measures regard fluid dynamics quantities and the efficiency was calculated starting from these parameters, the control effect is implicitly considered in the process. Analogously, turbine housing plays a fundamental role in turbine efficiency, however, also this aspect is implicitly considered in the process.

The following quantities were measured:

- Air and fuel mass flow rate ($\dot{m}_A$, $\dot{m}_F$);
- Temperature before and after compressor and turbine (T_1, T_2, T_7, and T_8);
- Pressure before and after compressor and turbine (p_1, p_2, p_7, and p_8);
- Torque to ICE shaft;
- Speed of ICE (rpm).

From these quantities it is possible to easily calculate:

- ICE power;
- Mean effective pressure (pme).

The quantities were measured for 20 different working points (i.e., different ICE speeds and mean effective pressures), which are reported in Table 1. For each working point, ICE power is reported, since the maximum load (i.e., the maximum pme) depends on ICE speed, maximum load is reported in the last line of the table.

Table 1. Internal combustion engine (ICE) power in the 20 analyzed points.

PME	1500 rpm	2000 rpm	2500 rpm	3000 rpm
6 bar	16.3 kW	20.7 kW	26.0 kW	33.5 kW
8 bar	21.8 kW	28.8 kW	35.6 kW	44.6 kW
11 bar	30.3 kW	40.0 kW	50.0 kW	60.8 kW
17 bar	44.5 kW	61.6 kW	77.3 kW	92.4 kW
Maximum load	55.3 kW (21.1 bar)	82.6 kW (22.9 bar)	101 kW (22.6 bar)	115 kW (21.6 bar)

The measurements were performed on a traditional ICE, where the turbocharger is self-sustaining, since no power is delivered outside.

The thermodynamic first law is shown in Equation (1), where the heat provided to the system and the work provided by the system are considered positive. The work performed by the compressor can be therefore evaluated from Equation (2), where the transformation was considered as adiabatic, since there is a negligible thermal exchange. dh is the enthalpy infinitesimal variation, dW is the infinitesimal workm and dQ_e is the infinitesimal exchange heat per mass unit.

$$dh + dW - dQ_e = 0, \tag{1}$$

$$dh_c + dW_c = 0. \tag{2}$$

Being the enthalpy gap defined as in Equation (3), the power delivered to the compressor (Pc) can be evaluated as in Equation (4). C_{pA} is the air constant pressure specific heat capacity, T_2 is the compressor output temperature, T_1 is the ambient temperature and $\dot{m}_A$ is the air flow.

$$\Delta h_c = C_{pA}(T_2 - T_1), \tag{3}$$

$$P_c = \dot{m}_A \cdot |\Delta W_c| = \dot{m}_A C_{pA}(T_2 - T_1). \tag{4}$$

To evaluate turbine work, the assumption of adiabatic transformation is not valid since there is a high difference between outflow temperature and ambient temperature. The enthalpy gas can therefore be evaluated, taking into account the external work and the thermal losses, as in Equation (5), with the same sign convention as Equation (1).

$$dW_T + dh_T - dQ_T = 0. \tag{5}$$

One can assume that the thermal losses are related to the temperature jump between the flow upstream of the turbine (i.e., the flow in the pipe between the ICE and the turbine) and the ambient temperature, since the turbine thermal resistance is almost constant in the various working conditions. As a matter of fact, turbine thermal resistance is related to thermal conduction through the component case, conveniently insulated, and through the turbocharger shaft. Therefore the power delivered by the turbine (P_T) can be evaluated as in Equation (6).

$$P_T = \dot{m}_t \cdot |\Delta W_T| = \dot{m}_t C_p(T_7 - T_8) - q_t, \tag{6}$$

where $C_p = C_p(T_7, T_8)$ and $q_t = q_t(T_7, T_a)$, q_t being the exchanged heat per time unit ($q_t = \dot{m}_t \cdot dQ_t$). Please note that the turbine does not have a dedicated cooling system; the terms q_t includes also the heat subtracted by the lubricating oil.

Since the turbocharger is self-sustaining, the turbine work has to equal the compressor work plus the mechanical losses; introducing the mechanical efficiency η_m, the relation between turbine power and compressor power is given by Equation (7).

$$P_T = \frac{P_c}{\eta_m},\tag{7}$$

where $\eta_m = \eta_m(P_C)$. η_m and q_T are not directly measured and therefore must be evaluated through appropriate relations with the measured quantities. Equations (6) and (7) have three unknown quantities (P_t, q_t, η_m), therefore a third relation is necessary.

Since $q_t = q_t(T_7, T_a)$, a linear relation $q_t = \alpha\, T_7 + \varphi$ was supposed, indeed the thermal resistance is almost constant in the various working conditions. The value of η_m should be in accordance with the values shown in detailed studies in the technical literature, which show a mechanical efficiency around 95%–97% at maximum load and around 40%–50% for very low power [37]. Therefore, the system of Equation (8) was solved.

$$\begin{cases} P_T = \dot{m}_t \cdot |\Delta W_T| = \dot{m}_t C_p(T_7 - T_8) - q_t \\ P_T = \frac{P_c}{\eta_m} \\ q_t = \alpha\, T_7 + \varphi \\ \eta_m(1\ \mathrm{kW}) = 0.5 \\ \eta_m(20.5\ \mathrm{kW}) = 0.97 \end{cases}.\tag{8}$$

The relation for the exchanged heat, evaluated from Equation (8) is reported in Equation (9) and plotted with the red line in Figure 2.

$$q_T = 1.7 \cdot 10^{-3} \cdot T_7 - 0.521.\tag{9}$$

The corresponding η_m values obtained combining Equations (6), (7) and (9) are reported with red rhombi in Figure 3. One can note that the η_m values obtained with the proposed process are in accordance with the values shown in [37]. For these points the fitting curve shown in Equation (10) can be used.

$$\eta_m = 0.525 \cdot P_c^{0.229 - 1.46 \cdot 10^{-3} \cdot P_c}.\tag{10}$$

Evaluating the exchanged heat q_t with the fitting of Equation (10) and using Equation (6) and (7), one obtained the blue rhombi in Figure 2. Observing the good correspondence between Equation (9) (red line in Figure 2) and the blue rhombi (obtained with the fitting in Equation (10)), one can verify that the proposed fittings represent the turbocharger well.

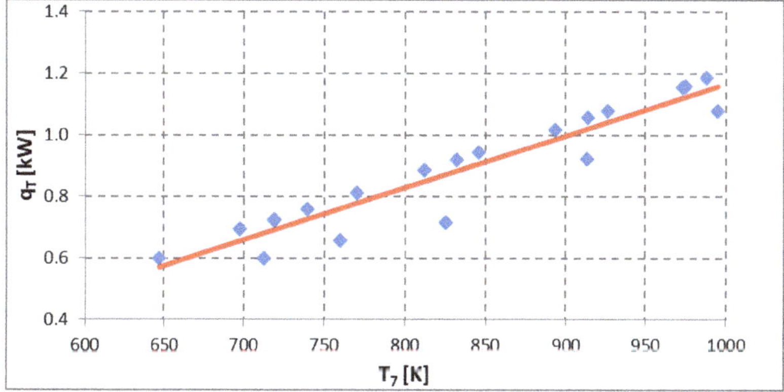

Figure 2. Exchanged heat as a function of turbine input gas temperature.

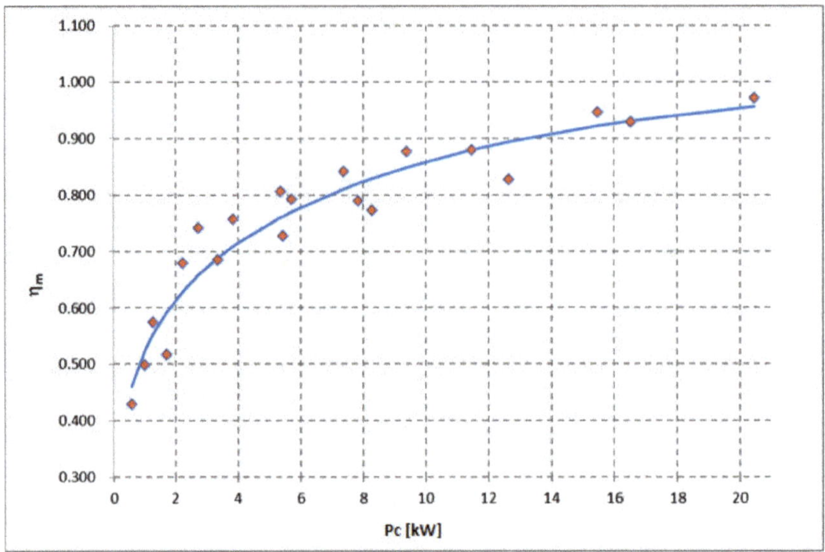

Figure 3. Turbocharger mechanical efficiency as a function of compressor power.

The turbine power P_T obtained from Equation (7), can be compared with the isentropic enthalpy gap related to the expansion rate in the turbine. In this way, one obtains an "apparent" efficiency value ($\widetilde{\eta}_T$), defined in Equation (11), significantly higher than the efficiency value experimentally measured for these types of turbines with stationary flow.

$$\widetilde{\eta}_T = \frac{P_T}{\dot{m}_t \cdot dh_T}. \tag{11}$$

The difference is related to the cyclic operation of the ICE, which produces a pulsating outflow.

The relation between the above "apparent" turbine efficiency $\widetilde{\eta}_t$ and the turbine efficiency which can be measured in stationary flow $\overline{\eta}$ is therefore given by Equation (12), where ε is the pulsating factor [34–39]:

$$\widetilde{\eta}_t = \overline{\eta}_t \cdot \varepsilon. \tag{12}$$

This phenomenon is related to the ratio between flow average speed in the turbine, which is related to the flow rate, and the amplitude of speed pulsation. Indeed, the energy provided to the turbine is related to the kinetic energy of the flow, which depends on speed quadratic value and is filtered by the turbine inertia. The mean kinetic energy is therefore the mean of quadratic speed, which is higher than the quadratic of the main speed (i.e., the speed measured during the experimental test), as summarized in Equation (13), where $tcycle$ is the period of a thermodynamic cycle:

$$\frac{1}{tcycle}\int_0^{tcycle} c^2 dt > \frac{1}{tcycle}\left(\int_0^{tcycle} cdt\right)^2. \tag{13}$$

The pulsating factor (ε) is therefore related to the ratio between the pulsation amplitude (A) and the main speed (c) and can be evaluated as in Equation (14):

$$\varepsilon = 1 + \tau \cdot (A/c)^2, \tag{14}$$

where:

$$A = c_{\max} - c_{\min}, \tag{15}$$

and τ is a coefficient which depends on the flux waveform (e.g., for a square wave it is 0.25).

In the pipe between the ICE and the turbine, taking away the work produced by the piston during the exhaust stroke and the irreversibility in the valve, the kinetic energy variation is obviously related to the pressure difference between the cylinder pressure (just before the exhaust valve opens) and the mean pressure in the exhaust pipe.

Indeed from energy conservation law in Equation (16), neglecting dW and dW_a, and integrating the Equation, one obtains Equation (17). dp is the pressure gap, v is the flow specific volume, dW is the infinitesimal work, and dW_a are the friction losses.

$$c \cdot dc + v \cdot dp + dW + dW_a = 0, \tag{16}$$

$$c_{\max}^2 - c_{\min}^2 = 2v \cdot \Delta p. \tag{17}$$

Therefore, the pulsating factor can be expressed as in Equation (18).

$$\varepsilon = 1 + \zeta \frac{\Delta p}{c^2}, \tag{18}$$

where:

$$\zeta = 2v \cdot \tau. \tag{19}$$

From Equations (18) and (20) one obtains Equation (21), where $\rho = \rho(T_7)$.

$$\dot{m} = \Omega \rho c, \tag{20}$$

$$\varepsilon = 1 + \frac{\zeta(p_6 - p_7)}{\dot{m}_t^2} \Omega^2 \rho^2. \tag{21}$$

From Equation (21) is it possible to obtain the ε value for each working point, which is reported in Table 2.

Table 2. ε as a function of ICE speed and mean effective pressure.

RPM	6 bar	8 bar	11 bar	17 bar	22 bar
1500	1.67	1.70	1.67	1.45	1.36
2000	1.28	1.35	1.34	1.25	1.20
2500	1.15	1.18	1.19	1.15	1.12
3000	1.07	1.08	1.11	1.09	1.08

Analyzing Equation (21), one can observe that ε depends on ρ (therefore on T_7), on p_6-p_7 and on $\dot{m}_t$, therefore the fitting proposed in Equation (22) can be considered.

$$\varepsilon = 1 + \frac{0.5 \cdot 10^{-3} - 0.75 \cdot 10^{-6} \cdot (T_7 - 647)}{\dot{m}_t^2} \cdot (p_6 - p_7) \cdot 10^{-5}. \tag{22}$$

In order to verify if the proposed analytical expressions properly model the turbocharger behavior, the efficiency in stationary flow ($\overline{\eta}_t$) obtained with Equations (9), (10) and (22), starting from the available measurements, was compared with the value of $\overline{\eta}_t$ available in the technical literature (which normally identify the combined effect of isentropic efficiency $\widetilde{\eta}_T$ and mechanical efficiency in stationary flow η_m). The combined efficiency obtained with Equations (10), (12) and (22) is shown in Table 3 for the 20 analyzed points.

Table 3. $\overline{\eta}_T \cdot \eta_m$ as a function of ICE speed and mean effective pressure.

RPM	6 bar	8 bar	11 bar	17 bar	22 bar
1500	0.35	0.35	0.36	0.54	0.58
2000	0.45	0.48	0.49	0.56	0.61
2500	0.58	0.57	0.59	0.64	0.66
3000	0.65	0.67	0.65	0.67	0.67

The combined efficiency $\overline{\eta}_T \cdot \eta_m$ in Table 3 can be plotted as a function of the expansion ratio β, obtaining the green rhombi in Figure 4. The rhombi can be compared with the results curve deriving from the fitting of experimental results obtained in [38], showing a good correspondence between them and therefore the model can be considered valid.

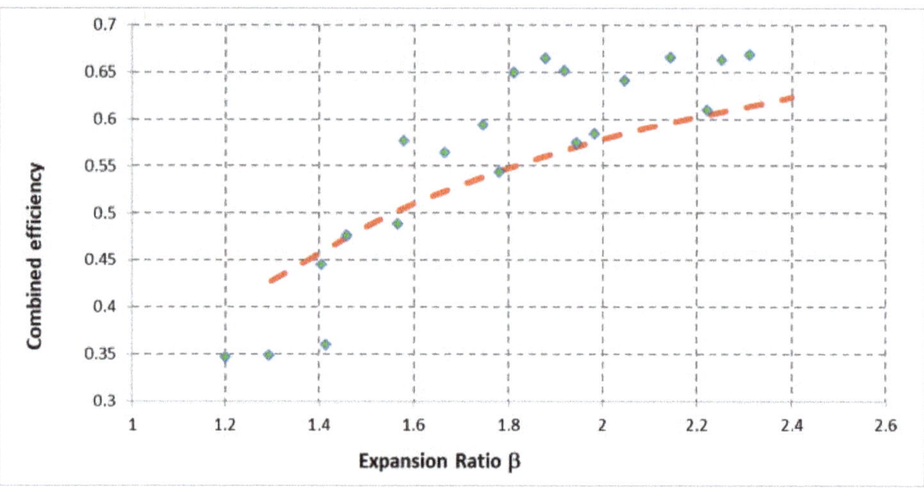

Figure 4. Combined efficiency $\overline{\eta}_T \cdot \eta_m$ as a function of expansion ratio β. Green rhombi are the value obtained with Equations (10), (12) and (22), whereas the red line shows the experimental results obtained in [38].

2.3. Turbocompound Modelling: An Extrapolation of Turbocharger Analytical Model

In order to extract power from the turbocharger, the power required from the compressor being the same as in self-sustaining, the expansion ratio of the radial mono-stage turbine should be increased; moreover, the turbine diameter should be increased and the cross-section reduced compared to the one in self-sustaining.

For each of the 20 working points for which experimental measurements were available, increasing values of p_7 were assigned. These values were given between the pressure related to the self-sustaining of the turbocharger and the pressure in the cylinder exactly before exhaust valve opening, which are the minimum and the maximum theoretical values that the pressure can reach. The increase of pressure upstream the turbine involves, for the same flow rate, an increase in turbine work, thanks to the increase of expansion ratio. The new value of the turbine work per cycle will be evaluated considering also an enthalpy increment caused by the piston, which has to oppose a higher pressure during the exhaust stroke.

For this reason, the power delivered to the ICE shaft will decrease of ΔP_p, with a power shift from the ICE to the turbocompound.

As aforementioned, also the pulsation amplitude of the flow will reduce, due to the reduction of the ratio between the pressure inside the cylinder immediately before exhaust valve opening and the mean pressure in the pipe upstream the turbine, causing a reduction in the pulsating factor. Therefore, the power generated by the turbine is given by Equation (23):

$$P_T = \dot{m}_t \, C_p \, T_7 \left(1 - \beta_T^{-\lambda}\right) \varepsilon \, \overline{\eta}_t, \tag{23}$$

where:

$$\beta_T = \frac{p_7}{p_8}. \tag{24}$$

For each of the 20 working points, P_T will be evaluated for each increasing value of p_7. For each increased value of p_7, the value of T_7, η_m, and ε will be recalculated with Equations (7), (9)–(11), (23) and (24), whereas $\overline{\eta_T}$ is considered constant for each of the 20 working points ($\overline{\eta_T}$ varies for each of the 20 points but is constant for increasing values of p_7). Indeed, it is hypothesized that a correct design allows the modified turbine to have the same $\overline{\eta_T}$ of the original component.

The power delivered externally from the turbocompound (i.e., the power to the electric generator shaft) is given by Equation (25):

$$P_{TC} = (P_T - P_C){\cdot}\eta_m. \tag{25}$$

As a consequence, the total power of the power unit is given by Equation (26), where ΔP_p is the incremental pumping loss (i.e., the difference between the pumping loss with the turbocharger in self-sustaining and the pumping loss with the turbocompound). This loss was evaluated taking into account the pression increase in turbocompound configuration, considering the engine displacement and the engine speed.

$$P_{PU} = P_{TC} + P_{ICE} - \Delta P_P. \tag{26}$$

The new efficiency of the power unit, for the same fuel power (fuel flow rate), for each working point will be given by Equation (27):

$$\eta_{PU} = \frac{P_{PU}}{F}. \tag{27}$$

Finally, the efficiency increase thanks to the introduction of turbocompound is given by Equation (28):

$$\psi = \frac{\eta_{PU}}{\eta_{ICE}}. \tag{28}$$

As an example, power unit efficiency η_{PU} is shown in Figure 5 for 2500 rpm and 17×10^5 Pa of pme, for the same fuel flow rate and as a function of pressure p_7.

2.4. Turbine Limits and Turbocompound Efficiency

The analytic analysis performed in the previous section showed that by increasing turbine counter-pressure, the power unit efficiency increases, thanks to the higher power recovered on the exhaust gas. However, conceptual and constructive constraints limit the maximum counter-pressure that can be imposed for each working point. These constraints have to be respected in the machine sizing and defining the behavior of the variable-geometry control system. The following main limit criteria were taken into account:

The main pressure in the pipe between the ICE and the turbine should be lower than the pressure in the cylinder at the exhaust valve opening.

The expansion ratio β should be compatible with the limits of single-stage radial turbines.

The pipe section reduction, for the same ICE speed and main effective pressure, should not exceed 50% in comparison to the pipe section of a traditional ICE with the turbocharger in self-sustaining.

In Section 2.3 the good correspondence between the analytical results and the experimental results of [38] was shown and therefore the model was validated. For this reason, the proposed model was used to extrapolate the results also for higher values of the expansion ratio β. These higher values are obtained when the turbocharger is no longer in self-sustaining but is producing work (turbocompound application).

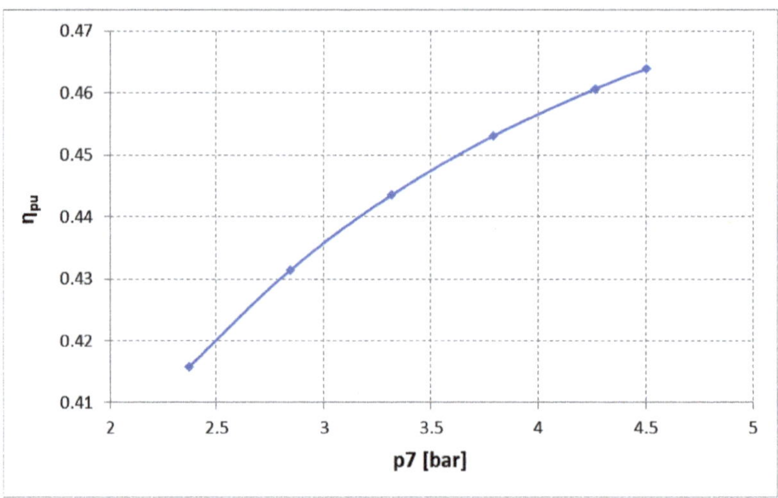

Figure 5. Power unit efficiency as a function of pressure *p7* for 2500 rpm and 17 bar pme.

In Figure 6 the combined efficiency $\overline{\eta}_T \cdot \eta_m$ obtained with the extrapolation process of the proposed model (red dots) are compared with the values available in [38] (green line), whereas the blue rhombi and the red line are the same as in Figure 4. Additionally, in the extrapolation area, the values obtained with the analytical approach show a good correspondence with the experimental results and therefore also in the extrapolation area the model is validated.

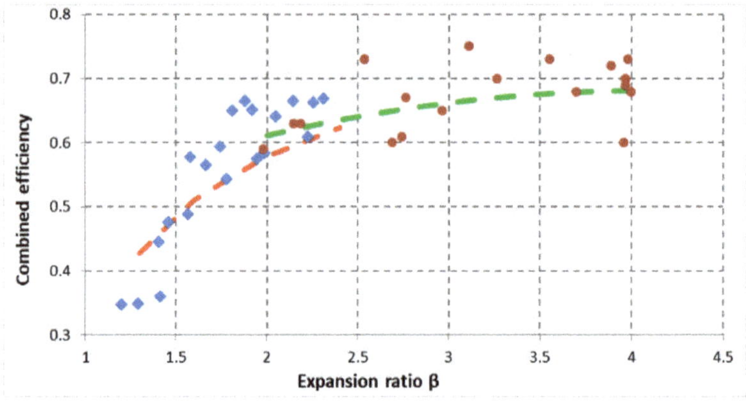

Figure 6. Combined efficiency $\overline{\eta}_T \cdot \eta_m$ as a function of expansion ratio β. Red dots are the value obtained with Equations (8), (11), (21), and the extrapolation process whereas the green line shows the results obtained in [38] for the same values of β.

With the abovementioned procedure and using Equations (24) and (27), power unit efficiency improvements and ICE power reduction (in comparison to a traditional solution with a self-sustaining turbocharger) were evaluated as a function of fuel rate per thermodynamic cycle. In Figure 7 the efficiency increase of the power unit is shown, as a function of the fuel flow rate per cycle. As it can be noticed from the green line, the influence of the ICE speed is negligible, and the efficiency increase is more related to the fuel flow rate.

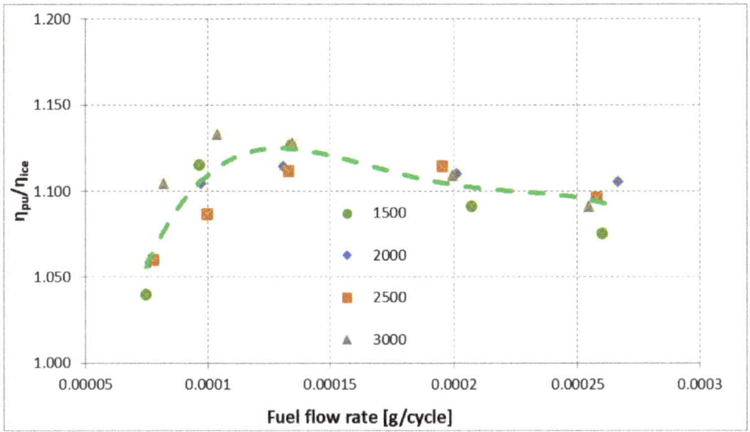

Figure 7. Power unit efficiency η_{PU} compared to ICE efficiency without turbocompound η_{ICE}, as a function of fuel flow rate (g/cycle). Marks correspond to ICE speed (rpm).

While working in power unit configuration (ICE + TC) a part of the power originally produced by the ICE is produced by the turbocompound. The shift of power from the ICE to the TC is shown in Figure 7 as a function of the fuel flow rate; again, the red line shows that the influence of ICE speed on this parameter is negligible.

From Figure 8 it is easy to notice that for the same fuel flow rate and same operating condition (i.e., ICE speed) there is a shift of power from the ICE to the turbine. This justifies the necessity of considering the two subsystems together to evaluate power unit performance.

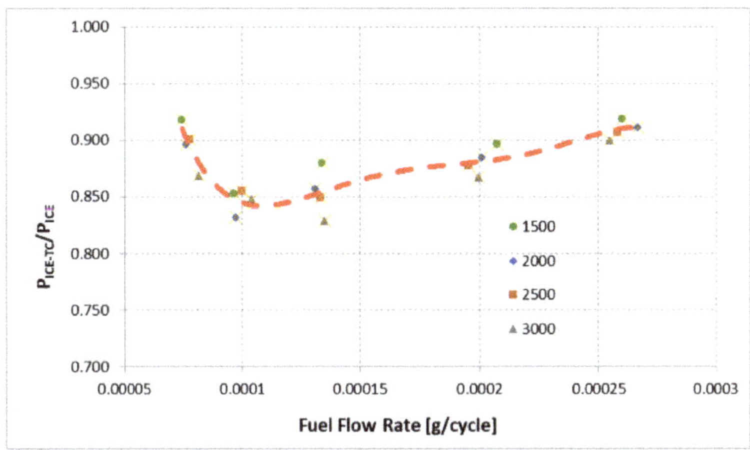

Figure 8. Ratio between ICE power with turbocompound (P_{ICE_TC}) and ICE power without turbocompound (P_{ICE}), as a function of fuel flow rate (g/cycle). Marks correspond to ICE speed (rpm).

The ICE contour map, which is easily calculated from measured fuel flow and power delivered to the ICE shaft, is reported in Figure 9a. The map regards the efficiency of the original ICE, without the turbocompound. The results in Figures 8 and 9a allow the efficiency contour maps for the turbocompound power unit to be obtained, which is reported in Figure 9b. The contour maps are obtained interpolating the efficiency of the 20 points analyzed in the study. These contour maps were used to perform vehicle fuel consumption simulations, as shown in the following sections.

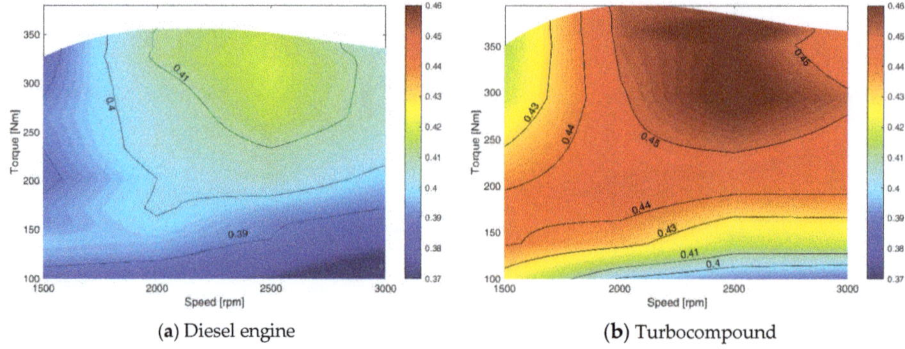

(a) Diesel engine   **(b)** Turbocompound

Figure 9. ICE efficiency contour map.

3. Turbocompound Application on Series Hybrid Vehicles

In the previous sections, starting from the measurements available for a turbocharged ICE, an analytical model of the turbocharger was carried out. The model was used to extrapolate the results for a turbocompound power unit, and the efficiency contour was map evaluated. These results are used in this section to evaluate the fuel economy of a series hybrid vehicle based on supercapacitor storage and equipped with a turbocompound diesel power unit. In Section 3.1 the road mission profiles used for simulations are described. An overview on series hybrid vehicles is shown in Section 3.2, whereas the vehicle modelling is reported in Section 3.3. Finally, the simulation results are shown in Section 3.4.

3.1. Simulated Road Missions

The different powertrain configurations were tested over six different road missions. At first, three American standard drive cycles were considered: Highway Fuel Economy Test (HWFET), Urban Dynamometer Driving Schedule (UDDS), and the Supplemental Federal Test Procedures (SFTP or US06) [39]. The advantage of using standard driving cycles is that they allow an easier comparison with other studies. However, since downhill are critical for the storage system (in this case supercapacitors), because a large quantity of potential energy should be stored, three additional real missions were simulated, whose data were experimentally acquired in the area of Genoa (Italy) [10,18,40]. In particular, the extra-urban mission is characterized by a long downhill road. The mission main features are reported in Table 4.

Table 4. Road mission features.

Road Mission	Average Speed (km/h)	Maximum Speed (km/h)	Length (km)	Change of Altitude	Maximum Road Slope (%)
US06	78	130	13	-	-
UDDS	31	90	12	-	-
HWFET	78	97	16.5	-	-
Urban	24	57	11.4	20	Negligible
Fast-urban	27	68	22	62	6
Extra-urban	45	80	36	300	9

3.2. Overview on Series Hybrid Vehicles

Using batteries as storage systems, the efficiency being low, ICE provided power has to be instantaneously close to the demanded power, in order to minimize the stored energy. Moreover, not only the stored energy has to be minimized from an efficiency point of view, but battery life in terms of number of cycles is quite low. Indeed, on medium sized car, parallel (or analogously series/parallel) architecture is primarily developed. Passalacqua et al. [22] proposed a series hybrid architecture for medium size cars based on supercapacitor as a stand-alone storage system. The efficiency of supercapacitors being high, the ICE can always work at high load, even if the average demanded power is low (e.g., in urban missions) since the exceeding power can be stored at high efficiency. Moreover, also power electronics efficiency has a significant effect on series architecture; as a matter of fact, supercapacitor storage is connected to the DC-link with a DC-DC converter, and all the power provided by the ICE is subjected to a double conversion (i.e., from mechanical to electrical and from electrical to mechanical), where an inverter is involved in both conversions. As a result, not only storage efficiency has a great influence on series architecture, but also power electronics play a fundamental role.

In Figure 10, average powertrain efficiency (obtained in [10,22] considering a spark ignition engine) over the simulated road missions is plotted as a function of storage and converter efficiency. Please note that the average storage and converter efficiency is the product of storage, generator inverter, motor inverter, and DC-DC converter average efficiency, whereas powertrain efficiency is defined as in Equation (29).

$$\eta_{\text{powertrain}} = \frac{\text{Energy to overcome friction}}{\text{Primary energy (fuel)}}. \tag{29}$$

From Figure 10 it can be noticed that low storage efficiency parallel and series/parallel architecture are more efficient than series architecture, this fact explains why today those structures are the most widespread; however, with high storage and converter efficiency, series architecture becomes competitive [22,41]. The break-even point is around 86%, which is a high value to be reached with batteries (as a matter of fact, the combined efficiency of storage and converters being 86%, battery efficiency should be higher than 92%–94%). Moreover, battery life is a critical issue, therefore from an economic point of view, break-even point could be at higher values. Nevertheless, high efficiency values can be achieved using supercapacitor storages and series architecture benefits can be further increase using Silicon Carbide components [18,22,41].

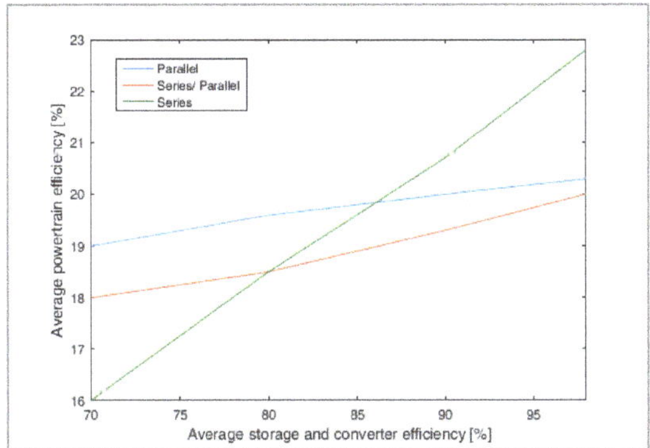

Figure 10. Average power train efficiency over the simulated road missions as a function of storage and converter efficiency (%) [22].

Although series architecture introduces additional losses, powertrain efficiency increases since ICE can work in optimal working conditions, as it can be noticed from Figure 11 [10,22] (again the results are shown on a spark-ignition engine) As a matter of fact, ICE working points are at high-load and in the maximum efficiency area in the last configuration. Please note that in the first two figure ICE maximum power is about 72 kW, whereas in the last figure ICE maximum power is about 40 kW. Indeed, as it will be shown after, ICE downsizing is possible in series architecture.

In addition, as observed in [22], the series architecture allows a relatively easy introduction of a diesel engine exploiting its strengths and softening its weaknesses. As a matter of fact, spark ignition engines are today widely employed in the automotive hybrid industry; the reason for this trend is given by their lower local pollutant emission and a higher flexibility in terms of on/off frequency. Nonetheless, firstly, the series architecture can reduce the number of ICE on/offs [22,40], secondly, by slowing down ICE transients, and by working in a limited working area, the local emissions are reduced. To sum up, not only using supercapacitor storage, series architecture becomes more competitive than parallel, but this gap can be even increased with the use of diesel engines.

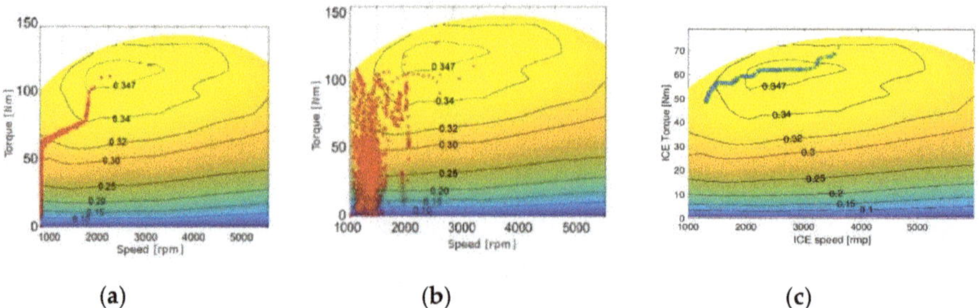

Figure 11. ICE work points in the parallel (**a**), series/parallel (**b**), and series (**c**) architecture [22].

In this scenario, not only fuel economy can increase using traditional ICEs, but series architecture can further benefit from turbocompound introduction, the benefits of which, in terms of efficiency, are relevant at high load, as observed by the modelling at Section 2 and by many works in the scientific literature [23–28,42].

3.3. Vehicle Modelling

In order to quantify the benefits of turbocompound introduction in a series hybrid vehicle, a MATLAB/Simulink model was created according to what was presented in [22]. The model is a quasi-stationary model and components were modelled as efficiency look-up tables. As a matter of fact, the aim of the study is to evaluate powertrain efficiency and fuel economy with different configurations, therefore vehicle dynamic was implemented, as well as components efficiency and energy management system (EMS). The efficiency contour map obtained with the process of Section 2 was used in the model, as well as efficiency map for a diesel engine without turbocompound (Figure 9a). In order to compare the powertrain efficiency with different engine configuration, spark-ignition engine efficiency map [43] was also used, which is reported in Figure 11, where also the operating points are shown. Regarding series architecture, a great downsizing of the ICE is possible, as shown in [18,22], and in this study a 40 kW ICE was considered. The efficiency contour maps shown in Figure 9 are considered valid also for a 40 kW ICE, where the speed is kept constant and the torque is conveniently scaled. The efficiency contour maps for a 40-kW diesel engine and 40-kW diesel + TC power unit are shown in Figure 12a,b, respectively, together with series architecture working points. As can be noticed from Figure 12, for each required power, the ICE always works in the maximum efficiency point for that specific power.

Initial and final storage state of charge (SOC) are generally different. However, to compare the fuel consumption of the different architectures, the initial SOC should be the same of final SOC, in this way electric consumption is zero. To avoid this problem every mission was repeated several times; indeed, the longer the road mission is, the lower is the SOC variation influence on fuel consumption. Two different simulations were performed for each road mission: one with maximum initial SOC and one with minimum initial SOC. When the difference between the fuel consumption obtained with the two SOC values was below 1%, the number of road mission repetitions was considered sufficient.

Vehicle longitudinal parameters are reported in Table 5.

Supercapacitor storage features are reported in Table 6; a 162-Wh supercapacitor module is used to perform simulations, according to what was presented in [40].

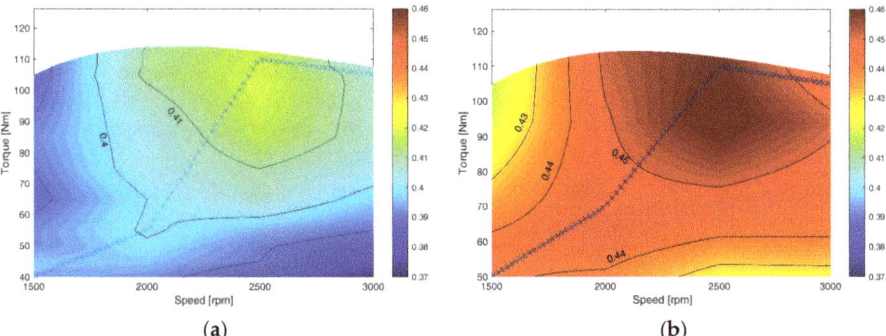

Figure 12. The 40-kW diesel (**a**) and 40-kW diesel + TC (**b**) efficiency contour map. ICE working point in series architecture are shown.

Table 5. Vehicle longitudinal parameters.

Parameters	Value
Mass [1] (kg)	1450
Rolling coefficient	0.01
Aerodynamic drag coefficient	0.25
Front section (m^2)	2.3
Wheel radius (m)	0.3
Final gear ratio	8
Differential gear efficiency	0.97
Gearbox efficiency (parallel)	0.95
Power-split gear efficiency (series/parallel)	0.95
Air density (kg/m^3)	1.22
ICE maximum power (parallel) (kW)	72
ICE maximum power [2] (series) (kW)	40
Mass [1] (kg)	1450
Rolling coefficient	0.01
Aerodynamic drag coefficient	0.25

[1] The vehicle mass includes the electric machines (about 80 kg), the storage system (85 kg), and the electric converter (about 15 kg). [2] The vehicle performance is related to the electric motor and it is independent of the ICE sizing in the series architecture; as a consequence, ICE downsizing is about 50%.

Table 6. Supercapacitor storage features.

Features	Value
Elementary module employed	BMOD0165 P048 C01
Series module number	4
Total mass (kg)	56
Rated capacitance, C (F)	41.2
Rated voltage, V_{max} (V)	192
Equivalent series resistance (ESR) (mΩ)	24
Rated power [1] (kW)	365
Usable energy [2] (Wh)	162

[1] The storage rated power is one order of magnitude higher than the converter nominal power, therefore, this is not a critical parameter for the storage. Indeed, the sizing criterion for supercapacitors is the stored energy. [2] Usable energy, calculated considering a working condition between 50% and 100% of the rated voltage.

3.4. Simulation Results

Different vehicle configurations are taken into account:

- Regular vehicle (RV): spark ignition engine, automatic transmission optimized for consumption minimization.
- Spark Ignition Series Hybrid Vehicle (SI SHV): spark ignition engine, series architecture, supercapacitor module, SiC power converters.
- Diesel Series Hybrid Vehicle (D SHV): diesel engine, series architecture, supercapacitor module, SiC power converters.
- Turbocompound Diesel Series Hybrid Vehicle (TCD SHV): diesel engine combined with the turbocompound technology, series architecture, supercapacitor module, SiC power converters.

Each configuration corresponds to a different MATLAB/Simulink model [10,18,22]. In order to provide a fair comparison, machine and inverter efficiency are the same for each configuration (according to what was presented in [10,18,22]) and vehicle longitudinal parameters (i.e., parameters in Table 5) are equal for all configurations. In Table 7, fuel economy is reported for various road missions for all the above-mentioned configurations.

Table 7. Fuel economy of different hybrid electric vehicle (HEVs) over different road missions (km/L).

Road Mission	RV	SI SHV	D SHV	TCD SHV
US06	16.1	20.7	27.5	29.1
UDDS	19.1	34.7	45.4	49.5
HWFET	23.2	27.6	36.2	38.9
Urban	19.5	35	45.2	50.2
Fast-urban	20	35.3	46	50
Extra-urban	25.2	36	46.9	50.6

It can be notice how turbocompound introduction guarantees a great fuel consumption reduction, as it can be highlighted in Table 8 where fuel saving (i.e., fuel reduction moving from D SHV to TCD SHV) is shown. The use of a series architecture allows a great improvement from turbocompound use also in urban missions (UDDS, Urban, and Fast-urban); as a matter of fact, although average demanded power is low (thus turbocompound would not allow significant fuel saving in traditional vehicle or parallel configurations [23]), ICE constantly works at high-load in series architecture. Please note that high-load does not mean maximum load, indeed turbocompound efficiency improvement in diesel engines is significant starting from 50% of full load, whereas it does not give substantial benefits for lower loads.

Table 8. Turbocompound fuel saving.

Road Mission	Turbocompound Fuel Saving
US06	10.2%
UDDS	11.6%
HWFET	11.0%
Urban	12.5%
Fast-urban	11.5%
Extra-urban	11.1%

4. Conclusions

In this paper an analytic study to determine efficiency map of a four cylinder, common rail, 2000 cm^3 diesel engine equipped with turbocompound is carried out. Once the efficiency contour map was obtained, it was used to evaluate benefits on a series hybrid architecture for a medium sized car. This architecture, using supercapacitors as storage system, which can store energy at high efficiency, allows the ICE to work in optimal conditions (i.e., high-load). Therefore, not only this architecture is more efficient with traditional ICE, but it can further benefit from TC introduction. As a matter of fact, ICE generally works at low-load in traditional vehicles and parallel hybrid vehicles, hence TC would lead to negligible advantages in these architectures.

The simulation results show a fuel saving from 10% to 12.5% in the diesel series architecture thanks to the introduction of the TC system. The proposed architecture shows a fuel consumption reduction from 44% to 61% compared to a traditional vehicle and from 43% to 55% compared to a series/parallel hybrid vehicle with battery storage and spark-ignition engine.

Author Contributions: M.R. and A.P.P. conceived the turbocompound modelling and managed the study of Section 2. M.P. and L.V. managed the study of Section 3. M.P. wrote the article. M.M. and A.P.P. revised the article and supervised the all work. All authors have read and agreed to the published version of the manuscript.

Funding: This research received no external funding.

Conflicts of Interest: The authors declare no conflict of interest.

Nomenclature

Symbol	Definition
h	Enthalpy for mass unit (kJ/kg)
W	Work for mass unit (kJ/kg)
Q	Exchanged heat for mass unit (kJ/kg)
p	Pressure (Pa)
T	Temperature (K)
c	Speed (m/s)
m	Mass (kg)
η	Efficiency
β	Expansion ratio
ε	Pulsating factor
A	Pulsation amplitude
λ	(k-1)/k where k is the exponent of Poisson adiabatic equation
P	Power (kW)
ICE	Internal Combustion Engine
TC	Turbocharger
PU	Power Unit (ICE + TC)
v	Specific volume (m^3/kg)
pme	Mean effective pressure (Pa)
C_p	Constant pressure specific heat capacity (kJ/K)
F	Fuel power (kW)
q	Exchanged heat per time unit (kW)
Ω	Pipe section (m^2)
ρ	Density (kg/m^3)
Ψ	Efficiency increase
α	Exchanged heat equation slope
φ	Exchanged heat equation intercept
Subscript	
e	External
t	Total
m	Mechanical
T	Turbine
max	Maximum
min	Minimum
a	Friction
c	Compressor
A	Air
F	Fuel

Superscript

$^-$	Stationary conditions
$^{\cdot}$	Time derivative
$^{\sim}$	Apparent

Thermodynamic Cycle Points

1	Ambient condition
2	Compressor output
3	Intercooler output
4	ICE end of compression
5	ICE End of combustion
6	ICE end of expansion
7	Turbine input
8	Turbine output

References

1. Salmasi, F.R. Control strategies for hybrid electric vehicles: Evolution, classification, comparison, and future trends. *IEEE Trans. Veh. Technol.* **2007**, *56*, 2393–2404. [CrossRef]
2. Uebel, S.; Murgovski, N.; Tempelhahn, C.; Bäker, B. Optimal energy management and velocity control of hybrid electric vehicles. *IEEE Trans. Veh. Technol.* **2017**, *67*, 327–337. [CrossRef]
3. Awadallah, M.; Tawadros, P.; Walker, P.; Zhang, N.; Tawadros, J. A system analysis and modeling of a HEV based on ultracapacitor battery. In Proceedings of the 2017 IEEE Transportation Electrification Conference and Expo, Asia-Pacific (ITEC Asia-Pacific), Harbin, China, 2–5 August 2017; pp. 792–798.
4. Millo, F.; Cubito, C.; Rolando, L.; Pautasso, E.; Servetto, E. Design and development of an hybrid light commercial vehicle. *Energy* **2017**, *136*, 90–99. [CrossRef]
5. Cheng, Y.-H.; Lai, C.-M. Control strategy optimization for parallel hybrid electric vehicles using a memetic algorithm. *Energies* **2017**, *10*, 305. [CrossRef]
6. Lanzarotto, D.; Passalacqua, M.; Repetto, M. Energy comparison between different parallel hybrid vehicles architectures. *Int. J. Energy Prod. Manag.* **2017**, *2*, 370–380. [CrossRef]
7. Burress, T.A.; Campbell, S.L.; Coomer, C.; Ayers, C.W.; Wereszczak, A.A.; Cunningham, J.P.; Lin, H.T. Evaluation of the 2010 Toyota Prius hybrid synergy drive system. In *Power Electronics and Electric Machinery Research Facility*; Technical Report ORNL/TM2010/253; Oak Ridge National Laboratory (ORNL): Oak Ridge, TN, USA, 2010.
8. Kim, N.; Cha, S.; Peng, H. Optimal control of hybrid electric vehicles based on Pontryagin's minimum principle. *IEEE Trans. Control Syst. Technol.* **2011**, *19*, 1279–1287.
9. Chung, C.-T.; Wu, C.-H.; Hung, Y.-H. Effects of electric circulation on the energy efficiency of the power split e-CVT hybrid systems. *Energies* **2018**, *11*, 2342. [CrossRef]
10. Bonfiglio, A.; Lanzarotto, D.; Marchesoni, M.; Passalacqua, M.; Procopio, R.; Repetto, M. Electrical-loss analysis of power-split hybrid electric vehicles. *Energies* **2017**, *10*, 2142. [CrossRef]
11. Kim, H.; Wi, J.; Yoo, J.; Son, H.; Park, C.; Kim, H. A Study on the fuel economy potential of parallel and power split type hybrid electric vehicles. *Energies* **2018**, *11*, 2103. [CrossRef]
12. Pei, H.; Hu, X.; Yang, Y.; Tang, X.; Hou, C.; Cao, D. Configuration optimization for improving fuel efficiency of power split hybrid powertrains with a single planetary gear. *Appl. Energy* **2018**, *214*, 103–116. [CrossRef]
13. Xiang, C.; Ding, F.; Wang, W.; He, W. Energy management of a dual-mode power-split hybrid electric vehicle based on velocity prediction and nonlinear model predictive control. *Appl. Energy* **2017**, *189*, 640–653. [CrossRef]
14. Yang, Y.; Pei, H.; Hu, X.; Liu, Y.; Hou, C.; Cao, D. Fuel economy optimization of power split hybrid vehicles: A rapid dynamic programming approach. *Energy* **2019**, *166*, 929–938. [CrossRef]
15. Chen, J.; Du, J.; Wu, X. Fuel economy analysis of series hybrid electric bus with idling stop strategy. In Proceedings of the 2014 9th International Forum on Strategic Technology (IFOST), Cox's Bazar, Bangladesh, 21–23 October 2014; pp. 359–362.
16. Kim, M.; Jung, D.; Min, K. Hybrid Thermostat strategy for enhancing fuel economy of series hybrid intracity bus. *IEEE Trans. Veh. Technol.* **2014**, *63*, 3569–3579. [CrossRef]

17. Zhao, Y.; Yao, J.; Zhong, Z.M.; Sun, Z.C. The research of powertrain for supercapacitor-based series hybrid Bus. In Proceedings of the 2008 IEEE Vehicle Power and Propulsion Conference, Harbin, China, 3–5 September 2008; pp. 1–4.

18. Passalacqua, M.; Lanzarotto, D.; Repetto, M.; Marchesoni, M. Advantages of Using Supercapacitors and Silicon Carbide on Hybrid Vehicle Series Architecture. *Energies* **2017**, *10*, 920. [CrossRef]

19. Xie, S.; Hu, X.; Liu, T.; Qi, S.; Lang, K.; Li, H. Predictive vehicle-following power management for plug-in hybrid electric vehicles. *Energy* **2019**, *166*, 701–714. [CrossRef]

20. Liu, T.; Wang, B.; Yang, C. Online Markov Chain-based energy management for a hybrid tracked vehicle with speedy Q-learning. *Energy* **2018**, *160*, 544–555. [CrossRef]

21. Passalacqua, M.; Lanzarotto, D.; Repetto, M.; Marchesoni, M. Conceptual design upgrade on hybrid pow ertrains resulting from electric improv ements. *Int. J. Transp. Dev. Integr.* **2018**, *2*, 146–154. [CrossRef]

22. Passalacqua, M.; Lanzarotto, D.; Repetto, M.; Vaccaro, L.; Bonfiglio, A.; Marchesoni, M. Fuel Economy and ems for a series hybrid vehicle based on supercapacitor storage. *IEEE Trans. Power Electron.* **2019**, *34*, 9966–9977. [CrossRef]

23. Jain, A.; Nueesch, T.; Naegele, C.; Lassus, P.M.; Onder, C.H. Modeling and control of a hybrid electric vehicle with an electrically assisted turbocharger. *IEEE Trans. Veh. Technol.* **2016**, *65*, 4344–4358. [CrossRef]

24. Mamat, A.M.I.B.; Martinez-Botas, R.F.; Rajoo, S.; Romagnoli, A.; Petrovic, S. Waste heat recovery using a novel high performance low pressure turbine for electric turbocompounding in downsized gasoline engines: Experimental and computational analysis. *Energy* **2015**, *90*, 218–234. [CrossRef]

25. Dimitriou, P.; Burke, R.; Zhang, Q.; Copeland, C.; Stoffels, H. Electric turbocharging for energy regeneration and increased efficiency at real driving conditions. *Appl. Sci.* **2017**, *7*, 350. [CrossRef]

26. Jye, A.T.S.; Pesiridis, A.; Rajoo, S. *Effects of Mechanical Turbo Compounding on a Turbocharged Diesel Engine*; SAE International: Warrendale, PA, USA, 2013.

27. Katsanos, C.; Hountalas, D.; Zannis, T. Simulation of a heavy-duty diesel engine with electrical turbocompounding system using operating charts for turbocharger components and power turbine. *Energy Convers. Manag.* **2013**, *76*, 712–724. [CrossRef]

28. Kant, M.; Romagnoli, A.; Mamat, A.M.; Martinez-Botas, R.F. Heavy-duty engine electric turbocompounding. *Proc. Inst. Mech. Eng. Part D J. Automob. Eng.* **2015**, *229*, 457 472. [CrossRef]

29. Zhao, R.; Zhuge, W.; Zhang, Y.; Yang, M.; Martinez-Botas, R.; Yin, Y. Study of two-stage turbine characteristic and its influence on turbo-compound engine performance. *Energy Convers. Manag.* **2015**, *95*, 414–423. [CrossRef]

30. Zhao, R.; Zhuge, W.; Zhang, Y.; Yin, Y.; Zhao, Y.; Chen, Z. Parametric study of a turbocompound diesel engine based on an analytical model. *Energy* **2016**, *115*, 435–445. [CrossRef]

31. Marelli, S.; Carraro, C.; Marmorato, G.; Zamboni, G.; Capobianco, M. Experimental analysis on the performance of a turbocharger compressor in the unstable operating region and close to the surge limit. *Exp. Therm. Fluid Sci.* **2014**, *53*, 154–160. [CrossRef]

32. Bozza, F.; de Bellis, V.; Marelli, S.; Capobianco, M. 1D simulation and experimental analysis of a turbocharger compressor for automotive engines under unsteady flow conditions. *SAE Int. J. Engines* **2011**, *4*, 1365–1384. [CrossRef]

33. Pasini, G.; Lutzemberger, G.; Frigo, S.; Marelli, S.; Ceraolo, M.; Gentili, R.; Capobianco, M. Evaluation of an electric turbo compound system for SI engines: A numerical approach. *Appl. Energy* **2016**, *162*, 527–540. [CrossRef]

34. Zinner, K.A. *Supercharging of Internal Combustion Engines: Additional*; Springer Science & Business Media: Berlin/Heidelberg, Germany, 2012.

35. Marelli, S.; Capobianco, M.; Zamboni, G. Pulsating flow performance of a turbocharger compressor for automotive application. *Int. J. Heat Fluid Flow* **2014**, *45*, 158–165. [CrossRef]

36. Confidential data related to test performed by the Department of Mechanical Engineering (DIME) of the University of Genova in collaboration with Centro Ricerche Fiat (CRF).

37. Deligant, M.; Podevin, P.; Descombes, G. Experimental identification of turbocharger mechanical friction losses. *Energy* **2012**, *39*, 388–394. [CrossRef]

38. Confidential data of the Department of Mechanical Engineering (DIME), regarding studies on turbocharged power unit for car racing application.

39. United States Environmental Protection Agency (EPA). Vehicle and Fuel Emissions Testing Dynamometer Drive Schedules. Available online: https://www.epa.gov/vehicle-and-fuel-emissions-testing/dynamometer-drive-schedules (accessed on 14 January 2020).

40. Passalacqua, M.; Carpita, M.; Gavin, S.; Marchesoni, M.; Repetto, M.; Vaccaro, L.; Wasterlain, S. Supercapacitor storage sizing analysis for a series hybrid vehicle. *Energies* **2019**, *12*, 1759. [CrossRef]

41. Lanzarotto, D.; Marchesoni, M.; Passalacqua, M.; Prato, A.P.; Repetto, M. Overview of different hybrid vehicle architectures. *IFAC PapersOnLine* **2018**, *51*, 218–222. [CrossRef]

42. Zhao, R.; Zhuge, W.; Zhang, Y.; Yin, Y.; Chen, Z.; Li, Z. Parametric study of power turbine for diesel engine waste heat recovery. *Appl. Therm. Eng.* **2014**, *67*, 308–319. [CrossRef]

43. Königstein, A.; Grebe, U.D.; Wu, K.-J.; Larsson, P.-I. Differentiated analysis of downsizing concepts. *MTZ Worldwide* **2008**, *69*, 4–11. [CrossRef]

Article

Analysis, Design, and Implementation of Improved LLC Resonant Transformer for Efficiency Enhancement

Zhenxing Zhao [1,2], **Qianming Xu** [1,*], **Yuxing Dai** [1,3] **and Hanhang Yin** [1]

[1] College of Electrical and Information Engineering, Hunan University, Changsha 410082, China; 22006@hnie.edu.cn (Z.Z.); daiyx@hnu.edu.cn (Y.D.); yinhanhang@hnu.edu.cn (H.Y.)

[2] College of Electrical Information, Hunan Institute of Engineering, Xiangtan 411100, China

[3] College of Physical and Electrical Engineering, Wenzhou University, Wenzhou 325000, China

* Correspondence: xqm@hnu.edu.cn; Tel.: +86-731-8882-3964

Received: 9 November 2018; Accepted: 22 November 2018; Published: 25 November 2018

Abstract: In battery charging applications, the charger changes its output voltage in a wide range during the charging process. This makes the design of LLC converters difficult to be optimized between the efficiency and the gain range. In this paper, an improved resonant transformer is presented for LLC resonant converter charger to improve the gain adjustment and charger efficiency. The resonant inductance and magnetizing inductance are integrated in the designed LLC transformer, and the magnetizing inductance can be adjusted dynamically with the change of output voltage and load, which is realized by a switch-controlled inductor (SCI) parallel to the secondary winding of transformer. The proposed transformer has 22.4% reduction in losses under full load conditions compared to conventional solutions. Moreover, the conduction loss and switching loss of LLC resonant tank are reduced by dynamically adjusting the magnetizing inductance, which improves the comprehensive efficiency of the whole charging process. The proposed transformer design is verified on a 720 W prototype.

Keywords: LLC resonant converter; resonant transformer; fringing effect; adjustable magnetizing inductance; efficiency

---

1. Introduction

LLC resonant converter has been widely used in electric vehicle battery chargers, flat panel television (TV), and photovoltaic (PV) system due to its high-power density and high conversion efficiency [1–4]. For now, it has become one of the most concerned DC/DC converters.

In constant output voltage applications, LLC resonant converter can achieve high efficiency. However, there are many challenges for LLC resonant converter in the charger applications which requires a wide output voltage adjustment range [5]. It requires a small magnetizing inductance to obtain a wide output voltage adjustment range, but it can lead to increased conduction loss and switching loss as well as efficiency reduction [6]. The magnetizing inductance is usually integrated in LLC resonant transformer. The usual structure of magnetic integrated LLC transformer is as shown in Figure 1, the leakage layer is set between the primary winding and secondary winding. The required resonant inductance is obtained by using stray flux and the magnetizing inductance is achieved by inserting an appropriate air-gap in the magnetic circuit [7,8]. The transformer with that structure has high integration and low cost and has been widely used in converters with power level from dozens to hundreds of watts.

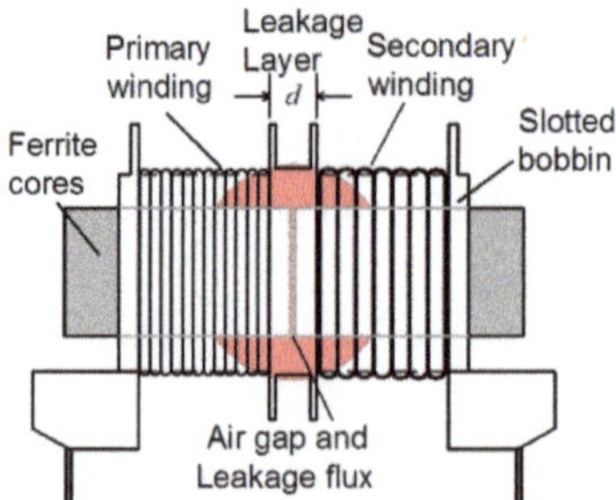

Figure 1. Structure of conventional LLC transformer and leakage flux effect on the windings.

As the core device of LLC resonant converter, the transformer plays a critical role in converter's efficiency, volume, power density, and reliability. However, the whole loss of the LLC resonant transformer has a much higher percentage than that of phase-shift full-bridge transformer in the same power [9]. One of the main reasons is that the air gap in the magnetic core of LLC resonant transformer causes fringing effect, leading to an increase in the equivalent resistance of windings near the air gap, and the transformer inner temperature rise distributed imbalance. The other reason is that the small magnetizing inductance leads to increased conduction loss and switching loss of the LLC resonant converter, so the light-load efficiency is lower.

In order to improve the efficiency of LLC transformer, different improvement designs are proposed. In [10], a novel shape magnetic core is used to achieve the integration of two LLC transformers. In [11], a magneto plated wire is used to effectively decrease the winding loss caused by the proximity effect when working in high frequency. In [12], the integration of resonant inductance required is achieved by inserting a layer of flexible magnetic material between the primary and secondary windings. The designs proposed by [10–12] have high integration, but without regard for the fringing effect, so they are more suitable for low power applications. In [13], the copper loss of the Litz wire caused by the air gap is analyzed, and the influence of the fringing effect is weakened by using multiple small air gaps in series, but this increases the cost and difficulty of transformer production. In [7], a design method of LLC resonant transformer is presented, and the current density in the conductor near the air gap is simulated. In [14], the optimal design method of the transformer in conventional LLC resonant converter is extended to the design and application of flat panel transformer, and a complete design scheme and a detailed application method are given. In [7,14], the fringing effect is mentioned, but the influence of the effect is not analyzed and the solution is not given. In [15], the relationship between the temperature rise and the switch frequency as well as the winding number of the LLC resonant transformer are discussed, and a combination of two smaller transformers is used instead of one larger transformer to reduce the temperature rise. Matrix transformer is used to achieve a higher output power in [16]. However, with the increase of power, the matrix transformer which consists of too many transformers will lead the system more complex. In [17,18], to reduce the influence of the fringing effect, the distance between the winding and the air gap is increased by making changes of the winding structure, which leads to the increase in difficulty of processing windings.

To improve the light-load efficiency, the transformers in [19,20] are designed with variable magnetizing inductance. In [19], by using utilizing a step-gap in the core column, a larger magnetizing inductance is obtained at light load, and it decreases at heavy load, but the variety of magnetizing inductance is non-linear and uncontrollable. In [20], a bidirectional power switch is used to achieve the parallel operation of two transformers. When under heavy load, the switch turn-on, two transformers primary side work in parallel, the magnetizing inductance of resonant tank is small, and can satisfy the gain demand required in the initial stage of charging. When under light load, the switch turns off, only one transformer works, the magnetizing inductance of resonant tank is large, and the conduction loss and switching loss of primary side decrease. However, the resonant inductance cannot be integrated into the transformer, the magnetizing inductance can only be changed between two fixed values, and when under light load, only one transformer works and the other is vacant.

Aiming to address the design problems of the conventional LLC transformer caused by the large influence of the fringing effect and the unchangeable magnetizing inductance, this paper presents a design for the integrated LLC resonant transformer. The magnetizing inductance does not need to be obtained by inserting air gap, so the fringing effect of the transformer is small. Moreover, the magnetizing inductance can be dynamically adjusted by a SCI parallel to the secondary winding of the transformer. This can not only improve the efficiency and the temperature rise of the transformer but also optimize the comprehensive efficiency of the LLC resonant converter during the whole charging process.

This paper includes following aspects. Section 2 analyzes the conventional integrated LLC resonant transformer. Section 3 presents the proposed design scheme, analysis of magnetic flux in the transformer core and the application circuit. Section 4 gives the calculation formula and design procedure of the transformer. Section 5 shows the experimental results of a 720 W LLC resonant converter. The conclusion is given in the last section.

2. Analysis of Conventional LLC Transformer

In high-power applications, the fringing effect near the air gap leads to a significant increase in the current density of the windings near the air gap, resulting in an apparent increase in the partial temperature rise and copper loss of the transformer [7]. To analyze the influence of fringing effect, finite element analysis (FEA) simulation is carried out by using Maxwell software. Figure 2 illustrates the simulation of fringing effect at 100 kHz. It can be seen the current density of the copper conductor near the air gap is about 70% higher than that away from the air gap, so the copper loss of this part is significantly increased. The increased copper loss results in the conspicuous regional temperature rise near the air gap. The simulation also shows that the fringing effect has a significant impact near the air gap, and the affected range increases with the increase of transformer power from the center of the air gap.

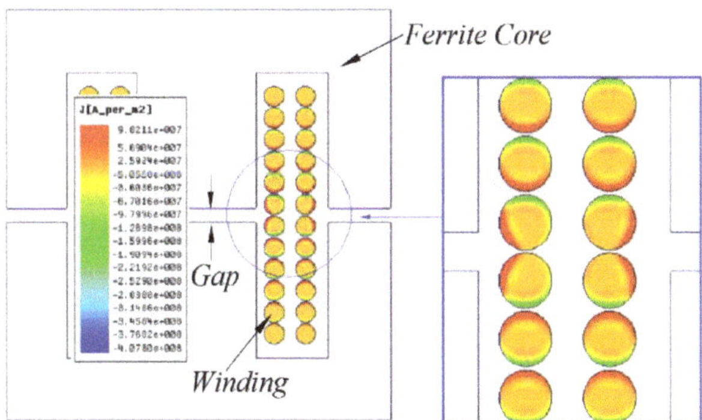

Figure 2. High frequency fringing effect near the air gap.

After the winding turns and air gap size of the transformer are determined, the transformer magnetizing inductance is a fixed value. Choosing small magnetizing inductance to obtain sufficient output voltage gain will result in increased conduction loss and switching loss of the LLC resonant converter.

The simulation results show that the induced magnetic flux increases in pace with the increase of the transformer power and input voltage, the area influenced by the air gap expand accordingly. In order to reduce the fringing effect, it is needed to increase distance d of leakage layer, as shown in Figure 1. However, the increasing d causes the available area of the core window decrease, under the same diameter and turns, the layer p of transformer winding is increase. According to the Dowell's formula as below, this effect can lead to a significant increase in proximity loss.

$$\frac{R_{ac}}{R_{dc}} = \Delta \left[\frac{\sinh 2\Delta + \sin 2\Delta}{\cosh 2\Delta - \cos 2\Delta} + \frac{2(p^2 - 1)}{3} \frac{\sinh \Delta - \sin \Delta}{\cosh \Delta + \cos \Delta} \right] \tag{1}$$

where, p is the number of transformer windings layers, Δ is the ratio of the winding layer thickness d of the skin depth δ_0.

To integrate the resonant inductance into the transformer, the primary winding and the secondary winding cannot use 'sandwich' method, which is used in full-bridge phase-shifting transformer to decrease the proximity loss. Therefore, at high frequency condition, use Litz wire to decrease the impact of the proximity effect and the skin effect.

In summary, the fringing effect, proximity effect, and the increase of the conduction loss and switching loss caused by small magnetizing inductance are main factors affect the transformer efficiency.

3. Analysis of Improved LLC Resonant Transformer

In order to promote the efficiency, an improved LLC transformer design is presented.

3.1. Main Structure

The structure of proposed LLC resonant transformer is shown in Figure 3, which has the following characteristics. (1) The skeleton adopts double groove structure, thus primary winding and secondary winding are wound respectively in two grooves. Therefore, the resonant inductance is still integrated in the transformer. (2) The air gap is not inserted in the magnetic circuit, so there is nearly no leakage flux, the intensive windings in the transformer is almost not affected by the fringing effect. (3) The thickness d of the middle leakage layer can be small, which does not occupy the space of the windings and improves the utilization rate of the core window, reducing the proximity effect.

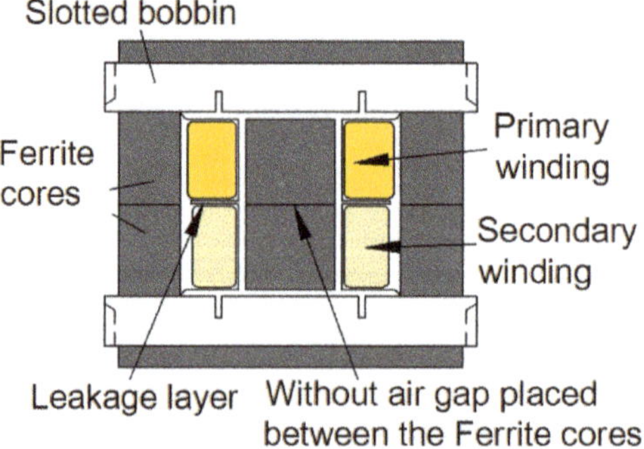

Figure 3. Main structure of proposed transformer.

The fringing effect simulations comparison is shown in Figure 4. It can be seen that the conductor current density near the air gap of the proposed transformer is greatly reduced, and the conduction loss is effectively decreased, so the internal temperature rise can be improved.

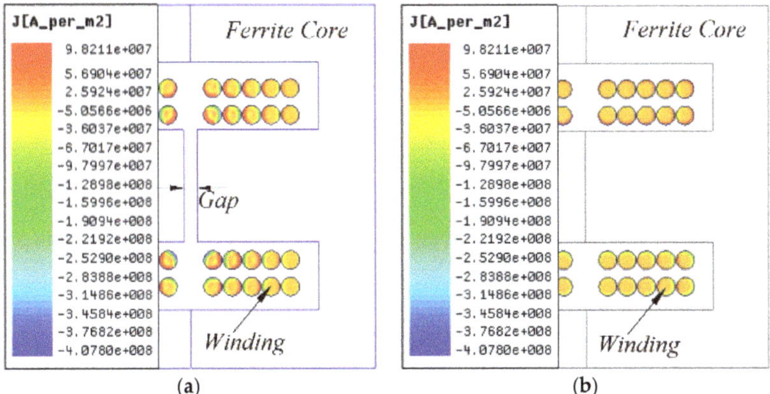

Figure 4. Fringing effect comparison: (**a**) Conventional scheme; (**b**) Proposed scheme.

However, the proposed transformer almost has no air gap, and the equivalent magnetic permeability of the magnetic core is large. Hence, its magnetizing inductance is much larger than that of conventional LLC transformer with air gap, which cannot satisfy the LLC resonant tank demand. For this reason, a SCI is parallel to the secondary winding to achieve the magnetizing inductance adjustment, as shown in Figure 5, Co is the output filter capacitor, and R_L is the load.

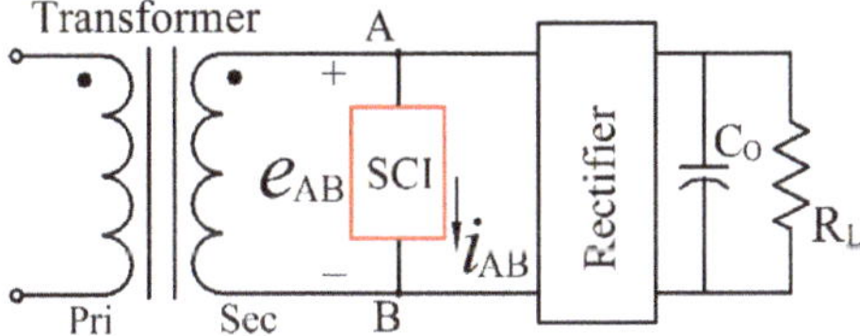

Figure 5. Switch controlled inductor parallel to secondary winding of transformer.

3.2. Switch Controlled Inductor (SCI)

The SCI in Figure 5 adopts series type [21], and its circuit structure and working waveforms are shown in Figure 6. The SCI consists of linear inductance La and two control switches Tu and Tb. The voltage e_{AB} added to the both sides of SCI is the transformer secondary winding voltage close to square wave. V_{Ta} and V_{Tb} are the drive signal of switch devices Ta and Tb respectively, and their frequencies are equal to the operating frequency of the LLC resonant converter. The current flowing through the SCI is i_{AB}. δ is the phase angle of the switch drive signal. The working principle of SCI in a switch period is analyzed as follows.

At t_0, when the secondary winding voltage e_{AB} is positive polarity and Ta is turn-off, the inductor current flow through SCI is zero. At t_1, the drive signal makes Ta turning on and Tb turning off, and i_{AB} increases linearly from zero. Therefore, Ta turns on at zero current and Tb turn off at zero current. At t_2, when the polar of e_{AB} reverses, the inductor current i_{AB} reaches a maximum and then begins to decrease linearly. At t_3, the inductor current i_{AB} decreases to zero. Because of the reverse blocking of Tb, i_{AB} remains zero until t_4. Therefore, Ta realizes zero current turn-off and Tb realizes zero current turn-on. The opposite direction is the same. Since the conduction resistance of Ta and Tb is very small, the switching loss and conduction loss of switch devices Ta and Tb are relatively small.

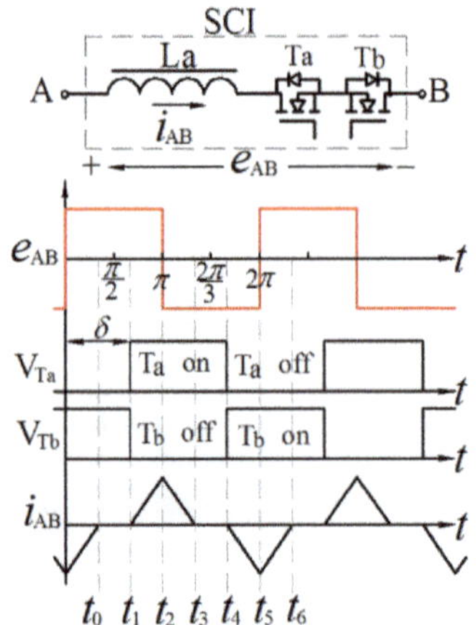

Figure 6. Topology and waveforms of SCI.

The equivalent inductance of SCI can be obtained from [22] as Equation (2)

$$L_{SCI} = \frac{L_a}{2 - (2\delta - \sin 2\delta)/\pi}$$

(2)

According to Equation (2), when the phase angle of switch drive signal ranges between $\pi/2$ and π, the equivalent switch-controlled inductance L_{SCI} ranges between L_a and ∞. The relationship curve between inductance ratio L_{SCI}/L_a and the phase angle of switch drive signal is shown in Figure 7. It can be found the circuit shown in Figure 5 can be equivalent to an inductance which can be adjusted from L_a to ∞.

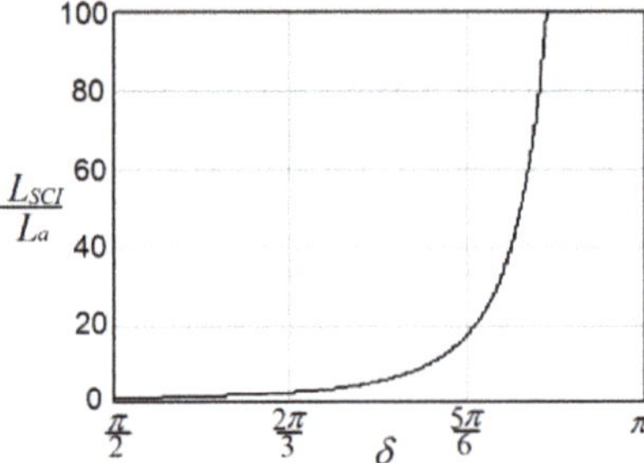

Figure 7. Equivalent inductance of SCI.

3.3. Equivalent Circuit of Resonant Tank and Magnetic Field Analysis

The equivalent circuit of LLC resonant tank for secondary winding parallel SCI is shown in Figure 8. where C_r is the resonant capacitance, L_r is the resonant inductance and $L_{m\text{-}ini}$ is the large magnetizing inductance integrated in transformer, as shown in Equation (3).

$$L_{m-ini} = \frac{\mu_{eff} \cdot N_p^2 \cdot A_c}{l_c}$$

(3)

where μ_{eff} is the equivalent magnetic permeability of the magnetic circuit approximately equal to the relative permeability of the core material μ_r. N_p is the turns of primary winding, A_c is the cross-section area of the magnetic core, l_c is the magnetic circuit length. Since $\mu_{eff} \approx \mu_r$, the value of L_{m-ini} is large as result of the large relative permeability of the core material.

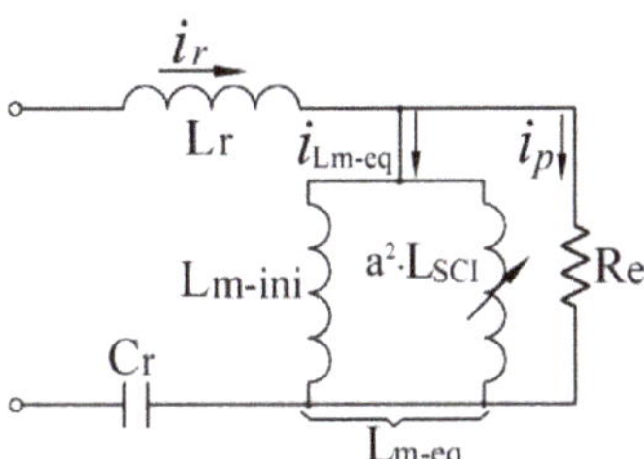

Figure 8. Equivalent circuit of the proposed LLC resonant tank.

Setting a is the ratio of the transformer, and $a^2 \cdot L_{SCI}$ is the primary equivalent inductance reflected by the SCI. Re is the equivalent load impedance, and R_{load} is the load resistance. The magnetizing inductance in the resonant tank can be expressed as Equation (4).

$$L_{m-eq} = L_{m-ini} \big\| a^2 \cdot L_{SCI} \tag{4}$$

$$R_e = \frac{8}{\pi^2} \cdot a^2 \cdot R_{load} \tag{5}$$

Due to the adjusted value of L_{SCI} driven by δ, so the inductance $a^2 \cdot L_{SCI}$ in Equation (4) can be represented as an adjustable inductance. Therefore, the equivalent magnetizing inductance L_{m-eq} can also be adjusted by changing δ.

As shown in Figure 8, when the resonant tank is transferring energy to secondary side, the resonant frequency is f_r.

$$f_r = \frac{1}{2\pi\sqrt{L_r C_r}} \tag{6}$$

Otherwise, the magnetizing inductance L_{m-eq} participates in resonant, and the resonant frequency is Equation (7).

$$f_r' = \frac{1}{2\pi\sqrt{(L_r + L_{m-eq})C_r}} \tag{7}$$

Inner magnetic flux analysis of the transformer is shown in Figure 9. The primary side is on the left, and the secondary side is on the right. The flux ϕ_p and ϕ_S in the magnetic core established respectively by the primary resonant current i_p and the secondary current i_s cancel each other. Meanwhile, the flux ϕ_{Lm-eq} and ϕ_{AB} in the magnetic core established respectively by the magnetizing current i_{Lm-eq} and the SCI current i_{AB} cancel each other. It follows that, the transformer does not store energy, and the inductive energy storage required by LLC half-bridge switch for zero voltage switching (ZVS) is completed by SCI.

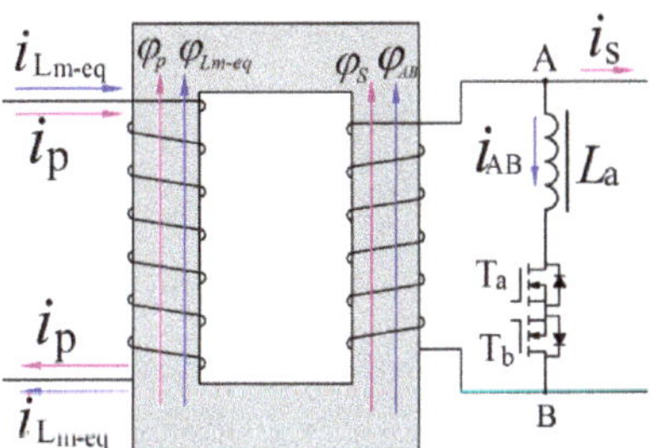

Figure 9. Magnetic flux in the core of transformer.

3.4. LLC Resonant Converter Charger Application

The application of the proposed transformer in LLC resonant converter charger is shown in Figure 10, and the secondary side adopts a full-bridge rectifier. The resonant tank consists of C_r, L_r, and $L_{m\text{-}eq}$.

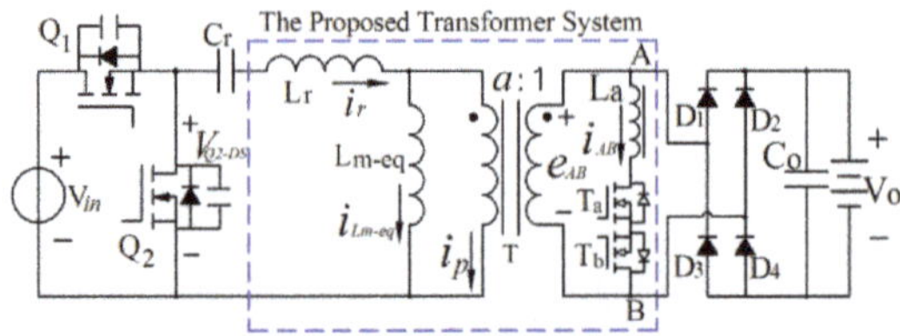

Figure 10. Proposed resonant transformer applied to LLC resonant converter charger.

The ratio of the equivalent magnetizing inductance to resonant inductance is defined as Equation (8).

$$k = \frac{L_{m-eq}}{L_r} \tag{8}$$

$$Q = \frac{\sqrt{L_r/C_r} \cdot \pi^2}{8 \cdot a^2 \cdot R_{load}} \tag{9}$$

In Equation (9), Q is the quality factor. Under constant L_r, C_r, and a, Q is only related to load. Under the same load (Q value), the voltage gain curves with different k are shown in Figure 11. The horizontal axis is the normalized frequency, which is the ratio of working frequency f_s to resonant frequency f_r, and the vertical axis is DC voltage gain. It can be seen that in the same frequency range, a resonant tank with smaller k can get a larger output voltage regulating range. Moreover, under the condition of fixed resonant inductance L_r, the proposed transformer can regulate $L_{m\text{-}eq}$ to adjust k, so that it can dynamically adjust the voltage gain of LLC resonant converter in the charging process according to the corresponding demand.

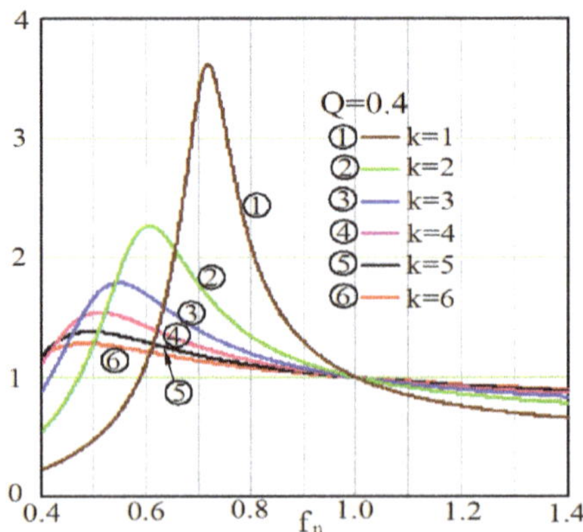

Figure 11. Gain curves with different k value.

Furthermore, the simplified typical battery charging curve is shown in Figure 12. The charging process can be divided to three stages: constant current charging, constant power charging, and constant voltage charging. The output power becomes very small at the charging end.

$$Q = \frac{\sqrt{L_r/C_r} \cdot \pi^2 \cdot I_o}{8 \cdot a^2 \cdot V_o} \tag{10}$$

At the initial stage of charging, the charger current I_o is large, and the output voltage V_o is relatively small. At the latter stage of charging, the output voltage V_o is relatively high and the output current I_o is small. According to Equation (10), the Q value should be large at the initial stage of charging. With the charging process carrying on, the Q value decreases. Figure 13 gives several gain curves of k value in the charging process. The shaded area (ZVS region) in Figure 13 is the work areas of LLC resonant converter. Usually, the operating frequency of LLC resonant converter is set as the resonant frequency with the maximum output power (point A in Figure 12), since the LLC converter can get highest efficiency [22].

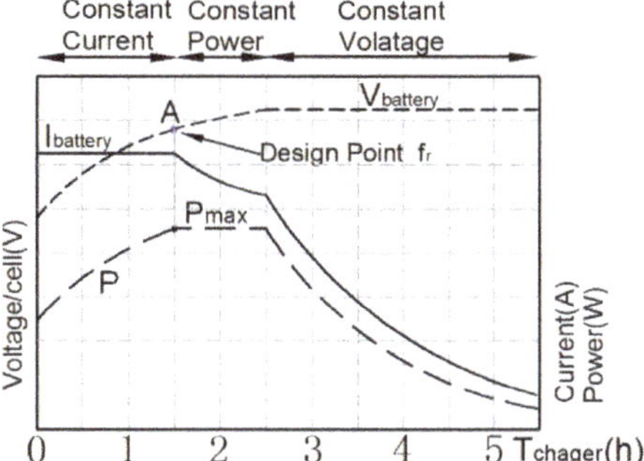

Figure 12. Simplified battery charging profile.

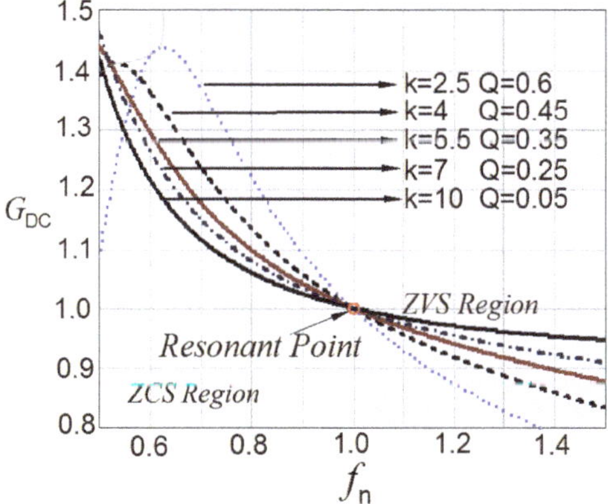

Figure 13. Change of gain curve during charging process.

Therefore, the operating frequency of the constant current charging stage in Figure 12 is located at the right of the resonant frequency ($f_n = 1$) in Figure 13. The primary switching device of LLC resonant converter works in the ZVS conduction state, but the turn-off loss increases with the increase of frequency and the secondary rectifier diode loses ZCS characteristic. Therefore, in this area, the operating frequency can be set as close as possible to the resonant frequency. As shown in Figure 13, the curve corresponding to a smaller K value is closer to the resonant frequency point (the frequency range is smaller with same gain range). When charging process enters constant power stage, the LLC resonant converter works at the left of the resonant frequency point. The primary switch of LLC resonant converter can realize ZVS conduction, and the secondary rectifier diode can realize ZCS turn-off, which can achieve better efficiency. With the decrease of operating frequency, the smaller K value leads to the increased resonant tank magnetizing current, conduction loss as well as turn-off loss. Especially, when at light load, the proportion of conduction loss in the whole loss increases and reduces overall efficiency. Therefore, with the decrease of output power and the increase of output voltage, the magnetizing inductance should gradually increase (k value gradually increases). In the conventional scheme, the K value is constant after the magnetizing inductance is selected. However, the proposed transformer can make the LLC resonant converter change its equivalent magnetizing inductance according to load and output voltage during the whole charging process, so that the gain characteristic, circulation loss, and switching loss of LLC resonant tank can be optimized via programming.

4. Design Methodology

4.1. Electrical Design Considerations

A: Selection of magnetizing inductance

The maximum magnetizing inductance to realize ZVS under idle condition is L_{m1}, and the maximum value to ensure the maximum gain at the lowest frequency is L_{m2}. The final value of the magnetizing inductance should satisfy the above two requirements at the same time. Hence, the smaller one between L_{m1} and L_{m2} can be chosen. If L_{m1} is greater than L_{m2}, the dead time t_{dead} should be appropriately reduced [23], as shown in Equation (11).

$$\begin{cases} L_{m1} = \dfrac{t_{dead} \cdot a \cdot V_{o(min)} \cdot \left(\frac{T_{s(min)}}{4} - \frac{t_{dead}}{2} \right)}{C_{HB} \cdot V_{in(max)}} \\[2ex] L_{m2} = L_r \cdot \dfrac{\pi^2}{4} \cdot \dfrac{1 - \frac{1}{G_{DC(max)}}}{f_r \cdot T_{s(max)} - 1} \\[2ex] L_{m-eq} = \min(L_{m1}, L_{m2}) \end{cases} \tag{11}$$

where $G_{DC(max)}$ is the maximum DC gain, and C_{HB} is the total equivalent capacitance of the H bridge.

B: Selection of minimum resonant inductance

The minimum resonant inductance should limit the maximum output current under short circuit when working at the highest frequency.

$$L_{r(min)} = \frac{a \cdot V_{in(norm)} \cdot V_{o(norm)}}{8 f_{s(max)} \cdot P_o} \tag{12}$$

C: Selection of resonant capacitance

After L_r is selected, the resonant capacitance can be obtained by Equation (13).

$$C_r = \frac{1}{(2\pi \cdot f_r)^2 L_r} \tag{13}$$

4.2. Transformer Loss and Thermal Design Considerations

The loss of transformer includes core loss and winding loss.

A: Core loss

The core loss and core loss can be calculated by Steinmetz formula.

$$P_{fe} = V_c K_c f^\alpha B_{max}{}^\beta \tag{14}$$

where V_c is the core volume, K_c is the typical value. α and β are provided by core manufacturer or obtained by loss curve.

B: Winding loss

The winding loss include DC loss and AC loss. The current through the winding can be calculated as below.

The current through the primary winding i_r is the sum of i_p and i_{Lm-eq}, and the current through secondary winding is the sum of i_s and i_{AB}. The expressions of the peak current through primary winding is I_{r-peak} with the RMS value I_{r-rms}, and the peak magnetizing current $I_{Lm-peak}$ are [7]

$$I_{r-peak} = \sqrt{\left(\frac{\pi \cdot I_0}{2af_n}\right)^2 + \left(\frac{aV_0}{4f_r L_{m-eq}}\right)^2} \tag{15}$$

$$I_{Lm-peak} = \frac{aV_0}{4L_{m-eq}f_r} \tag{16}$$

$$I_{r-rms} = \sqrt{\frac{a^2 V_0{}^2 T_r{}^2 (2T_s - T_r)}{32 L_m{}^2 T_s} + \frac{\pi^2 I_0{}^2 T_s{}^2}{8a^2 T_r{}^2}} \tag{17}$$

where T_s and T_r are the switching period and resonant period respectively, V_o and I_o are the output voltage and output current, respectively.

By using Equation (18), the RMS value of the current though secondary winding can be obtained.

$$i_{s\_rms} = \sqrt{\frac{2 \cdot \int_0^{\frac{T_r}{2}} (aI_{r\_peak}\sin[\omega_r t + \phi] + \frac{a^2 V_0}{4L_m f_r} - \frac{a^2 V_0}{L_m}t)^2 dt}{T_r}} \tag{18}$$

where $\omega_r = 2\pi f_r$, $\phi = arctan\left(-\frac{a^2 R_L f_s}{\omega_r L_m f_r}\right)$.

The DC copper loss is calculated as below.

$$P_{cu} = R_{p-rms}I^2{}_{r-rms} + R_{s-rms}I^2{}_{s-rms} \tag{19}$$

The AC impedance of winding can be got from Equation (1).

C: Temperature rise consideration

The maximum loss is determined by the core thermal resistance and the permissible temperature increase.

$$P_{L\_max} \approx \frac{\Delta T}{R_\theta} \tag{20}$$

where R_θ is the thermal resistance of the core provided by manufacturer or obtained from empirical data, h_c is thermal conductivity, and A_t is the surface area of transformer.

$$R_\theta = \frac{1}{h_c A_t} \tag{21}$$

4.3. Transformer Design Considerations

Under the premise of minimize core loss and winding loss, the design purpose of a transformer is to transfer energy from input side to output side by electromagnetic induction. The optimization result can be boiled down to one conclusion: iron loss is equal to copper loss [23]. The transformer design method is related to the best magnetic induction intensity and temperature rise of magnetic core and is limited by the maximum permissible power loss.

A: Core selection

The appropriate core is up to A_p value. A_p is the product of core window area W_a and core cross-section area A_c, as Equation (22).

$$A_p = \left(\frac{\sqrt{2}\sum VA}{K_v f_s B_o k_f K_t \sqrt{k_u \Delta T}} \right)^{\frac{8}{7}} \tag{22}$$

where, $\sum VA$ is the sum of each windings rated VA values, $K_v = 4.44$, f_s is the operating frequency, B_o is the best magnetic induction intensity value, $K_t = 48.2 \times 103$, k_f is the core lamination factor, A_m is the effective sectional area of magnetic circuit.

The best magnetic induction intensity B_o is given by Equation (23).

$$B_o = \frac{(h_c k_a \Delta T)^{\frac{2}{3}}}{2^{\frac{2}{3}} (\rho_w k_w k_u)^{\frac{1}{12}} (k_c K_C f^\alpha)^{\frac{7}{12}}} \left(\frac{K_v f k_f k_u}{\sum VA} \right)^{\frac{1}{6}} \tag{23}$$

where, h_c is the thermal convection transfer coefficient with typical value 10, and k_a, k_c, and k_w are dimensionless constants with the typical values $k_a = 40$, $k_c = 5.6$, $k_w = 10$. K_C, α are the material parameters, ρ_w is the wire resistivity.

B: Calculation of transformer winding turns and turns ratio

The winding turns can be calculated by Equation (24).

$$N = \frac{V_{rms}}{K_v f B_{max} A_m} \tag{24}$$

where V_{rms} is the wingding terminal voltage, A_m is the core cross-section area, B_{max} is the smaller one between B_o and B_{sat}. B_o is usually smaller than B_{sat} at the high frequency condition.

The turns ratio can be calculated by Equation (25). where, V_d is the conduction voltage drop of the secondary rectifier diode.

$$a = \frac{V_{in(norm)}}{2(V_{o(min)} + V_d)} \tag{25}$$

C: Transformer wire diameter selection

The current density J_0 in the wire should satisfy the temperature rise requirement under the whole power loss.

$$J_0 = K_t \frac{\sqrt{\Delta T}}{\sqrt{k_u(1+\gamma)} \ \sqrt[8]{A_p}} \tag{26}$$

The calculation of wire sectional area A_w is Equation (27).

$$A_w = \frac{I_{rms}}{J_0} \tag{27}$$

D: Determination of inductance La

According to Equation (2) and Figure 7, the switch-controlled inductance L_{SCI} can be adjusted in a wide range. When $\alpha = \pi/2$, $L_{SCI} = La$. While the k value of LLC resonant tank obtains a minimum value k_{min} as Equation (28) According to Equation (4).

$$L_a = \frac{k_{min} \cdot L_r \cdot L_{m-ini}}{a^2 (L_{m-ini} - k_{min} \cdot L_r)} \tag{28}$$

5. Experimental Verification

5.1. Design Specification

An LLC resonant DC/DC converter with 720 W output power is designed, its input voltage is 390 VDC and output voltage is 60 VDC–96 VDC with maximum output current is 8 A. The resonant frequency is 85 kHz. According to Equation (12), the resonant inductance L_r should be greater than 44 μH and set as 50 μH. According to Equation (13), the resonant capacitance C_r is 70 nF. According to Equation (24), the transformer turns ratio a = 2:1. According to Equation (28), k_{min} is 3, L_{min} is 1 mH and La = 44 μH.

The design requirement is shown in Table 1, the RMS value of resonant current, the peak magnetizing current and the RMS value of secondary current are calculated by Equations (15)–(18). The core window utilization factor k_u is 0.55.

Table 1. Design Specification.

Symbol	Description	Value
P_o	Output power	760 W
f_s	Switching frequency	50 k–300 kHz
f_{r1}	Resonant frequency	85 kHz
L_a	Magnetizing inductance	44 μH
L_r	Resonant inductance	50 μH
a	Turns ratio	2:1
ΔT	Temperature rise	70 °C
T_a	Ambient temperature	40 °C
I_{r-rms}	Resonant rms current	4.53 A
$I_{Lm-peak}$	Magnetizing peak current	1.26 A
I_{s-rms}	Secondary rms current	9.1 A
I_{AB-rms}	SCI current	2 A
$\sum VA$	The total rated VA values of windings	1466.3 VA

5.2. Transformer Parameters

The ferrite core is used for the high switch frequency. B_o = 0.1 T is calculated by Equation (23), A_p = 2.5 cm^4 is calculated by Equation (22), the ETD44 magnetic core is chosen. The magnetic material is the N87 from EPCOS company, the parameters of this material are given in Table 2, and the parameters of core and winding are given in Table 3. The saturation flux density of this material is 0.4 T. By Equation (24), the transformer secondary turns can be calculated as Ns = 11.9 ≈ 12. The primary turns number is 24.

Table 2. Material Specifications (Epcos N87).

Symbol	Description	Value
K_c	Steinmetz parameter	16.9
α	Steinmetz parameter	1.25
β	Steinmetz parameter	2.35
B_{sat}	Saturation magnetic flux	0.4 T
k_u	Window utilization factor	0.55
A_m	Effective magnetic circuit sectional area	213 mm^2
γ	Ratio of core loss to winding loss	1

Table 3. Core and Winding Parameters.

Symbol	Description	Value
A_c	Cross-section area	1.73 cm^2
l_c	Magnetic path length	10.3 cm
W_a	Window area	2.78 cm^2
A_p	Area product parameter	4.81 cm^4
V_c	Volume of core	17.70 cm^3
MLT	Mean length of a turn	7.77 cm
k_f	Core stacking factor	1.0
ρ_{20}	Copper resistivity (20 °C)	1.72 μΩ·cm
α_{20}	Constant	0.00393

The current density can be obtained by Equation (26) as 344 A/cm^2, and the wire sectional area of the primary winding is 0.0132 cm^2 from Equation (27). Considering the wire skin effect at high frequency, the skin depth is

$$\lambda_0 = \frac{66}{\sqrt{f_s}} \tag{29}$$

It is set as λ_0 = 0.226 mm and the specification parameters are shown in Table 4. The DC resistance of primary winding is 120.3 μΩ/cm (20 °C).

Table 4. Core and Winding Parameters.

AWG	Sectional Area	Resistivity	Diameter
24	0.2047 mm^2	842.1 μΩ/cm	0.51 mm

To reduce the secondary proximity effect, the Litz wire of 0.15 mm × 110 mm is chosen. The equivalent electric conduction area is 1.95 mm^2, the current density is 464 A/cm^2, and the primary DC internal resistance is 89.2 μΩ/cm (20 °C).

The DC internal resistance of each winding can be obtained with temperature correction as

$$R_{cu-dc} = N \cdot MLT \cdot \rho_w \cdot [1 + \alpha_{20}(T_{max} - 20\,°C)] \tag{30}$$

where T_{max} is the maximum temperature.

Considering the wire skin effect at high frequency, the skin effect factor is calculated by

$$\frac{R_{ac-pri}}{R_{dc-pri}} = 1 + \frac{(r_o/\delta_0)^4}{48 + 0.8(r_o/\delta_0)^4} = 1.019 \tag{31}$$

where, r_o. is the radius of wire and r_0 = 0.255 mm.

The influence of the proximity effect of the secondary winding is calculated by

$$\Delta_s = \frac{d_s}{\delta_0} = \frac{0.15}{0.26} = 0.577, \quad p_s = 4 \tag{32}$$

where, d_s is the diameter of wire. The ratio of secondary AC resistance and DC resistance is R_{ac_S}/R_{dc_S} = 1.2. Power loss and the efficiency of transformer are shown in Table 5. It is worth noting the efficiency of the transformer is calculated under the condition of resonant frequency and full power.

Table 5. Loss Calculations.

Symbol	Description	Value
$\dot{P}_{max}$	Allowed Maximum power loss	4.9 W
P_{cu_dc}	DC winding loss	1.56 W
P_{cu_ac}	AC winding loss	1.74 W
P_{fe}	Core loss	1.94 W
P_{La}	The loss of SCI	1.1 W
P_{tol}	Total loss	4.78 W
η	Transformer efficiency	99.37%

5.3. Experimental Results

The designed transformer is applied to 720 W LLC resonant converter charger. The experimental waveforms under different operating conditions are given in Figure 14. The operating waveform of LLC resonant converter at the initial stage of constant current charging is given in Figure 14a. The output power is 608 W and the operating frequency is 98.1 kHz, which is higher than the resonant frequency. The phase-shifting angle of the SCI is $\alpha = \pi/2$, and the current of the SCI branch i_{AB} is triangular wave. At this time, the magnetizing inductance of the resonant tank is small, the K value is also small, which helps reducing the switching frequency range.

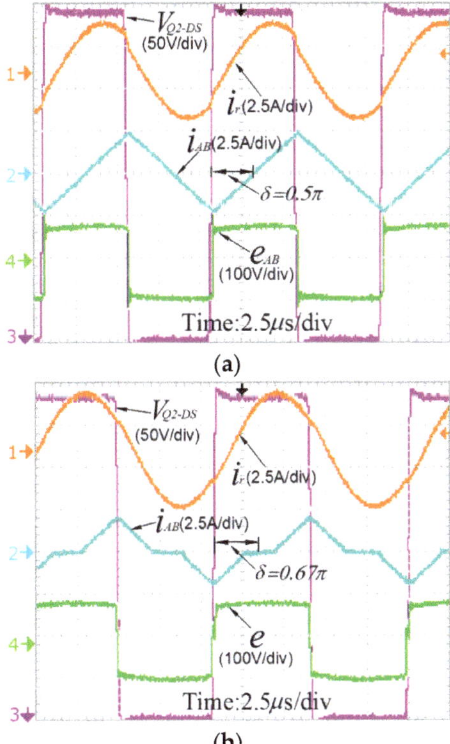

Figure 14. *Cont.*

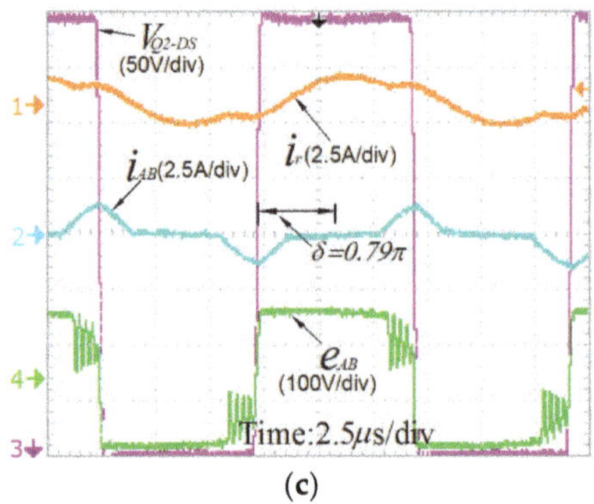

(c)

Figure 14. Experimental waveforms. (a) Constant Current Stage with $V_o = 76$ V and $I_o = 8$ A ($\alpha = 0.5\pi$); (b) Constant Power Stage with $V_o = 90$ V and $I_o = 8$ A ($\alpha = 0.67\pi$); (c) Constant Voltage Stage with $V_o = 96$ V and $I_o = 2.5$ A ($\alpha = 0.79\pi$).

Figure 14b shows the working waveform of LLC resonant converter at the constant power charging stage. The output power is 720 W and the operating frequency is 84.8 kHz, near the resonant frequency. The phase-shifting angle of the switch-controlled inductor is $\alpha \approx 0.67\pi$, and the magnetizing current of the transformer is relatively decreased, which reduces the conduction loss.

Figure 14c shows the working waveform of LLC resonant converter at constant voltage charging stage. The output power is 240 W, the operating frequency is 70.2 kHz, which is lower than the resonant frequency, and the output power is smaller. The phase-shifting angle of the switch-controlled inductor is $\alpha \approx 0.79\pi$, the magnetizing current of the transformer is decreased further and the light load efficiency is improve effectively.

The experiment results comparison of the transformer temperature after operating under a full load for 1 h is shown in Figure 15. The temperature rise of the proposed transformer is 65.2 °C. Experiments show that the transformer works in an ideal temperature rise range. In addition, the conventional transformer with air gap is applied to the same resonator circuit, the transformer retaining wall interval d is 4.3 mm, the temperature rise is measured as 83.8 °C. Moreover, the temperature rise of the secondary winding near the air-gap is the highest, which is 14.9 °C higher than the proposed transformer.

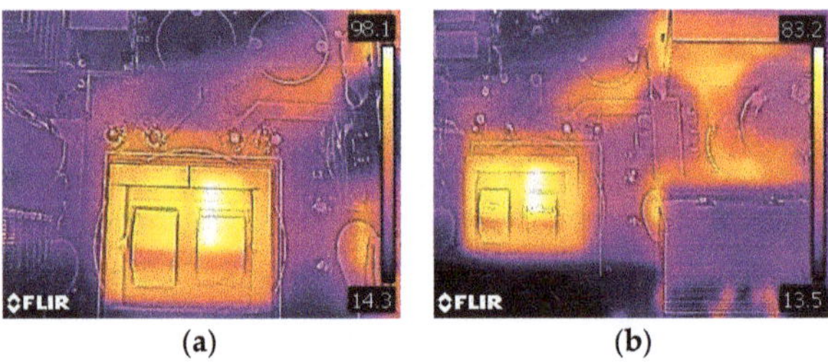

Figure 15. Temperature rise comparison. (a) The conventional scheme; (b) The proposed scheme.

Corresponding to Figure 14, the measured loss of the proposed transformer under 100%, 80%, and 33% loads are compared with those of the conventional LLC transformer, as shown in Figure 16. Experimental results show that the proposed design scheme is more efficient to reduce the temperature rise of magnetic components.

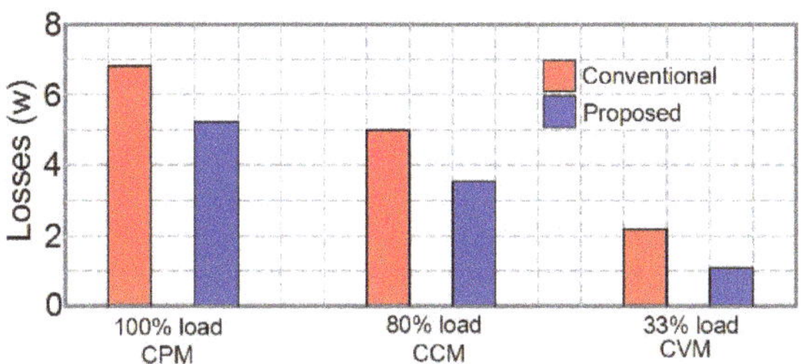

Figure 16. Measured loss comparison.

Figure 17 shows the efficiency comparison between the proposed LLC resonant converter charger and the conventional LLC resonant converter charger ($k = 5$) in the whole charging process. At the left of point A, the operating frequency is higher than resonant frequency. The proposed transformer magnetizing inductance can be dynamically adjusted, which not only improves the efficiency of the transformer itself, but also effectively reduces the conduction loss and switching loss of the LLC resonant converter. Hence, the comprehensive efficiency can be improved.

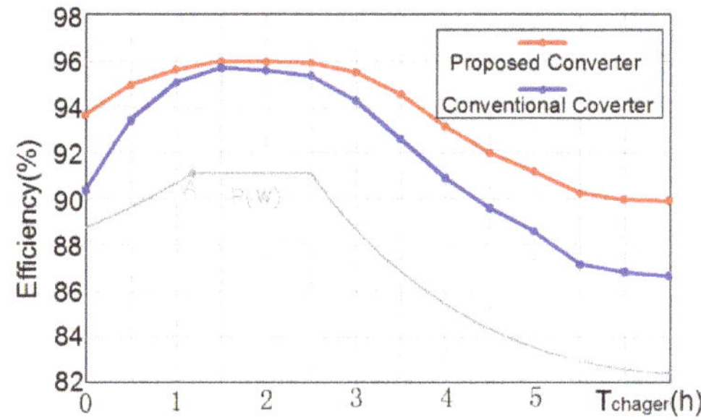

Figure 17. Efficiency comparison in the whole charging process.

6. Conclusions

Aiming at the relatively low efficiency problem of the conventional integrated transformer in LLC resonant converter, a controlled magnetizing inductance resonant transformer suitable for high power LLC resonant converter is proposed. The proposed transformer mainly removes the influence of fringing effect, improving the efficiency of the transformer, and the temperature rise of transformer. The experimental results show that the temperature rise of the transformer is reduced by 14.9 °C compared with the conventional scheme. Moreover, according to the change of charging process (output voltage and equivalent load impedance changes), the magnetizing inductance of proposed transformer can be dynamically adjusted to reduce the conduction loss and switching loss of the LLC resonant converter. Therefore, the comprehensive efficiency of the whole charging process can be improved further, especially at the initial charging stage and at the charging end (light load).

Author Contributions: All the authors conceived and designed the study. Z.Z., Q.X. performed the simulation and the experiment and wrote the manuscript with guidance from Y.D., Z.Z., Q.X., Y.D. and H.Y. reviewed the manuscript and provided valuable suggestions.

Funding: This work was supported by the National Natural Science Foundation of China (NSFC) under Grant No. 51807056.

References

1. Deng, J.; Mi, C.C.; Ma, R.; Li, S. Design of LLC resonant converters based on operation-mode analysis for level two PHEV battery chargers. *IEEE/ASME Trans. Mechatron.* **2015**, *20*, 1595–1606. [CrossRef]
2. Demirel, I.; Erkmen, B. A Very Low-Profile Dual Output LLC Resonant Converter for LCD/LED TV Applications. *IEEE Trans. Power Electron.* **2014**, *29*, 3514–3524. [CrossRef]
3. Uno, M.; Kukita, A. Two-Switch Voltage Equalizer Using an LLC Resonant Inverter and Voltage Multiplier for Partially Shaded Series-Connected Photovoltaic Modules. *IEEE Trans. Ind. Appl.* **2015**, *51*, 1587–1601. [CrossRef]
4. Yang, B.; Lee, F.C.; Zhang, A.J.; Huang, G. LLC resonant converter for front end DC/DC conversion. In Proceedings of the Seventeenth Annual IEEE Applied Power Electronics Conference and Exposition, Dallas, TX, USA, 10–14 March 2002; pp. 1108–1112.
5. Wang, H.; Li, Z. A PWM LLC Type Resonant Converter Adapted to Wide Output Range in PEV Charging Applications. *IEEE Trans. Power Electron.* **2018**, *33*, 3791–3801. [CrossRef]
6. Jeong, Y.; Moon, G.W.; Kim, J.K. Analysis on half-bridge LLC resonant converter by using variable inductance for high efficiency and power density server power supply. In Proceedings of the 2017 IEEE Applied Power Electronics Conference and Exposition (APEC), Tampa, FL, USA, 26–30 March 2017; pp. 170–177.
7. Zhang, J.; Hurley, W.G.; Wölfle, W.H. Gapped Transformer Design Methodology and Implementation for LLC Resonant Converters. *IEEE Trans. Ind. Appl.* **2016**, *52*, 342–350. [CrossRef]
8. Saket, M.A.; Shafiei, N.; Ordonez, N. LLC Converters with Planar Transformers: Issues and Mitigation. *IEEE Trans. Power Electron.* **2017**, *32*, 4524–4542. [CrossRef]
9. Watanabe, T.; Kurokawa, F. Efficiency comparison between phase shift and LLC converters as power supply for information and communication equipment. In Proceedings of the 2015 IEEE International Telecommunications Energy Conference (INTELEC), Osaka, Japan, 18–22 October 2015; pp. 1–5.
10. Kim, E.S.; Kang, C.H.; Hwang, I.G.; Lee, Y.S.; Huh, D.Y. LLC Resonant Converter Using A Planar Transformer with New Core Shape. In Proceedings of the 2014 IEEE Applied Power Electronics Conference and Exposition—APEC 2014, Fort Worth, TX, USA, 16–20 March 2014; pp. 3374–3377.
11. Yamamoto, T.; Bu, Y.; Mizuno, T.; Yamaguchi, Y.; Kano, T. Loss Reduction of Transformer for LLC Resonant Converter Using a Magneto plated Wire. In Proceedings of the 2016 19th International Conference on Electrical Machines and Systems (ICEMS), Chiba, Japan, 13–16 November 2016; pp. 1–6.
12. Kang, B.G.; Park, C.S.; Chung, S.K. Integrated transformer using magnetic sheet for LLC resonant converter. *Electron. Lett.* **2014**, *50*, 770–771. [CrossRef]
13. Stadler, A.; Gulden, C. Copper Losses of Litz-Wire Windings Due to an Air Gap. In Proceedings of the 2013 15th European Conference on Power Electronics and Applications (EPE), Lille, France, 2–6 September 2013; pp. 1–7.
14. Zhang, J.; Hurley, W.G.; Wölfle, W.H. Design of the Planar Transformer in LLC Resonant Converters for Micro-grid applications. In Proceedings of the 2014 IEEE 5th International Symposium on Power Electronics for Distributed Generation Systems (PEDG), Galway, Ireland, 24–27 June 2014; pp. 1–7.
15. Yang, S.; Abe, S.; Shoyama, M. Design Consideration of Two Flat Transformers in a Low-Profile LLC Resonant Converter. In Proceedings of the 8th International Conference on Power Electronics—ECCE Asia, Jeju, Korea, 30 May–3 June 2011; pp. 854–859.
16. Yang, S.; Abe, S.; Shoyama, M. Design Consideration of Flat Transformer in LLC Resonant Converter for Low Core Loss. In Proceedings of the 2010 International Power Electronics Conference (IPEC), Sapporo, Japan, 21–24 June 2010; pp. 343–348.
17. Alabakhshizadeh, A.; Midtgård, O.M. Winding Loss Analysis and Optimization of an AC Inductor for a Galvanically Isolated PV Inverter. In Proceedings of the 2012 International Conference and Exposition on Electrical and Power Engineering, Iasi, Romania, 25–27 October 2012; pp. 705–708.
18. Alabakhshizadeh, A.; Midtgård, O.M. Air Gap Fringing Flux Reduction in a High Frequency Inductor for a Solar Inverter. In Proceedings of the 2013 IEEE 39th Photovoltaic Specialists Conference (PVSC), Tampa, FL, USA, 16–21 June 2013; pp. 2849–2852.
19. Huang, W.N.; Lee, S.H.; Chen, C.G. Light-Load Efficiency Improvement Strategy for LLC Resonant Converter Utilizing a Step-Gap Transformer. In Proceedings of the 2014 International Power Electronics Conference, Hiroshima, Japan, 18–21 May 2014; pp. 1734–1737.

20. Hua, C.C.; Fang, Y.H.; Lin, C.W. LLC resonant converter for electric vehicle battery chargers. *IET Power Electron.* **2016**, *9*, 2369–2376. [CrossRef]
21. Gu, W.J.; Harada, K. A new method to regulate resonant converters. *IEEE Trans. Power Electron.* **1998**, *3*, 430–439.
22. Musavi, F.; Craciun, M.; Gautam, D.S.; Eberle, W.; Dunford, W.G. An LLC Resonant DC–DC Converter for Wide Output Voltage Range Battery Charging Applications. *IEEE Trans. Power Electron.* **2013**, *28*, 5437–5445. [CrossRef]
23. Hurley, W.G.; Wölfle, W.H. *Transformers and Inductors for Power Electronics. Theory, Design and Applications*; Wiley: London, UK, 2013.

Article

An Analysis of Non-Isolated DC-DC Converter Topologies with Energy Transfer Media

Se-Un Shin

Electrical Engineering and Computer Science, University of Michigan, Ann Arbor, MI 48109, USA; seuns@umich.edu

Received: 30 March 2019; Accepted: 15 April 2019; Published: 18 April 2019

Abstract: As miniaturized mobile devices with various functionalities are highly desired, the current requirement for loading blocks is gradually increasing. Accordingly, the efficiency of the power converter that supports the current to the loading bocks is a critical specification to prolong the battery time. Unfortunately, when using a small inductor for the miniaturization of mobile devices, the efficiency of the power converter is limited due to a large parasitic DC resistance (R_{DCR}) of the inductor. To achieve high power efficiency, this paper proposes an energy transfer media (ETM) that can make a switched inductor capacitor (SIC) converter easier to design, maintaining the advantages of both a conventional switched capacitor (SC) converter and a switched inductive (SI) converter. This paper shows various examples of SIC converters as buck, boost, and buck-boost topologies by simply cascading the ETM with conventional non-isolated converter topologies without requiring a sophisticated controller. The topologies with the ETM offer a major advantage compared to the conventional topologies by reducing the inductor current, resulting in low conduction loss dissipated at R_{DCR}. Additionally, the proposed topologies have a secondary benefit of a small output voltage ripple owing to the continuous current delivered to the load. Extensions to a multi-phase converter and single-inductor multiple-output converter are also discussed. Furthermore, a detailed theoretical analysis of the total conduction loss and the inductor current reduction is presented. Finally, the proposed topologies were simulated in PSIM, and the simulation results are discussed and compared with conventional non-isolated converter topologies.

Keywords: switched inductor capacitor converter; a power converter; energy transfer media; ripple voltage; efficiency; conduction loss

---

1. Introduction

The use of high-performance, power-hungry mobile devices has increased recently, prompting the need for longer battery life [1,2]. Accordingly, power management integrated circuits (PMICs) for mobile devices are becoming important. PMICs consist of a linear regulator, a switched capacitor (SC) converter, and a switched inductor (SI) converter [3,4]. Although linear regulators offer the advantage of low output voltage ripple, they have low power efficiency [5–8]. In contrast, SC converters have high power density with better power efficiency than linear regulators, but they suffer from severe degradation of efficiency when the conversion ratio of the SC converter differs from a pre-defined value [9–12]. Furthermore, when the load current (I_{LOAD}) increases, which is referred to as a heavy load condition, SC converters require many large external capacitors. Therefore, neither linear regulators nor SC converters are good candidates for powering high performance loading blocks that require a large I_{LOAD}. On the other hand, a SI converter with an external inductor is an efficient solution in heavy load conditions [13–16]. Buck, boost, and buck-boost SI converters exist for generating lower, higher, and lower or higher output voltage (V_O), respectively, compared with the battery voltage (V_{IN}).

In the SI converter, there are two representative power losses, as shown in Figure 1. One is the switching loss (P_{SW}) which is proportional to the switching frequency. When the switching frequency is fixed, the P_{SW} is a constant independent of the I_{LOAD}. The other is the conduction loss (P_{cond}). Since the P_{cond} is proportional to the square of the current, the P_{cond} is dominant in heavy load conditions. Therefore, reducing P_{cond} is important for improving power efficiency when I_{LOAD} is large.

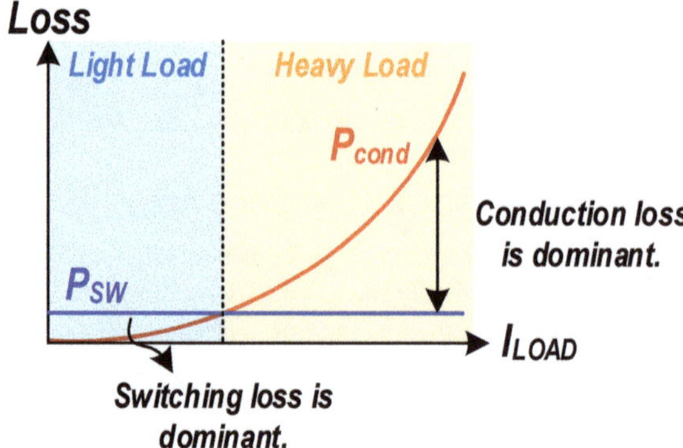

Figure 1. Graph of the switching loss and conduction loss in different conditions.

However, due to a large inductor current (i_L), under heavy load conditions, these efficient SI converters also suffer from significant conduction loss (P_{DCR}) dissipated at a parasitic DC resistance (R_{DCR}) of a small inductor for size-limited mobile devices as shown in Figure 2. This large P_{DCR} causes a severe thermal problem as well as low power efficiency in heavy load conditions. P_{DCR} is expressed as follows:

$$P_{DCR} = i_{L,RMS}{}^2 R_{DCR} = (I_L{}^2 + \frac{\Delta i_L{}^2}{12})R_{DCR} \tag{1}$$

where $i_{L,RMS}$, I_L, and Δi_L are the root-mean-square value, the DC value, and the ripple of the i_L, respectively. Since the small inductor for the miniaturized mobile device can have much larger R_{DCR} than the on-resistance of switches, reducing P_{DCR} can achieve a significant improvement in power efficiency. To minimize the P_{DCR}, reducing $i_{L,RMS}$ is the only solution when a large R_{DCR} of the small inductor is used as shown in Equation (1). In particular, as the inductor with larger R_{DCR} than the on-resistance of the switch is adopted, the efficiency improvement due to low $i_{L,RMS}$ is significantly increased.

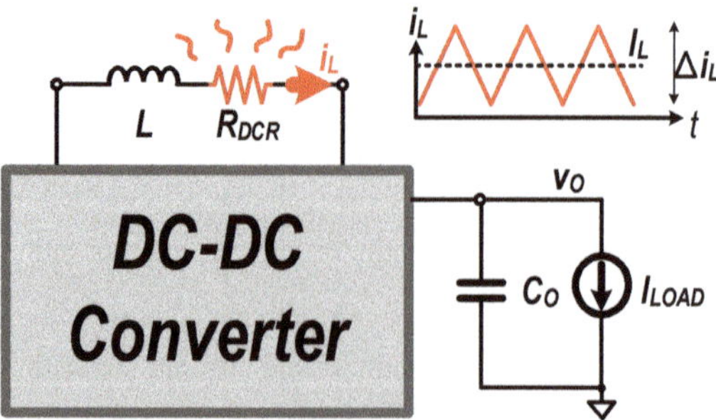

Figure 2. Conduction loss of a parasitic DC resistance (R_{DCR}) of the inductor in heavy load conditions.

There are some alternative topologies that can reduce $i_{L,RMS}$ in a SI converter. For example, Figure 3a shows a multi-level structure with an additional flying capacitor that reduces Δi_L, thereby improving power efficiency in light or medium load conditions [17,18]. However, under heavy loads, since I_L is much larger than Δi_L, the reduction of P_{DCR} is limited. Alternatively, Figure 3b shows a multi-phase structure that can reduce I_L and can result in increased power efficiency compared with the multi-level structure in heavy load conditions. However, it requires an additional inductor that is larger and more expensive than other passive components [19–21]. Furthermore, both the multi-level and the multi-phase structures require complex balancing circuits, as shown in Figure 3.

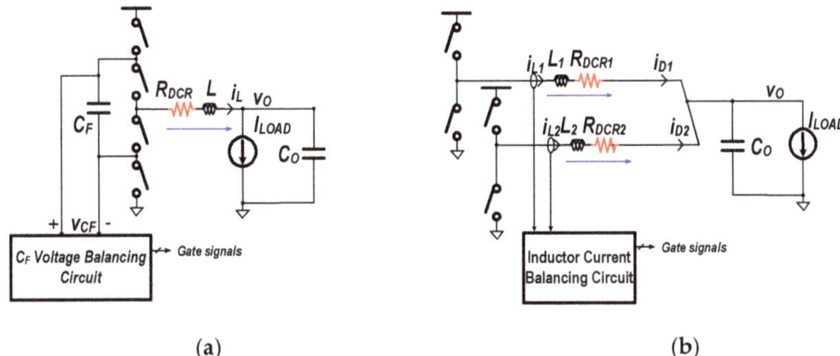

Figure 3. Alternative topologies for reducing conduction loss in R_{DCR}: (**a**) multi-level buck converter; (**b**) multi-phase buck converter.

To resolve these issues, this paper proposes and analyzes a new type of hybrid switched inductor capacitor (SIC) converter with energy transfer media (ETM) using an additional flying capacitor. The topologies with the ETM provide improved efficiency by lowering I_L owing to an additional current path in heavy load conditions.

The topologies with the proposed ETM are introduced in Section 2. In Sections 3 and 4, the operation principle and a detailed conduction loss analysis of both the buck and buck-boost topologies are provided. In Section 5, different examples of extension to other topologies are explained and discussed. The simulation results for verification are presented in Section 6. Finally, a brief concluding summary is given in Section 7.

2. Energy Transfer Media

A hybrid SIC converter that possesses the advantages of both a SC converter and a SI converter is an attractive solution [22–30]. However, it is complicated to design because of the many complex combinations of power switches and flying capacitors. Also, various factors such as conversion ratios, balancing circuits, and power loss should be considered. To make it easy to design, and to reduce P_{DCR} at the same time, this paper proposes an ETM that can be easily implemented with all types of non-isolated converters, such as buck, boost, and buck-boost topologies, with high efficiency under heavy load conditions. An ETM has been used previously to reduce only Δi_L [29,30]. However, similar to a multi-level converter, this structure is not effective at improving power efficiency under heavy load conditions. Therefore, we propose an ETM that uses an additional flying capacitor to obtain high efficiency in heavy load conditions, as shown in Figure 4.

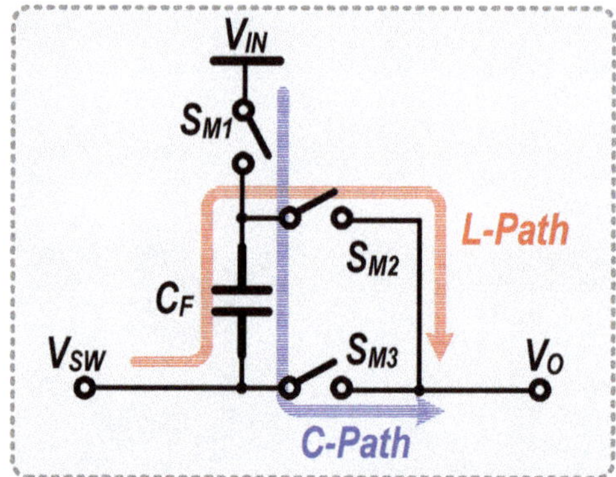

Figure 4. Energy transfer media with a flying capacitor.

The ETM uses one external flying capacitor (C_F) and three power switches (S_{M1}–S_{M3}). This approach offers the advantage of inserting the capacitor current path into the output current path (C-path) as well as the inductor current path (L-path). This ETM with dual current paths can be applied to conventional non-isolated topologies by simply cascading the ETM, as shown in Figure 5. Figure 5a is a conceptual structure showing that the ETM can be applied to conventional converter topologies. Figure 5b–d show examples of applying the ETM to buck, boost, and buck-boost converters, respectively. This paper analyzes buck type and buck-boost type topologies with ETMs as examples and discusses possible extensions to other topologies.

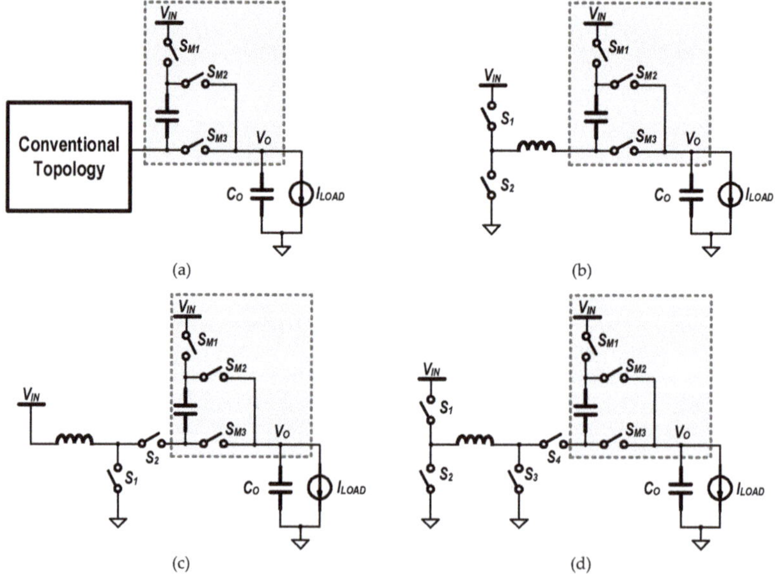

Figure 5. (**a**) Conventional topology with an energy transfer media; (**b**) buck converter; (**c**) boost converter; (**d**) buck-boost converter.

3. Buck Converter with Energy Transfer Media

Figure 5b shows a buck converter with ETM (BKETM), which is composed of power switches S_1–S_2 and S_{M1}–S_{M3}, one inductor (L), one flying capacitor (C_F), and one output capacitor (C_O). The BKETM uses two operating modes (Φ_1, Φ_2), as shown in Figure 6a. The operation waveforms of the BKETM are shown in Figure 6b.

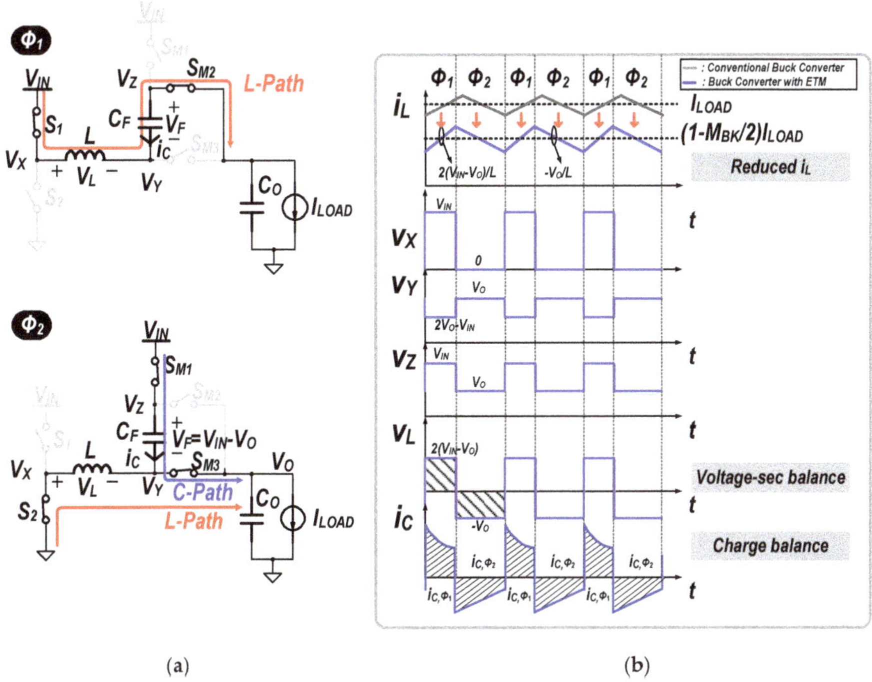

Figure 6. Operating mode (**a**) and waveforms (**b**) of buck converter with energy transfer media (BKETM).

In Φ_1, S_1 and S_{M2} are turned on, and S_2, S_{M1}, and S_{M3} are turned off. At this time, i_L is built up with a slope of $2(V_{IN} - V_O)/L$ and is delivered to the output. In Φ_2, S_2, S_{M1}, and S_{M3} are turned on, and S_1 and S_{M2} are turned off. While i_L is de-energized with a slope of $-V_O/L$, it is also delivered to the output. In the meantime, the capacitor current i_C of C_F flows to the output, while the voltage of C_F is charged to $V_{IN} - V_O$. To derive a conversion ratio (M_{BK}), applying the voltage sec balance to the inductor with duty cycle D is expressed as follows:

$$D(2V_{IN} - 2V_O) - (1-D)V_O = 0. \tag{2}$$

Simplifying Equation (2), M_{BK} is given by:

$$M_{BK} = \frac{V_O}{V_{IN}} = \frac{2D}{1+D} \quad 0 < D < 1. \tag{3}$$

The M_{BK} of the BKETM from Equation (3) has a value between 0 and 1 as D varies from 0 to 1, which is the same as the range of a conventional buck converter (CBK). Therefore, in spite of the SIC converter, the BKETM behaves like a CBK without the limit of the conversion ratio.

To obtain the average value of the C-path current ($I_{C,\Phi2}$) delivered to the output in Φ_2, we also apply charge balance to C_F as shown below:

$$DI_L - (1-D)I_{C,\Phi2} = 0. \tag{4}$$

Simplifying Equation (4), $I_{C,\Phi2}$ is given by

$$I_{C,\Phi2} = \frac{D}{1-D}I_L.$$

(5)

Applying the charge balance to the output capacitor C_O,

$$D(I_L - I_{LOAD}) + (1-D)(I_L + I_{C,\Phi2} - I_{LOAD}) = 0.$$

(6)

Substituting Equation (5) into Equation (6), I_L can be expressed with I_{LOAD} as shown below:

$$I_L = \frac{1}{1+D}I_{LOAD} = (1 - \frac{M_{BK}}{2})I_{LOAD}.$$

(7)

For the CBK, I_L is always the same as I_{LOAD} [3,4,13–16]. In contrast, for the proposed BKETM, I_L is I_{LOAD} divided by $(1 + D)$. As M_{BK} increases, I_L decreases. Therefore, I_L always has a smaller value than I_{LOAD} due to the two current paths (L-path and C-path). As I_L decreases, P_{DCR} also is reduced compared to that of the CBK. To compare the total conduction loss with that of the CBK, we assume that since the parasitic resistance (R_{ESR}) of the flying capacitors is typically much smaller than other resistances, the loss of R_{ESR} can be ignored for simplicity. Also, the on-resistance of each switch is assumed to be the same as R_{ON}. Thus, the total conduction loss ($P_{cond,CBK}$) of the CBK is expressed as follows:

$$P_{cond,CBK} = DI_L{}^2R_{ON} + (1-D)I_L{}^2R_{ON} + I_L{}^2R_{DCR} = I_L{}^2(R_{ON} + R_{DCR}) = I_{LOAD}{}^2(R_{ON} + R_{DCR}). \quad (8)$$

On the other hand, the total conduction loss ($P_{cond,BKETM}$) of the BKETM is as follows:

$$P_{cond,BKETM} = 2DR_{ON}I_L{}^2 + (1-D)R_{ON}(I_L{}^2 + I_{C,\Phi2}{}^2 + (I_L + I_{C,\Phi2})^2) + I_L{}^2R_{DCR}$$

(9)

$$= I_L{}^2[(1 + D + \frac{1}{1-D} + \frac{D^2}{1-D})R_{ON} + R_{DCR}]$$

(10)

$$= I_L{}^2(\frac{2-M_{BK}}{1-M_{BK}}R_{ON} + R_{DCR})$$

(11)

$$= \frac{(2-M_{BK})^2}{4}I_{LOAD}{}^2(\frac{2-M_{BK}}{1-M_{BK}}R_{ON} + R_{DCR}).$$

(12)

For the relative comparison with CBK, the ratio between $P_{cond,CBK}$ and $P_{cond,BKETM}$ is expressed as:

$$\frac{P_{cond,proposed}}{P_{cond,conv}} = \frac{\frac{(2-M_{BK})^2}{4}(\frac{2-M_{BK}}{1-M_{BK}}R_{ON} + R_{DCR})}{R_{ON} + R_{DCR}}.$$

(13)

Figure 7 depicts the value calculated by Equation (13) versus the conversion ratio M_{BK} for different R_{DCR}. It shows that $P_{cond,BKETM}$ is lower than $P_{cond,CBK}$ across a wide range of M_{BK} values. As described by Equation (7), the total conduction loss decreases because I_L is reduced as M_{BK} increases. Also, the larger the R_{DCR}, the lower the $P_{cond,BKETM}$ is compared with $P_{cond,CBK}$. Therefore, BKETM is a useful topology for step-down when a small inductor with a large R_{DCR} is used in heavy load conditions.

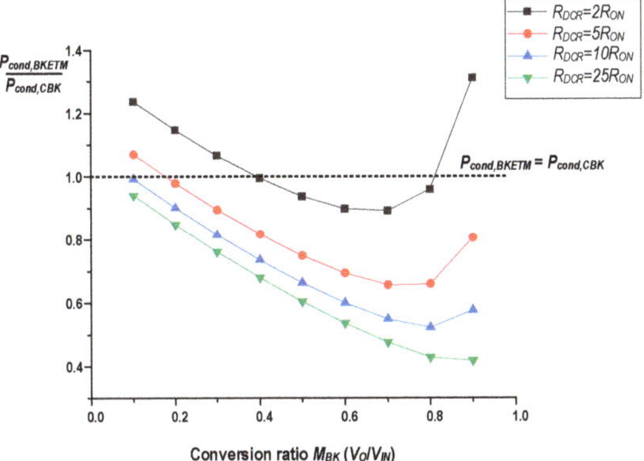

Figure 7. Conduction loss comparison for the buck converter with energy transfer media (BKETM) and conventional buck converter (CBK) with different R_{DCR} values.

4. Buck-Boost Converter with Energy Transfer Media

Since the boost converter with the ETM shown in Figure 5c has been previously described in detail in [27], this paper focuses on the buck-boost converter with ETM (BBETM) shown in Figure 5d. It is composed of power switches S_1–S_4, S_{M1}–S_{M3}, one inductor (L), one flying capacitor (C_F), and one output capacitor (C_O). The BBETM also uses two operating modes (Φ_1, Φ_2), as shown in Figure 8a. Its operation waveforms are shown in Figure 8b.

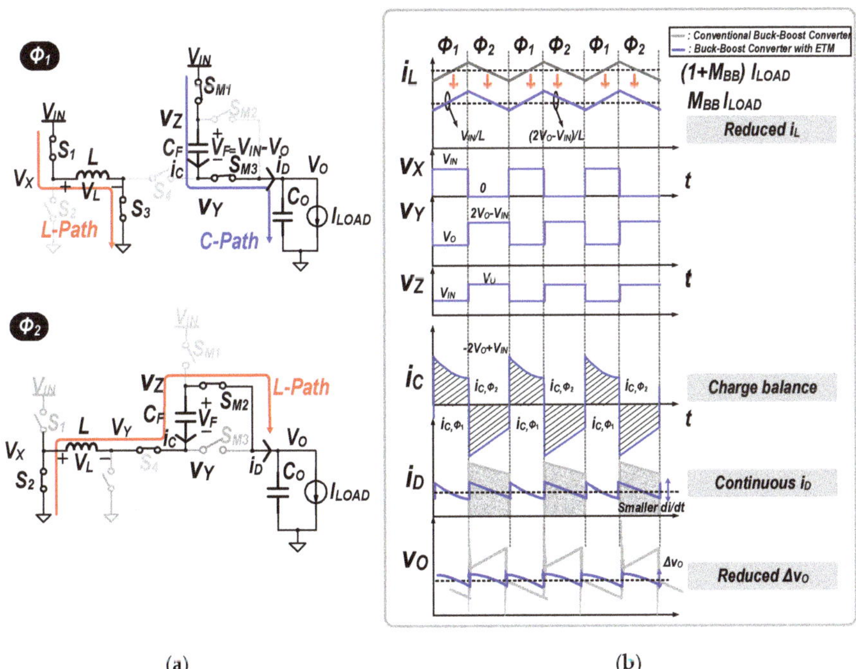

Figure 8. Operating modes (a) and waveforms (b) of buck-boost converter with ETM.

In Φ_1, S_1 and S_3 are turned on while i_L increases with a slope of V_{IN}/L and is delivered to the output. At the same time, S_{M1} and S_{M3} turn on, and i_C flows to the output through the C-path, while the voltage of C_F is charged to $V_{IN} - V_O$. With a conventional buck-boost (CBB) converter, current cannot be delivered to the output while i_L is building up. The ability of the BBETM to transfer the energy to the output during i_L build-up is one of the main differences between the BBETM and a CBB converter. Owing to this operation, the output delivery current (i_D) in the BBETM is continuous, resulting in a small output voltage ripple (ΔV_O). In Φ_2, S_2, S_{M2}, and S_{M3} turn on, and i_L is delivered to the output.

For the BBETM conversion ratio (M_{BB}), applying the voltage sec balance to the inductor based on the operation is expressed as follows:

$$DV_{IN} + (1 - D)(-2V_O + V_{IN}) = 0. \tag{14}$$

Simplifying Equation (14), M_{BB} is given by:

$$M_{BB} = \frac{V_{OUT}}{V_{IN}} = \frac{1}{2(1 - D)} \qquad 0 < D < 1. \tag{15}$$

M_{BB} of the BBETM from Equation (15) has a value between 0.5 and infinity as D varies from 0 to 1. This means that the BBETM can operate for step-up and step-down output voltages. Since the conversion ratio is limited to less than 0.5, this approach is not appropriate for applications with a low conversion ratio. However, it offers several advantages compared with a CBB converter.

First, similar to the buck type converter, the BBETM I_L is reduced compared with that of a CBB converter. To analyze this, the average value of the C-path current ($I_{C,\Phi 1}$) in Φ_1 can be obtained by applying charge balance to the C_F as shown below:

$$DI_{C,\Phi 1} - (1 - D)I_L = 0. \tag{16}$$

Simplifying Equation (16), $I_{C,\Phi 1}$ is given by

$$I_{C,\Phi 1} = \frac{1 - D}{D}I_L. \tag{17}$$

Applying the charge balance to C_O,

$$D(I_{C,\Phi 1} - I_{LOAD}) + (1 - D)(I_L - I_{LOAD}) = 0. \tag{18}$$

Substituting Equation (17) into Equation (18), I_L can be expressed with load current (I_{LOAD}) as shown below:

$$I_L = \frac{1}{2(1 - D)}I_{LOAD} = M_{BB}I_{LOAD}. \tag{19}$$

Due to the two current paths in the ETM, the BBETM I_L is as low as $M_{BB}I_{LOAD}$, while the CBB I_L is $(1 + M_{BB})I_{LOAD}$ [4]. Therefore, the BBETM P_{DCR} can be reduced. To compare the total conduction loss with that of the CBB, the on-resistance of each switch is assumed to be the same as R_{ON}, and the total conduction loss ($P_{cond,CBB}$) of the CBB is expressed as follows:

$$P_{cond,CBB} = 2I_L{}^2R_{ON} + I_L{}^2R_{DCR} = (1 + M_{BB})^2I_{LOAD}{}^2(2R_{ON} + R_{DCR}). \tag{20}$$

In contrast, the total conduction loss ($P_{cond,BBETM}$) of the BBETM is expressed as

$$P_{cond,BBETM} = I_L{}^2R_{ON} + I_L{}^2DR_{ON} + 2I_{C,\Phi 1}{}^2DR_{ON} + 2I_L{}^2(1 - D)R_{ON} + I_L{}^2R_{DCR} \tag{21}$$

$$= I_L{}^2[(3 - D + \frac{2(1 - D)^2}{D})R_{ON} + R_{DCR}] \tag{22}$$

$$= M_{BB}{}^2I_{LOAD}{}^2[(\frac{1}{M_{BB}(2M_{BB} - 1)} - \frac{2M_{BB} - 1}{2M_{BB}} + 3)R_{ON} + R_{DCR}]. \tag{23}$$

For the relative comparison with the CBB, the ratio between $P_{cond,CBB}$ and $P_{cond,BBETM}$ for the BBETM is expressed as

$$\frac{P_{cond,BBETM}}{P_{cond,CBB}} = \frac{M_{BB}{}^2[(\frac{1}{M_{BB}(2M_{BB}-1)} - \frac{2M_{BB}-1}{2M_{BB}} + 3)R_{ON} + R_{DCR}]}{(1 + M_{BB})^2(2R_{ON} + R_{DCR})}. \tag{24}$$

Figure 9 depicts the values of Equation (24) versus the conversion ratio M_{BB} for different R_{DCR} values, showing that the $P_{cond,BBETM}$ is lower than the $P_{cond,CBB}$ for a wide range of M_{BB} values. This finding demonstrates that the BBETM is more efficient than CBB due to the dual current paths. Also, it shows that the larger the R_{DCR}, the lower $P_{cond,BBETM}$ is, compared with $P_{cond,CBB}$. Therefore, the BBETM is useful for step-up and step-down applications when a small inductor with a large R_{DCR} is used in heavy load conditions.

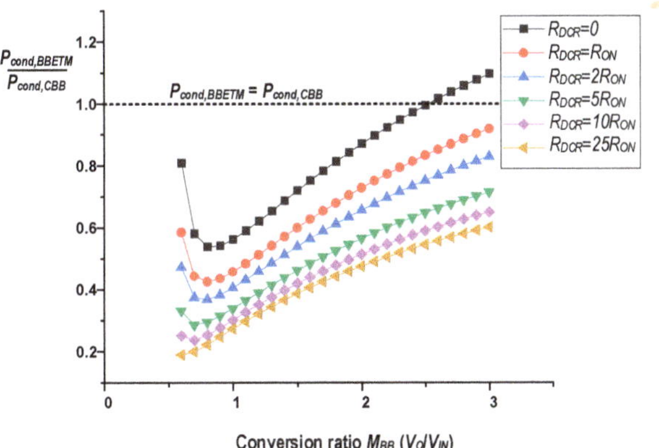

Figure 9. Conduction loss comparison of buck-boost converter with energy transfer media (BBETM) and conventional buck-boost (CBB) for different R_{DCR} values.

5. Extension to Other Topologies

5.1. Multi-Phase Buck Converter with ETM

The multi-phase converter is a structure that allows multiple inductors to transfer energy to the output when it is difficult to support sufficient energy for the output with only a single converter. The proposed ETM can also easily implement a SIC converter with a multi-phase structure. As shown in Figure 10a, the multi-phase buck converter can be designed with two ETMs for a single output with heavy I_{LOAD}. The proposed multi-phase buck converter with ETM (MBKETM) consists of one inductor and two flying capacitors, because two ETMs are used. Figure 10b shows the operation principle of the MBKETM. The advantage of the MBKETM over other topologies with an ETM is that the input frequency (f_{IN}) for the input duty (Φ_1, Φ_2) and the output switching frequency (f_{OUT}) for the output duty (Φ_{O1}, Φ_{O2}) can be independently controlled. As an example, in this paper, the input duty is controlled to regulate the output voltage, and the output duty is always fixed at 0.5 so that the C-path currents ($I_{C,\Phi O1}$, $I_{C,\Phi O2}$) can be maintained at the inductor current I_L. Then, applying charge balance to the output capacitor C_O in this condition,

$$I_L = I_{C,\Phi O1} = I_{C,\Phi O2} = \frac{1}{2}I_{LOAD}. \tag{25}$$

From Equation (25), I_L can also be maintained at half of I_{LOAD}, the same value as I_L of the conventional multi-phase buck converter (CMBK) with two inductors [19]. Thus, the MBKETM can generate triple current paths (one L-path, two C-paths) with a single inductor and two low-cost, small flying capacitors, thus reducing cost and size compared with a CMBK. Moreover, owing to these triple current paths, the ripple of the output delivery current i_D is reduced such that the output voltage ripple ΔV_O is smaller than that of the BKETM. Furthermore, by adopting a higher output frequency (f_{OUT}) for the output duty (Φ_{O1}, Φ_{O2}) than the input frequency (f_{IN}) for the input duty (Φ_1, Φ_2), ΔV_O can be further reduced. These characteristics are verified with simulation results in the next section.

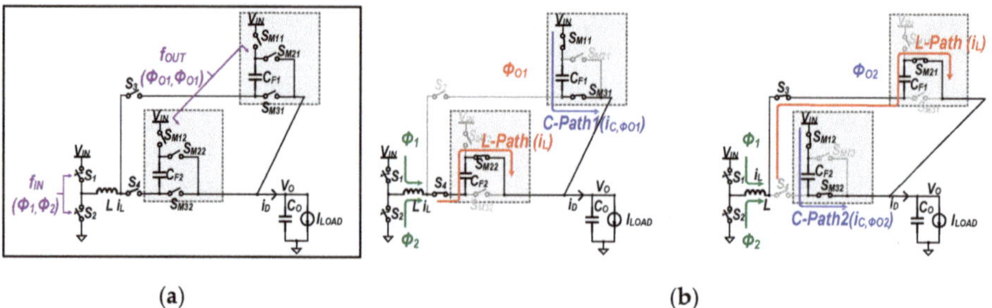

Figure 10. Topology (**a**) and operation (**b**) of a multi-phase buck converter with ETM.

5.2. Single-Inductor Multiple-Output Buck Converter with ETM

As the name suggests, a single-inductor multiple-output (SIMO) converter can regulate many outputs with one inductor [31–33]. As shown in Figure 11a as an example, it is also easy to make a dual-output structure with two ETMs. More outputs can be generated by increasing the number of ETMs. Figure 11b shows the operation of a single-inductor dual-output (SIDO) buck converter with two ETMs (SIDOETM). Because of the ETMs, I_L is reduced compared with that of a conventional SIMO converter (CSIMO). Moreover, since the SIDOETM has triple paths (one L-path and two C-paths), the currents (i_{D1}, i_{D2}) delivered to each output (V_{O1}, V_{O2}) are continuous. In contrast, with the CSIMO, the currents delivered to the output are discontinuous because a single inductor must be used to distribute the energy to each output during different time slots. Therefore, the CSIMO has the disadvantage of significant large voltage ripple at each output. In contrast, since the proposed SIDOETM has continuous i_{D1} and i_{D2}, the output voltage ripples (ΔV_{O1}, ΔV_{O2}) can be significantly reduced. Also, the CSIMO typically uses a comparator-based control for regulation of the outputs, which is very vulnerable to spike noise at the outputs [32,33]. Because the discontinuous output delivery currents can generate large spikes at every output due to the parasitic inductance that is connected to the output capacitors in series, it can cause a malfunction on the regulation control of the CSIMO. Therefore, the SIDOETM has an additional advantage of being able to alleviate the spike noise due to continuous output delivery current.

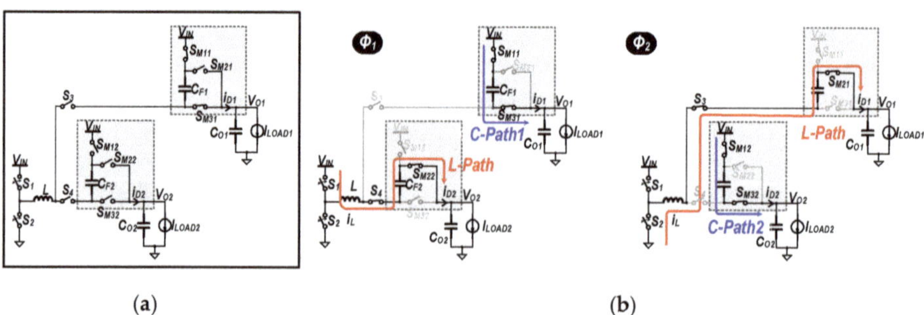

Figure 11. Topology (**a**) and operation (**b**) of a single-inductor dual-output buck converter with ETM.

6. Simulation Results and Discussion

6.1. Buck Converter with ETM

Table 1 shows the simulation conditions for the proposed BKETM. To obtain accurate simulation results, the switching loss model is included as well as the conduction loss in the simulation. C_{gate} and C_{oss} in Table 1 are the gate capacitance and the output capacitance of the power switch, respectively, for considering the switching loss. Figure 12 shows the simulated waveforms to confirm the operation of the converter.

Table 1. Simulation conditions for buck converter with ETM.

V_{IN}	V_{OUT}	I_{LOAD}	f_{IN}	L	R_{DCR}
5 V	2.8 V	1 A	1 MHz	4.7 µH	0.2 Ω

C_F	C_O	R_{ON}	R_{ESR}	C_{gate}	C_{oss}
4.7 µF	4.7 µF	50 mΩ	20 mΩ	250 pF	100 pF

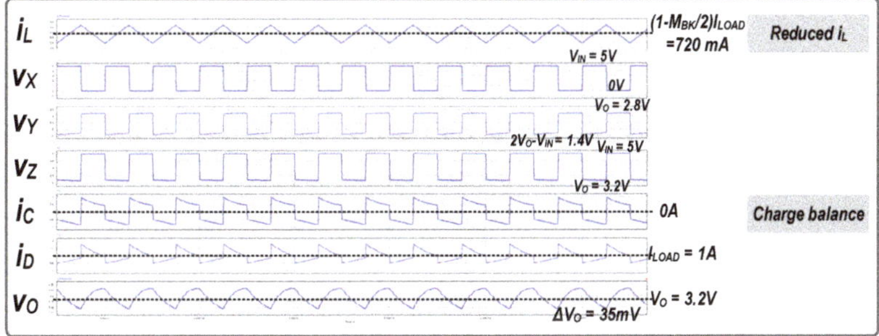

Figure 12. Simulation waveforms of a buck converter with ETM.

The results show a I_L of 720 mA, lower than the I_{LOAD} of 1 A. Also, the charge balance of the capacitor is satisfied through ic, and the voltage and current values of each node match the calculation in Section 3. Owing to the lowered I_L, high efficiency can be achieved even when using a small inductor, reducing P_{DCR}. This type of converter can also solve the heat problems associated with high performance mobile devices. However, since the BKETM has a pulsating i_D due to the C-path, a large ΔV_O of 35 mV is observed, as shown in Figure 12.

Figure 13 shows the simulated efficiency plots for both the BKETM and CBK with different conversion ratios (M_{BK}). From Equation (7), the larger the value of M_{BK}, the larger the reduction in I_L. Thus, the efficiency of the BKETM is higher than that of the CBK when M_{BK} is high. However, as M_{BK} approaches 1, the C-path current, $I_{C,\Phi2}$, rapidly increases according to Equation (5). Then, the total conduction loss increases again, resulting in the degradation of the efficiency of the BKETM.

Figure 14 shows the efficiency plots with different values of I_{LOAD} when M_{BK} is 0.6 or 0.3. When M_{BK} is high, as I_{LOAD} increases, the efficiency improves compared with that of the CBK. However, when M_{BK} is low, the reduction effect of I_L is not significant. Then, even if I_{LOAD} becomes large, the increment in efficiency is negligible. Therefore, the BKETM is an efficient topology when M_{BK} is high under heavy load conditions.

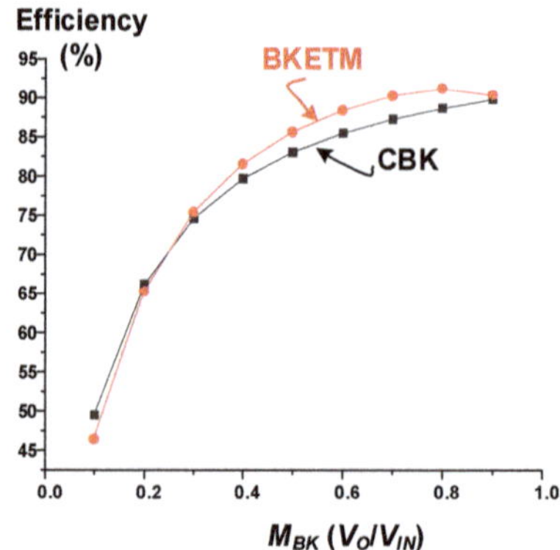

Figure 13. Simulated efficiency plot of a buck converter with ETM at different conversion ratios.

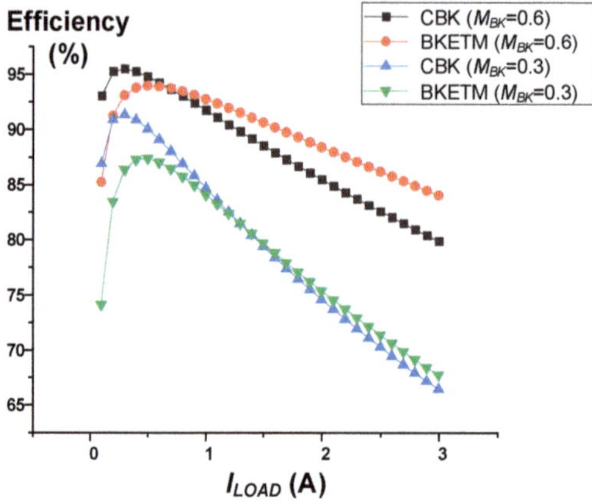

Figure 14. Simulated efficiency plot for a buck converter with ETM at different load currents.

6.2. Buck-Boost Converter with ETM

Table 2 shows the simulation conditions for the proposed BBETM. Figure 15 shows the simulation results to confirm the operation of the converter.

Table 2. Simulation conditions for buck-boost converter with ETM.

V_{IN}	V_{OUT}	I_{LOAD}	f_{IN}	L	R_{DCR}
5 V	6 V	1 A	1 MHz	4.7 μH	0.2 Ω
C_F	C_O	R_{ON}	R_{ESR}	C_{gate}	C_{oss}
4.7 μF	4.7 μF	50 mΩ	20 mΩ	250 pF	100 pF

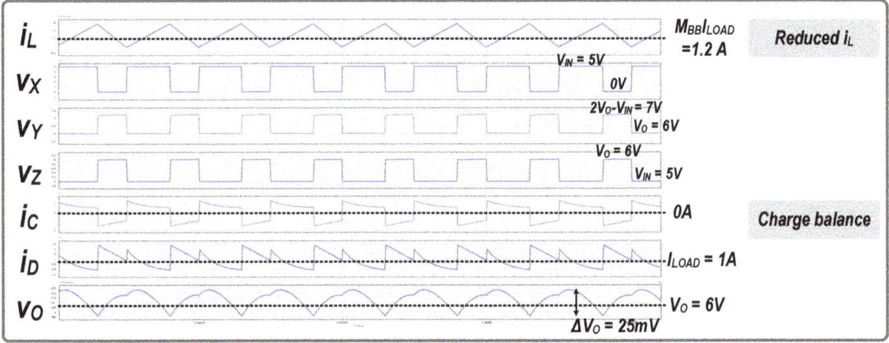

Figure 15. Simulation waveforms of buck-boost converter with ETM.

The simulation results show that the BBETM achieves a lower I_L (1.2 A) than that of the CBB (2.2 A). Also, the charge balance of C_F is satisfied through ic, and the voltage and current values of each node are matched with the calculation in Section 4. Due to reduced inductor current, the conduction loss can be decreased even when a small inductor is used. Moreover, in contrast to the CBB i_D, the BBTEM i_D is continuous due to the addition of the C-path, resulting in a smaller ΔV_O of 25 mV compared with the 130 mV observed for the CBB under the same operating conditions, as shown in Figure 16.

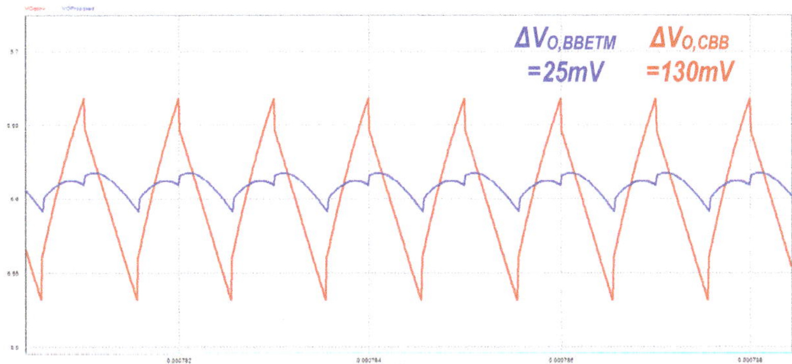

Figure 16. Simulated output voltage ripple of BBETM compared with CBB.

Figure 17 shows the simulated efficiency plots for both the BBETM and CBB with different conversion ratio values, M_{BB}. The BBETM has a much higher efficiency than the CBB across a wide range of M_{BB} values because the buck-boost topology generates a much larger I_L than the buck-type topology due to a structural characteristic.

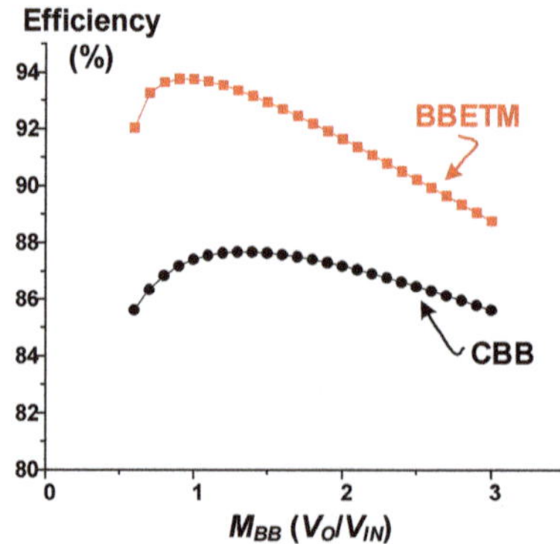

Figure 17. Simulated efficiency plot of a buck-boost converter with ETM at different conversion ratios.

Figure 18 shows the efficiency plots with different I_{LOAD} values when M_{BB} is 0.8 or 1.3. Based on Equation (19), a lower M_{BB} is associated with a larger reduction in I_L for the BBETM compared to the CBB. Thus, with a low M_{BB}, as I_{LOAD} increases, the increment in efficiency for BBETM compared with CBB becomes significant.

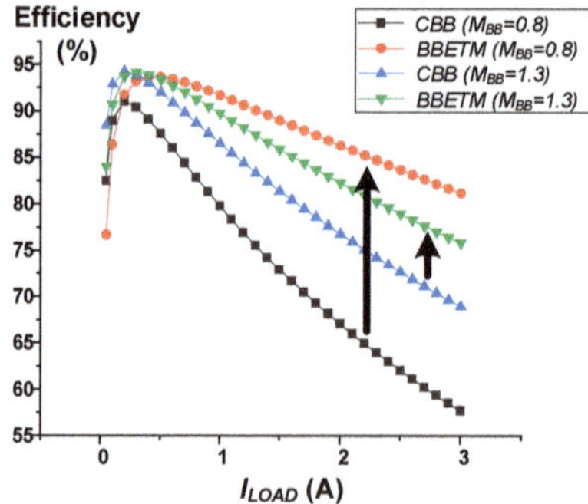

Figure 18. Simulated efficiency plot of a buck-boost converter with ETM at different load currents.

The BBETM has another benefit of a small ΔV_O. Figure 19 is a plot comparing the ΔV_O of the proposed BBETM and the CBB under the same operating conditions. The BBETM ΔV_O is lower than that of the CBB across a wide range of M_{BB} values. However, when M_{BB} is very low, which means there is a small duty cycle D, the C-path current rapidly increases as shown in Equation (17), resulting in a large ΔV_O again.

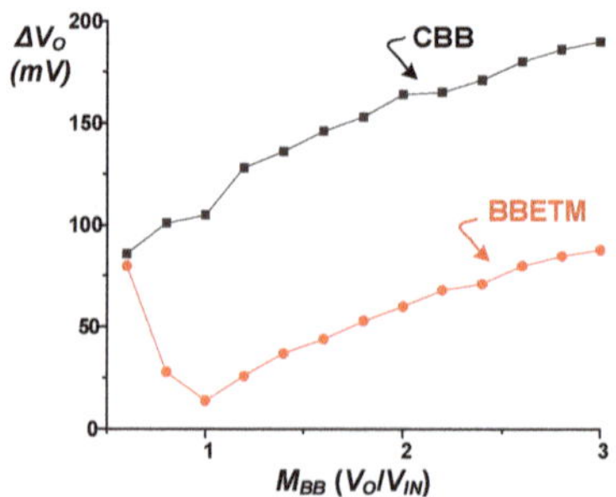

Figure 19. Simulated output voltage ripple plot of the BBETM and CBB across different conversion ratios.

6.3. Multi-Phase Buck Converter with ETM

The MBKETM is simulated under the operating conditions as shown in Table 3.

Table 3. Simulation conditions for multi-phase buck converter with ETM.

V_{IN}	V_{OUT}	I_{LOAD}	f_{IN}	L	R_{DCR}
5 V	3.7 V	1 A	1 MHz	4.7 µH	0.2 Ω
C_F	C_O	R_{ON}	R_{ESR}	C_{gate}	C_{oss}
4.7 µF	4.7 µF	50 mΩ	20 mΩ	250 pF	100 pF

It is possible to separate f_{IN} and f_{OUT} as mentioned in Section 5. Figures 20–22 show the simulation waveforms of the MBKETM with different values of f_{OUT}. Figure 20 shows the waveforms when f_{IN} and f_{OUT} are both equal to 1 MHz. I_L is reduced to half of I_{LOAD} because the output duty is fixed at 0.5, as shown in Equation (25). Moreover, ΔV_O is as small as 25 mV because i_D is continuous due to the presence of both the L-path and the C-path. Figure 21 shows the waveforms when f_{OUT} is two times higher than f_{IN}. Under these conditions, ΔV_O is further reduced to 15 mV because the effective frequency seen at the output is increased. Figure 22 shows the waveforms when f_{OUT} is triple the value of f_{IN}. Under these conditions, ΔV_O is further reduced to 10 mV. However, since high f_{OUT} can increase the switching loss in the converter, causing degradation of efficiency, there is a trade-off between ΔV_O and power efficiency.

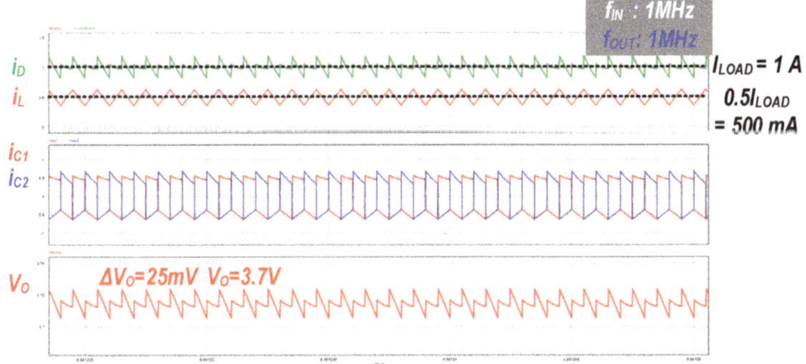

Figure 20. Simulated waveforms of the multi-phase buck converter with energy transfer media (MBKETM) with f_{IN} = 1 MHz and f_{OUT} = 1 MHz.

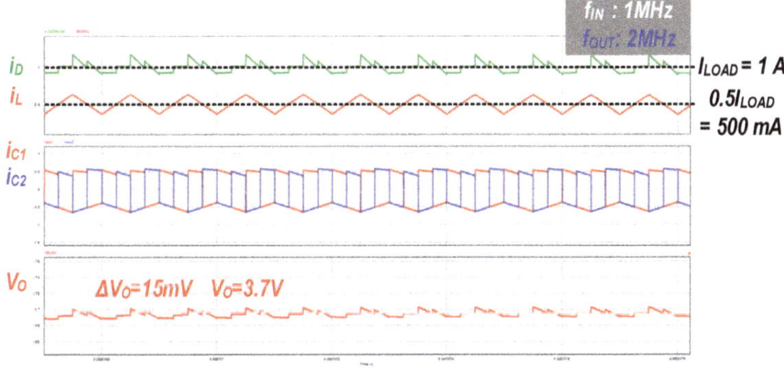

Figure 21. Simulated waveforms of the MBKETM with f_{IN} = 1 MHz and f_{OUT} = 2 MHz.

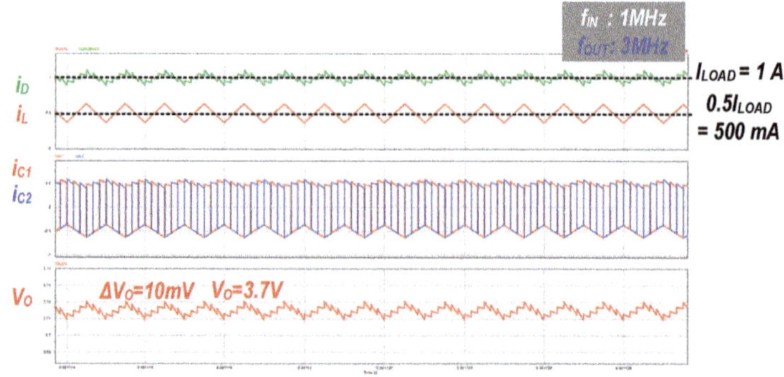

Figure 22. Simulated waveforms of the MBKETM with f_{IN} = 1 MHz and f_{OUT} = 3 MHz.

In summary, the proposed MBKETM, using one inductor and two flying capacitors, achieves a similar reduction of I_L to half of I_{LOAD} as the CMBK does with two large, expensive inductors, thereby reducing P_{DCR} significantly. Furthermore, because f_{IN} and f_{OUT} can be controlled independently, ΔV_O can be further reduced. Also, unlike the CMBK, the proposed MBKETM does not require a complex current balancing controller in spite of the multi-phase operation.

6.4. Single-Inductor Multiple-Output Buck Converter with ETM

Table 4 shows the simulation conditions of the proposed SIDOETM. Figure 23 shows the simulated i_L, i_{C1}, i_{C2}, i_{D1}, and i_{D2} of the proposed converter. Due to the dual current paths, I_L is lower than the sum of I_{LOAD1} and I_{LOAD2}.

Table 4. Simulation conditions for single-inductor multiple-output (SIDO) converter with ETM.

V_{IN}	V_{OUT1}/V_{OUT2}	I_{LOAD1}/I_{LOAD2}	f_{IN}	L	R_{DCR}
5 V	2.8 V/2 V	0.7 A/0.5 A	1 MHz	4.7 μH	0.2 Ω
C_F	C_O	R_{ON}	R_{ESR}	C_{gate}	C_{oss}
4.7 μF	4.7 μF	50 mΩ	20 mΩ	250 pF	100 pF

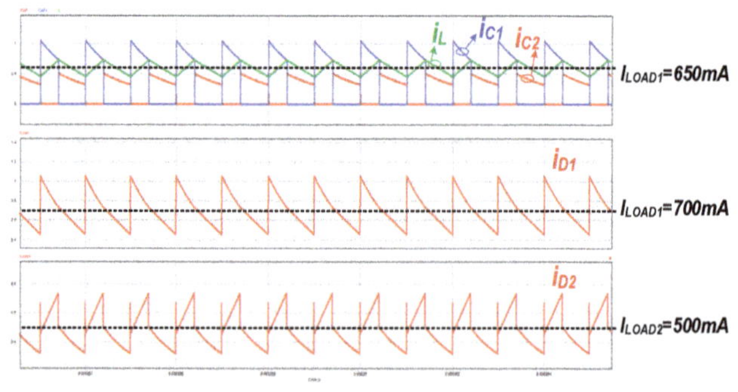

Figure 23. Simulated current waveforms of the proposed single-inductor dual-output converter with ETM (SIDOETM).

Moreover, since i_{D1} and i_{D2} do not drop to zero, the continuous current (i_{D1}, i_{D2}) flows to the respective output (V_{O1}, V_{O2}). Figure 24 shows that the output voltage is well regulated to 2.8 V and 2 V. The ripples of each output are 15 mV and 13 mV, respectively. It has a lower ΔV_O than the CSIMO [31–33].

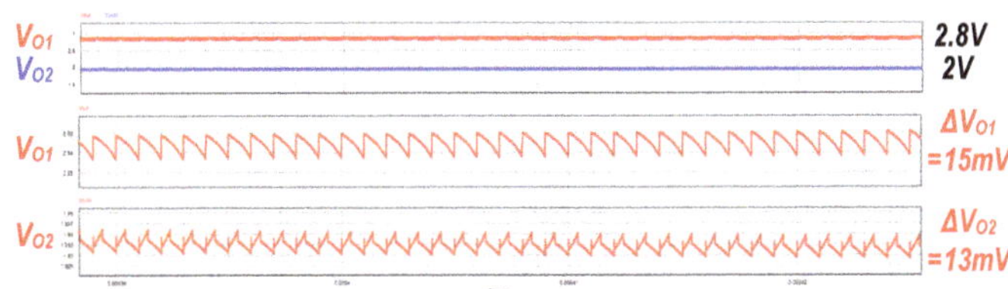

Figure 24. Simulated voltage waveforms of the proposed SIDOETM.

Tables 5 and 6 summarize the advantages of the proposed topologies with the ETM. The common advantage is the reduction of I_L, resulting in low total conduction loss in heavy load conditions. Moreover, unlike conventional topologies that have discontinuous i_D, such as the CBB and CSIMO topologies, the proposed ETM topologies have very low ΔV_O because of the continuous i_D. The MBKETM can control f_{IN} and f_{OUT} independently, resulting in further reduction of ΔV_O.

Table 5. Summary table with buck and buck-boost type converters.

Advantages	BKETM	CBK	BBETM	CBB
Reduction of I_L	O	×	O	×
Reduction of ΔV_O	×	×	O	×
Continuous i_D	O	O	O	×

O: Yes; ×: No.

Table 6. Summary table with multi-phase buck and single-inductor dual-output (SIDO) buck type converters.

Advantages	MBKETM	CMBK	SIDOETM	CSIMO
Reduction of I_L	O	×	O	×
Reduction of V_O	O	×	O	×
Continuous i_D	O	O	O	×
Separated frequency	O	×	×	×

O: Yes; ×: No.

7. Conclusions

In this paper, an ETM was proposed to make a promising hybrid switched inductor capacitor converter easier to design for heavy load conditions. New topologies with ETM, which generate dual current paths, were analyzed and compared with conventional topologies that have a single current path. Owing to the dual current paths (*L*-path and *C*-path), all of the topologies with the ETM shared the common advantage of reduced inductor current. Since it significantly decreases conduction loss dissipated at a considerable parasitic DC resistance of the inductor, the heating issue can be resolved at the same time as the power efficiency is improved, which was discussed with buck and buck-boost converters with ETMs as examples. Moreover, the buck-boost converter with ETM has continuous output delivery current, resulting in much smaller output voltage ripple than that of a conventional buck-boost converter. Also, a multi-phase converter and single-inductor multiple-output converter with several ETMs were proposed and simulated. The multi-phase converter with ETM offered the additional advantage of separating the switching frequency between the input frequency and the output frequency to further reduce the output ripple voltage. Additionally, the SIMO converter with ETM achieved a small output voltage ripple, similar to that of a buck-boost converter with ETM, due to the continuous output delivery current. In summary, the ETM can be implemented easily by combining with conventional topologies, and it has several merits such as reduced inductor current, small output voltage ripple, and independent frequency control. The proposed ETM can be applied to various non-isolated topologies as a promising solution for use in heavy load conditions with a small inductor.

Energies **2019**, *12*, 1468

Funding: This research received no external funding.

Conflicts of Interest: The author declares no conflicts of interest.

References

1. Carroll, A.; Heiser, G. An analysis of power consumption in a smartphone. In Proceedings of the 2010 USENIX Conference on USENIX Annual Technical Conference, Boston, MA, USA, 23–25 June 2010.
2. Lee, I.; Lee, Y.; Sylvester, D.; Blaauw, D. Battery Voltage Supervisors for Miniature IoT Systems. *IEEE J. Solid-State Circuits* **2016**, *51*, 2743–2756. [CrossRef]
3. Hella, M.M.; Mercier, P.P. *Power Management Integrated Circuits*, 1st ed.; CRC Press Publishers: Boca Raton, FL, USA, 2016.
4. Erickson, R.W.; Maksimovi´c, D. *Fundamentals of Power Electronics*, 2nd ed.; Kluwer Academic Publishers: Norwell, MA, USA, 2001.
5. Park, J.; Ko, W.-J.; Kang, D.-S.; Lee, Y.; Chun, J.-H. An Output Capacitor-Less Low-Dropout Regulator with 0–100 mA Wide Load Current Range. *Energies* **2019**, *12*, 211. [CrossRef]
6. Hazucha, P.; Karnik, T.; Bloechel, B.A.; Parsons, C.; Finan, D.; Borkar, S. Area-Efficient Linear Regulator with Ultra-Fast Load Regulation. *IEEE J. Solid-State Circuits* **2005**, *40*, 933–940. [CrossRef]
7. Milliken, R.J.; Silva-Martinez, J.; Sanchez-Sinencio, E. Full On-Chip CMOS Low-Dropout Voltage Regulator. *IEEE Trans. Circuits Syst. I* **2007**, *54*, 1879–1890. [CrossRef]
8. Guo, J.; Leung, K.N. A 6-_WChip-Area-Efficient Output-Capacitorless LDO in 90-nm CMOS Technology. *IEEE J. Solid-State Circuits* **2010**, *45*, 1896–1905. [CrossRef]
9. Seeman, M.D.; Sanders, S.R. Analysis and optimization of switched-capacitor DC-DC converters. *IEEE Trans. Power Electron.* **2008**, *23*, 841–851. [CrossRef]
10. Le, H.P.; Sanders, S.R.; Alon, E. Design Technique for fully integrated Switched-Capacitor DC-DC Converters. *IEEE J. Solid-State Circuits* **2011**, *46*, 2120–2131. [CrossRef]
11. Bang, S.; Blaauw, D.; Sylvester, D. A Successive-Approximation Switched-Capacitor DC–DC Converter with Resolution of VIN/2N for a Wide Range of Input and Output Voltages. *IEEE J. Solid-State Circuits* **2016**, *51*, 543–556.
12. Saif, H.; Lee, Y.; Lee, H.; Kim, M.; Khan, M.B.; Chun, J.-H.; Lee, Y. A Wide Load Current and Voltage Range Switched Capacitor DC–DC Converter with Load Dependent Configurability for Dynamic Voltage Implementation in Miniature Sensors. *Energies* **2018**, *11*, 3092. [CrossRef]
13. Chiang, C.; Chen, C. Zero-Voltage-Switching Control for a PWM Buck Converter Under DCM/CCM Boundary. *IEEE Trans. Power Electron.* **2009**, *24*, 2120–2212. [CrossRef]
14. Suh, J.; Seok, J.; Kong, B. A Fast Response PWM Buck Converter with Active Ramp Tracking Control in Load Transient period. *IEEE Trans. Circuits Syst. II Express Briefs* **2018**, *66*, 467–471. [CrossRef]
15. Calderón, A.; Vinagre, B.; Feliu, V. Fractional order control strategies for power electronic buck converters. *Signal Process.* **2006**, *86*, 2803–2819. [CrossRef]
16. Suh, J.-D.; Yun, Y.-H.; Kong, B.-S. High-Efficiency DC–DC Converter with Charge-Recycling Gate-Voltage Swing Control. *Energies* **2019**, *12*, 899. [CrossRef]
17. Abdulslam, A.; Mohammad, B.; Ismail, M.; Mercier, P.; Ismail, Y. A 93% Peak Efficiency Fully-Integrated Multilevel Multistate Hybrid DC–DC Converter. *IEEE Trans. Circuits Syst. I* **2018**, *65*, 2617–2630. [CrossRef]
18. Kim, W.; Brooks, D.; Wei, G.-Y. A fully-integrated 3-level DC-DC converter for nanosecond-scale DVSF. *IEEE J. Solid-State Circuits* **2012**, *47*, 206–219. [CrossRef]
19. Li, P.; Xue, L.; Hazucha, P.; Karnik, T.; Bashirullah, R. A delay-locked loop synchronization scheme for high-frequency multiphase hysteretic DC-DC converters. *IEEE J. Solid-State Circuits* **2009**, *44*, 3131–3145. [CrossRef]
20. Abedinpour, S.; Bakkaloglu, B.; Kiaei, S. A Multistage Interleaved Synchronous Buck Converter with Integrated Output Filter in 0.18μm SiGe Process. *IEEE Trans. Power Electron.* **2007**, *22*, 2164–2175. [CrossRef]
21. Huang, C.; Mok, P.K.T. A 100 MHz 82.4% Efficiency Package-Bondwire Based Four-Phase Fully-Integrated Buck Converter with Flying Capacitor for Area Reduction. *IEEE J. Solid-State Circuits* **2013**, *48*, 2977–2988. [CrossRef]
22. Rodi˘c, M.; Milanovi˘c, M.; Trunti˘c, M.; Ošlaj, B. Switched-Capacitor Boost Converter for Low Power Energy Harvesting Applications. *Energies* **2018**, *11*, 3156. [CrossRef]

23. Tran, V.-T.; Nguyen, M.-K.; Choi, Y.-O.; Cho, G.-B. Switched-Capacitor-Based High Boost DC-DC Converter. *Energies* **2018**, *11*, 987. [CrossRef]
24. Ju, Y.; Shin, S.; Huh, Y.; Park, S.; Bang, J.; Kim, K.; Choi, S.; Lee, J.; Cho, G. A hybrid inductor-based flying-capacitor-assisted step-up/step-down DC-DC converter with 96.56% efficiency. In Proceedings of the 2017 IEEE International Solid-State Circuits Conference (ISSCC), San Francisco, CA, USA, 5–9 February 2017; pp. 184–185.
25. Liu, W.; Assem, P.; Lei, Y.; Hanumolu, P.K.; Pilawa-Podgurski, R. A 94.2%-peak-efficiency 1.53A direct-battery-hook-up hybrid Dickson switched-capacitor DC-DC converter with wide continuous conversion ratio in 65nm CMOS. In Proceedings of the 2017 IEEE International Solid-State Circuits Conference (ISSCC), San Francisco, CA, USA, 5–9 February 2017; pp. 182–183.
26. Ko, M.; Kim, K.; Woo, Y.; Shin, S.; Han, H.; Huh, Y.; Kang, G.; Cho, J.; Lim, S.; Park, S.; et al. A 97% high-efficiency 6µs fast-recovery-time buck-based step-up/down converter with embedded 1/2 and 3/2 charge-pumps for li-lon battery management. In Proceedings of the 2018 IEEE International Solid-State Circuits Conference (ISSCC), San Francisco, CA, USA, 11–15 February 2018; pp. 428–430.
27. Shin, S.; Huh, Y.; Ju, Y.; Choi, S.; Shin, C.; Woo, Y.; Choi, M.; Park, S.; Sohn, Y.; Ko, M.; et al. A 95.2% efficiency dual-path DC-DC step-up converter with continuous output current delivery and low voltage ripple. In Proceedings of the 2018 IEEE International Solid-State Circuits Conference (ISSCC), San Francisco, CA, USA, 11–15 February 2018; pp. 430–432.
28. Huh, Y.; Shin, S.; Hong, S.; Woo, Y.; Ju, Y.; Choi, S.; Cho, G. A Hybrid Dual-Path Step-Down Converter with 96.2% Peak Efficiency Using a 250mΩ Large-DCR Inductor. In Proceedings of the 2018 IEEE Symposium on VLSI Circuits, Honolulu, HI, USA, 18–22 June 2018; pp. 225–226.
29. Wang, S.; Woo, Y.; Yuk, Y.; Lee, B.; Cho, G.; Cho, G. Efficiency enhanced Single-Inductor Boost-Inverting Flyback converter with Dual Hybrid Energy transfer media and a Bifurcation Free Comparator. In Proceedings of the 2010 Proceedings of ESSCIRC, Seville, Spain, 14–16 September 2010; pp. 450–453.
30. Wang, S.; Woo, Y.; Yuk, Y.; Cho, G.; Cho, G. High efficiency Single-Inductor Boost/Buck Inverting Flyback converter with hybrid energy transfer media and multi level gate driving for AMOLED panel. In Proceedings of the 2010 Symposium on VLSI Circuits, Honolulu, HI, USA, 16–18 June 2010; pp. 59–60.
31. Dongsheng, M.; Wing-Hung, K.; Chi-Ying, T. A pseudo-CCM/DCM SIMO switching converter with freewheel switching. *IEEE J. Solid-State Circuits* **2003**, *38*, 1007–1014. [CrossRef]
32. Lu, D.; Qian, Y.; Hong, Z. 4.3 An 87%-peak-efficiency DVS-capable single-inductor 4-output DC-DC buck converter with ripple-based adaptive off-time control. In Proceedings of the 2014 IEEE International Solid-State Circuits Conference Digest of Technical Papers (ISSCC), San Francisco, CA, USA, 9–13 February 2014; pp. 82–83.
33. Goh, T.Y.; Ng, W.T. Single Discharge Control for Single-Inductor Multiple-Output DC–DC Buck Converters. *IEEE Trans. Power Electron.* **2018**, *33*, 2307–2316. [CrossRef]

Article

Miniature DC-DC Boost Converter for Driving Display Panel of Notebook Computer

Seok-Hyeong Ham [1] and Hyung-Jin Choe [2],*

[1] Department of Electrical Engineering, Pohang University of Science and Technology, Pohang, Kyungpook 37673, Korea
[2] Department of Electrical Engineering, Tongmyong University, 428 Sinseon-ro, Nam-gu, Busan 48520, Korea
* Correspondence: hjchoe@tu.ac.kr; Tel.: +82-51-629-1318

Received: 26 June 2019; Accepted: 29 July 2019; Published: 30 July 2019

Abstract: This paper proposes a miniature DC-DC boost converter to drive the display panel of a notebook computer. To reduce the size of the circuit, the converter was designed to operate at a switching frequency of 1 MHz. The power conversion efficiency improved using a passive snubber circuit that consisted of one inductor, two capacitors, and two diodes; it reduced the switching losses by lowering the voltage stress of the switch and increased the voltage gain using charge pumping operations. An experimental converter was fabricated at 2.5 cm × 1 cm size using small components, and tested at input voltage $5\,V \leq V_{IN} \leq 17.5\,V$ and output current $30\,mA \leq I_O \leq 150\,mA$. Compared to existing boost converters, the proposed converter had ~7.8% higher power conversion efficiency over the entire range of V_{IN} and I_O, only ~50% as much voltage stress of the switch and diodes, and a much lower switch temperature $T_{SW} = 49.5\,°C$. These results indicate that the proposed converter is a strong candidate for driving the display panel of a notebook computer.

Keywords: backlight; DC-DC converter; passive snubber; voltage stress

---

1. Introduction

Notebook or tablet computers are powered by a Li ion battery pack [1–4]. Typically, the battery pack is composed of 2–4 Li ion battery cells. Each Li ion cell has a terminal voltage of 2.6–4.2 V depending on the charging state, so the voltage of the battery pack changes in the range of 5–17 V [5–7]. To supply power to each part of the computer, the pack voltage is converted to several voltages by DC-DC converters.

Many computers use a light-emitting diode (LED) backlight unit (BLU) for the liquid crystal display (LCD) panel. For a 17 inch panel, the voltage required to drive the LED backlight can be as high as 40 V [8]. The DC-DC boost converter for the LED BLU in a notebook or tablet computer should have a circuit topology that can reduce the overall size and weight [9–16]. The conventional boost converter (Figure 1) has the simplest structure, and can be implemented in small sizes by operating at a high switching frequency f_S. However, the switching loss increases as f_S increases, and thereby decreases the power conversion efficiency η_e and can cause overheating of the switch. This problem has been solved using soft-switching converters [17–21].

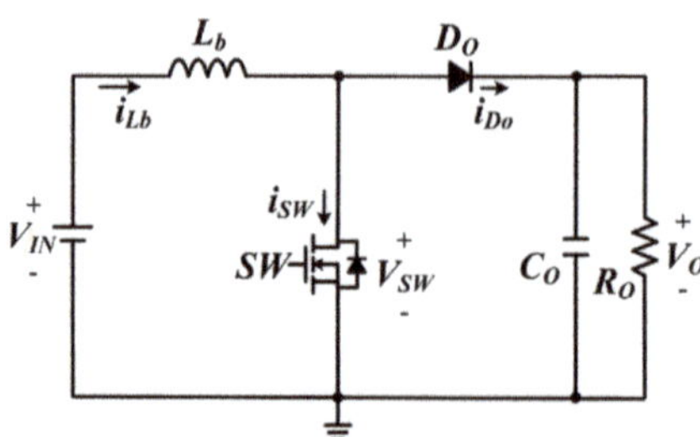

Figure 1. Schematic of conventional DC-DC boost converter.

Soft-switching converters [17–21] use active snubbers to reduce switching loss. These snubbers are implemented using an auxiliary switch, diodes, and energy storage elements. The auxiliary switch requires a separate gate drive circuit that is controlled by a sophisticated algorithm. The snubber of Reference [17] discharges the drain-source capacitor C_{DS} of the main switch before the switch turns on, so zero-voltage (ZV) turn-on of the switch is possible. The snubbers of References [18,19] force the switch current to zero before the switch turns off, so zero-current (ZC) turn-off of the switch is possible. The snubber of Reference [20] achieves ZV turn-off of the switch by reducing the rate of voltage increase during the turn-off transient, and achieves ZC turn-on of the switch by reducing the rate of current increase during the turn-on transient. However, the converters of References [17–20] have some limitations when they operate at high f_S; they use a resonance between an inductor (L) and a capacitor (C) in the active snubber, the inductance L of L should be small enough to achieve ZV or ZC switching in a short time; the current peak of the switch or diodes increases as L decreases, therefore, this change increases the conduction loss and reduces the maximum output power. In the converter of Reference [21], the capacitance C_{DS} discharges through L of the snubber when the auxiliary switch turns on. The ZV turning on of the main switch is achieved by discharging C_{DS} completely. The L required to achieve the ZV turning on of the switch decreases as f_S increases; as a consequence, the switching loss in the auxiliary switch increases. Therefore, simultaneous reduction of switching losses in the main and auxiliary switches is very difficult at high f_S. The converter of Reference [9] uses a passive snubber to reduce the switching loss. The switching loss is reduced by lowering the turn-on slope of the switch current; the inductor L_S that is connected in series with the switch controls the turn-on slope. However, the energy stored in L_S is released as a voltage spike because the current path of L_S is blocked when the switch is turned off. To ensure that this voltage spike does not exceed the maximum voltage limit of the switch, L_S should be less than 50 nH at high switching frequencies; This value of L_S was calculated using Equation (9) of the design guide in Reference [9]. The low value of the chip type inductor has a low maximum current limit. This low inductance also causes undesired resonance due to the parasitic inductance or capacitance of the circuit. Because of this, the converter of Reference [9] is difficult to use at high switching frequencies. Additionally, the switch current significantly increases the conduction loss in L_S because the components of small size have non-negligible series resistance.

To improve upon the aforementioned drawbacks in the existing converters, this paper proposes a miniature DC-DC boost converter for the display panel of a notebook computer. The proposed converter (Figure 2) uses a passive snubber to reduce the switching loss; the capacitor C_2 is connected in parallel with the switch, and the inductor L_1 is connected in series with C_2, and the value of L_1 and C_2 are large enough to prevent resonance at high f_S. This lowers V_{SW} to V_O-V_{C1} and eliminates the switching loss. In addition, L_1 is placed apart from the main current path to reduce the conduction loss in the passive snubber. The switch operates under a hard-switching condition to achieve fast switching, but the switching loss is reduced by lowering V_{SW} (Figure 3). The switching loss P_{SW} is calculated using the waveforms of i_{SW} and V_{SW} as:

$$P_{sw} = \frac{1}{T_s} \int_{t_0}^{t_2} V_{sw}(t) i_{sw}(t) dt + \frac{1}{T_s} \int_{t_3}^{t_5} V_{sw}(t) i_{sw}(t) dt$$

Compared with the conventional converter, the proposed converter has a similar current peak, but has $V_{SW}(t_1)$ and $V_{SW}(t_4)$ lowered by ~1/2, so P_{SW} is reduced.

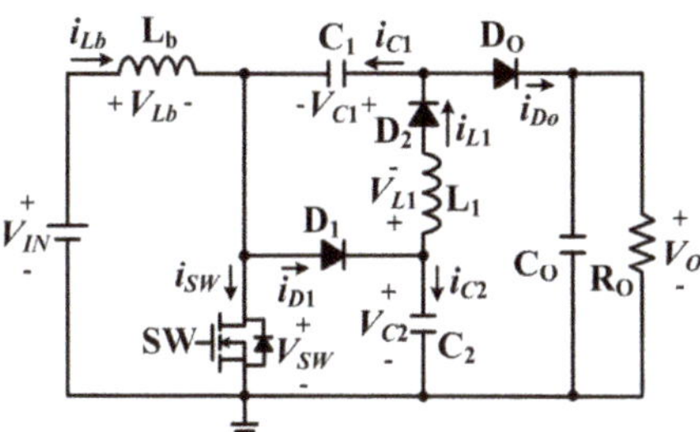

Figure 2. Schematic of proposed DC-DC boost converter.

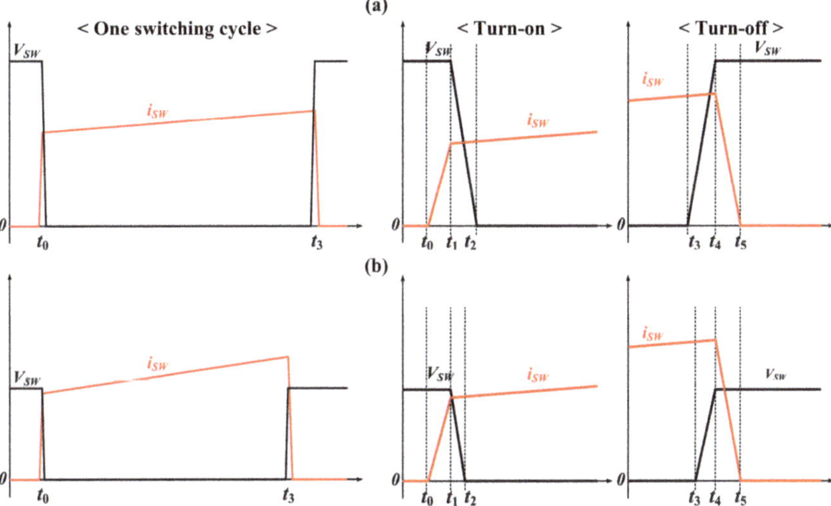

Figure 3. Voltage and current waveforms of the switch: (**a**) conventional boost converter, (**b**) proposed converter.

The proposed converter has a higher voltage conversion ratio than the conventional boost converter, and therefore has a wide range of output voltage for a given range of input voltage. Section 2 describes the circuit structure and operating principle of the proposed converter, Section 3 gives the design guidelines, Section 4 shows the experimental results, and Section 5 concludes the paper.

2. Circuit Structure and Operating Principles of the Proposed Converter

The proposed converter consists of two capacitors C_1 and C_2, two diodes D_1 and D_2, and an inductor L_1, in addition to the conventional boost components L_b, D_O, C_O, and a switch (SW). C_1 is located between boost inductor L_b and output diode D_O, and acts as a charge pump capacitor to increase the voltage gain. L_1 and C_2 reduce the voltage stress of SW by dividing the output voltage V_O. D_1 and D_2 provide a charging or discharging path for C_2.

The operating modes (Figure 4) and key waveforms (Figure 5) of the proposed converter include three sequential modes of operation.

Mode 1 ($t_0 \sim t_1$): The first operation (Figure 4a) starts at $t = t_0$ by turning ON SW. During this operation, D_2 turns ON, and D_1 and D_O turn OFF. $v_{Lb} = V_{IN}$ and $v_{L1} = V_{C2} - V_{C1}$, so

$$i_{Lb}(t) = \frac{V_{IN}}{L_b}(t - t_0) + i_{Lb}(t_0) \tag{1}$$

$$i_{L1}(t) = i_{C1}(t) = -i_{C2}(t) = \frac{V_{C2} - V_{C1}}{L_1}(t - t_0) \tag{2}$$

C_1 is charged and C_2 is discharged when SW is turned ON. *Mode* 1 ends when SW is turned OFF.

Mode 2 ($t_1 \sim t_2$): The second operation (Figure 4b) starts at $t = t_1$ when SW turns OFF. During *mode* 2, D_2 stays ON, and D_1 and D_O are turned ON. When SW is turned OFF, the anode voltage of D_1 is V_{C2} because $V_{Lb} = V_{C2} - V_{IN}$ according to the voltage second valance law. The cathode voltage of D_1 is $V_{C2} - \Delta V_{C2}$ because C_2 was discharged in *Mode* 1, so D_1 is turned ON. After D_1 is turned ON, the anode voltage of D_O is $V_{C1} + V_{C2}$ and cathode voltage is $V_O - \Delta V_O$ because C_O was discharged in *Mode* 1. Therefore, D_O is turned ON because $V_{C1} + V_{C2} = V_O$ in steady state. In the same manner, after D_1 and D_O are turned ON, D_2 is turned ON because the anode voltage of D_2 is $V_{C1} + V_{C2}$ and the cathode voltage is $V_O - \Delta V_O$. $v_{Lb} = V_{IN} - V_{C2}$ and $v_{L1} = -V_{C1}$, so

$$i_{Lb}(t) = \frac{V_{IN} - V_{C2}}{L_b}(t - t_1) + i_{Lb}(t_1), \; i_{L1}(t) = -\frac{V_{C1}}{L_1}(t - t_1) + i_{L1}(t_1).$$

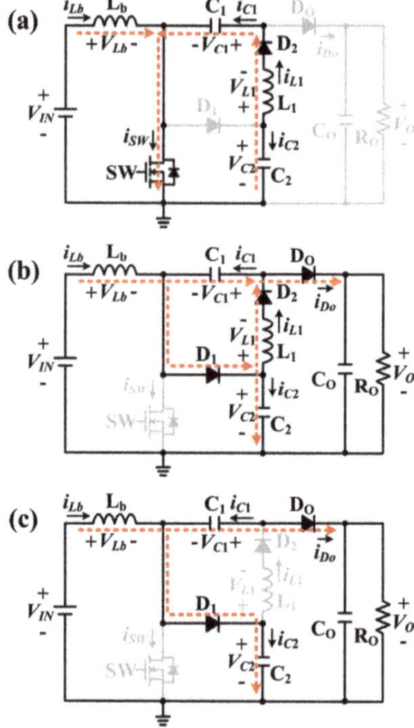

Figure 4. Modes of operation: (**a**) *Mode* 1 ($t_0 < t < t_1$), (**b**) *Mode* 2 ($t_1 < t < t_2$), (**c**) *Mode* 3 ($t_2 < t < t_3$).

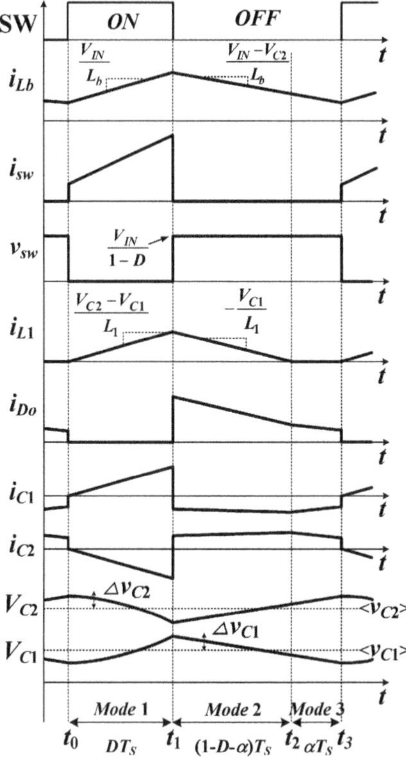

Figure 5. Operating waveforms of the proposed DC-DC boost converter.

The voltage stress of SW is V_{C2} because D_1 is turned ON when SW is turned OFF. The output voltage V_O is divided into V_{C1} and V_{C2}, so the voltage stress of SW is smaller than V_O, which is the voltage stress of the conventional boost converter.

Mode 3 $(t_2 \sim t_3)$: The last operation (Figure 4c) starts at $t = t_2$ when D_2 turns OFF. D_1 and D_O stay ON during this mode. $v_{Lb} = V_{IN} - V_{C2}$, so

$$i_{Lb}(t) = \frac{V_{IN} - V_{C2}}{L_b}(t - t_1) + i_{Lb}(t_1)$$

In this mode, $v_{L1} = 0$ because $i_{L1} = 0$. *Mode* 3 ends when SW is turned ON for the next switching cycle.

The voltage–second balance law of inductances L_b and L_1 yields

$$V_{IN}D + (V_{IN} - V_{C2})(1 - D) = 0 \tag{3}$$

$$(V_{C2} - V_{C1})D - V_{C1}(1 - D - \alpha) = 0 \tag{4}$$

where D is the duty of switching and αT_S is the duration of *Mode* 3. Solving Equations (3) and (4) for V_{C1} and V_{C2} yields

$$V_{C2} = \frac{V_{IN}}{1 - D}, \quad V_{C1} = \frac{D}{1 - \alpha}\frac{V_{IN}}{1 - D}. \tag{5}$$

Equation (5) and $V_O = V_{C1} + V_{C2}$ then give the voltage conversion ratio as

$$\frac{V_O}{V_{IN}} = \frac{1}{1-D}\frac{1-\alpha+D}{1-\alpha} \tag{6}$$

The average current of L_1, I_{L1} for one switching period T_S is equal to the average output current I_O, so

$$I_O = \frac{V_{C2}-V_{C1}}{2L_1}D^2T_S + \frac{V_{C1}}{2L_1}(1-D-\alpha)^2T_S \tag{7}$$

Inserting Equation (5) into Equation (7) and solving for α yields

$$\alpha = \frac{(V_{IN}DT_S - 2I_OL_1)(1-D)}{V_{IN}DT_S} \tag{8}$$

Combining Equations (6) and (8) yields the output voltage

$$V_O = \frac{2V_{IN}(V_{IN}D^2T_S + I_OL_1(1-D))}{(1-D)(V_{IN}D^2T_S + 2I_OL_1(1-D))} \tag{9}$$

V_O versus D (Figure 6) for $L_1 = 10\ \mu\text{H}$, $f_S = 1\ \text{MHz}$, $I_O = 150\ \text{mA}$, $V_{IN} = 5\ \text{V}$, and $0.1 \leq D \leq 0.9$ was calculated using Equation (9) for the proposed converter, but the conventional boost converter was calculated using their voltage gain equations. To have $V_O = 40\ \text{V}$ for given V_{IN}, $D = 0.775$ was applied to the proposed converter, whereas $D = 0.875$ was applied to the conventional boost converter. At given V_{IN} and D, the proposed converter had higher V_O than the conventional boost converters.

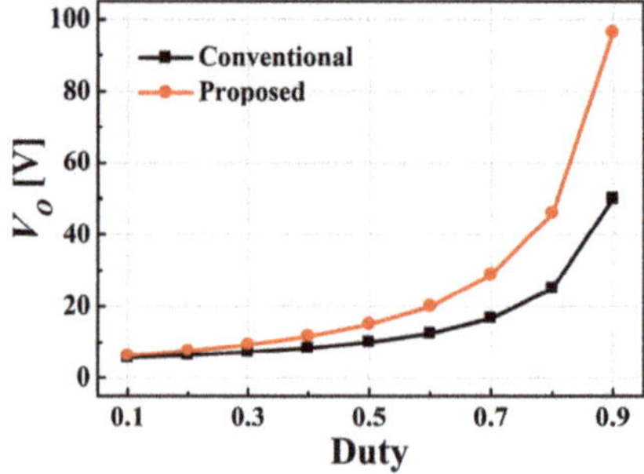

Figure 6. V_O versus duty D for the experimental boost converters; $V_{IN} = 5\ \text{V}$, $0.1 \leq D \leq 0.9$.

3. Design Guideline

3.1. Boost Inductor L_b

The inductance L_b of L_b should be determined so that the proposed converter operates in a continuous conduction mode (CCM). The current through L_b is minimum at t_0 and maximum at t_1. The following equation is obtained using the average of i_{Lb} and Equation (1),

$$i_{Lb}(t_1) - i_{Lb}(t_0) = \frac{V_{IN}DT_S}{L_b}, \quad i_{Lb}(t_1) + i_{Lb}(t_0) = 2I_{IN}. \tag{10}$$

Solving Equation (10) for $i_{Lb}(t_0)$ yields

$$i_{Lb}(t_0) = I_{IN} - \frac{V_{IN}DT_S}{2L_b}$$

The condition $i_{Lb}(t_0) > 0$ should be satisfied to operate the proposed converter in CCM. This condition yields

$$L_b > \frac{V_{IN}DT_S}{2I_{IN}} \tag{11}$$

For $5 \leq V_{IN} \leq 17.5$ V, $V_O = 40$ V, $f_S = 1$ MHz, and $30 \leq I_O \leq 150$ mA, $L_b = 33$ µH was determined using Equations (9) and (11).

3.2. Snubber Inductor L_1

A small value of L_1 causes a drastic current decrease of D_2 in mode 2. Therefore, the limit of L_1 is calculated using the following condition for a soft turn-off of D_2.

$$\frac{V_{C1}}{L_1} << \frac{I_{D2,max}}{t_{D2,fall}}$$

where $I_{D2,max}$ is the highest diode current and t_{fall} is the diode turn-off time; $L_1 \gg 0.2$ µH for $I_{D2,max} = 1$ A and $t_{fall} = 10$ ns. The diode turn-off loss increases as L_1 decreases, so $L_1 = 10$ µH was chosen to provide an operating margin and to reduce turn-off loss of D_2.

3.3. Snubber Capacitors C_1 and C_2

The capacitance of C_1 is determined by the condition that $\Delta V_{C1} << V_{C1}$ because V_{C1} should be almost constant for one switching period T_S, where ΔV_{C1} is the voltage ripple of C_1. During *Mode 1*, C_1 charges through L_1; this results in a voltage ripple ΔV_{C1} of C_1; $\Delta V_{C1} = (V_{C2} - V_{C1})(DT_S)^2 / (2C_1 L_1)$ because the current through C_1 decreases linearly from 0 A at t_0 to $(V_{C2} - V_{C1})DT_S / L_1$ at t_1.

Therefore,

$$C_1 >> \frac{V_{C2} - V_{C1}}{2V_{C1}L_1}(DT_S)^2. \tag{12}$$

For $V_{IN} = 5$ V, $V_O = 40$ V, $f_S = 1$ MHz, and $L_1 = 10$ µH, $C_1 = 1$ µF was determined using Equations (5) and (12). C_2 has been chosen to be the same as C_1 because $\Delta V_{C1} = \Delta V_{C2}$ and $i_{C1} = -i_{C2}$ in *Mode 1*.

4. Experimental Results

The proposed boost converter (Figure 7) was designed to operate at $V_{IN} = 5$ V, $V_O = 40$ V, $30 \leq I_O \leq 150$ mA, and $f_S = 1$ MHz. The inductances were set at $L_b = 33$ µH and $L_1 = 10$ µH. Capacitances C_1 and C_2 were both set at 1 µF. The experimental circuit was implemented using the following miniature components (Table 1): TPS55340 DC-DC controller (Texas Instruments Inc., Dallas, TX, USA), which has an internal n-MOS switch (40 V, 5A), RB160M-60TR diodes (Rohm Co.), IFSC-1515AH-01 (Vishay Inc.) and SSMC252008R47SC (SST Inc.) chip inductors, and multilayer ceramic chip capacitors of a size 3.2 mm × 1.6 mm. The chip inductors and capacitors have series resistances given in the parentheses. To ensure a fair comparison, the conventional boost converter used the same components.

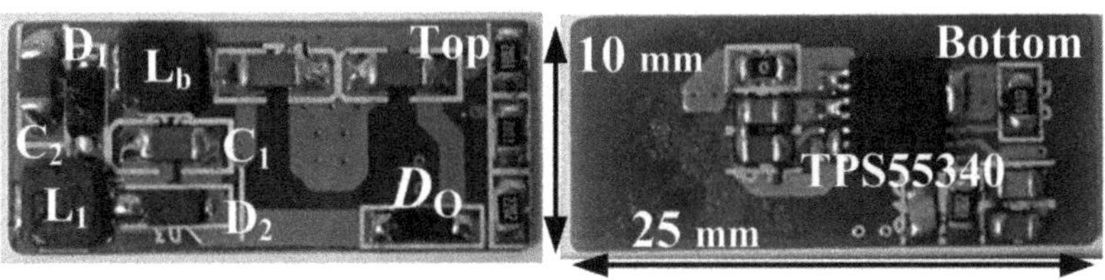

Figure 7. Photograph of the proposed DC-DC boost converter.

Table 1. Component values for the experimental boost converters.

Components	Conventional Boost Converter	Proposed Converter
L_b	33 µH (450 mΩ)	33 µH (450 mΩ)
L_1	-	10 µH (150 mΩ)
C_1	-	1 µH (12.5 mΩ)
C_2	-	1 µH (12.5 mΩ)
C_O	4.7 µF (10 mΩ)	4.7 µF (10 mΩ)
D_1	-	RB160M-60TR
D_2	-	RB160M-60TR
D_O	RB160M-60TR	RB160M-60TR
Driver IC (SW)	TPS55340 ($R_{on} = 0.1$ Ω)	TPS55340 ($R_{on} = 0.1$ Ω)

The TPS55340 regulates the output voltage with current mode PWM control, and has an internal oscillator. The pulse width modulation of the gate pulse V_g is achieved in the control circuit for the proposed converter (Figure 8) as follows; each clock pulse resets the flip-flop and the ramp generator, which sets V_g to the 'high' state; the output V_c of the error amplifier increases as V_O increases; the comparator output changes from 'low' to 'high' when the output voltage of the ramp generator exceeds V_c, which changes the inverter output V_g to the 'low' state. When the output current I_O increases/decreases abruptly, V_O decreases/increases instantly, which increases/decreases V_c. Thus, the pulse width of V_g increases/decreases to keep V_O constant.

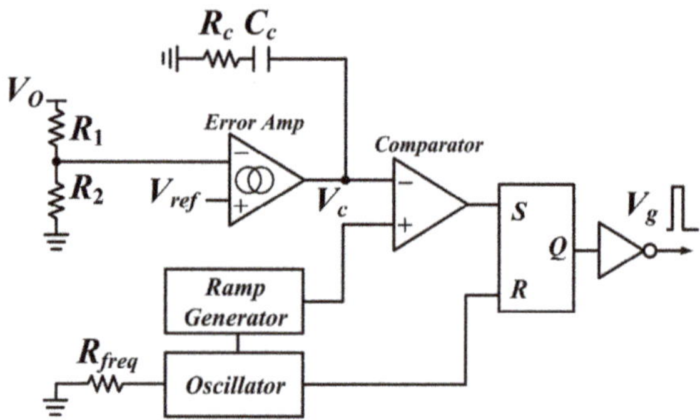

Figure 8. Block diagram of the control circuit for the proposed converter.

The voltage and current waveforms of SW (Figure 9) were measured at $V_{IN} = 5$ V, $V_O = 40$ V, $I_O = 150$ mA, and $f_S = 1$ MHz. The voltage stress of the proposed converter (Figure 9a) was lower than that of the conventional boost converter (Figure 9b) because the voltage stress was $V_O - V_{C1}$ in the proposed converter but V_O in the conventional boost converter. The current stresses of the proposed converter and the conventional boost converter were similar. The current waveforms of the diodes are shown in Figure 10. The diodes current stresses of the proposed converter was lower than the conventional boost converter. The time-averaged values $<i_{D1}>$, $<i_{D2}>$, and $<i_{Do}>$ were all equal to I_O.

The curves of η_e on I_O (Figure 11a) were measured at $30 \leq I_O \leq 150$ mA, $V_{IN} = 5$ V, $V_O = 40$ V, $f_S = 1$ MHz. At $I_O = 30$ mA, η_e was 84.0% for the proposed converter and 76.9% for the conventional boost converter. At $I_O = 150$ mA, η_e was 80.4% for the proposed converter and 72.6% for the conventional boost converter. The curves of η_e on V_{IN} (Figure 11b) were measured at $5 \leq V_{IN} \leq 17.5$ V, $I_O = 150$ mA, $V_O = 40$ V, $f_S = 1$ MHz; η_e at $V_{IN} = 17.5$ V was 89.1% for the proposed converter and 88.6% for the conventional boost converter. The η_e of the proposed converter was up to 7.8% higher than the conventional boost converter at $V_{IN} = 5$ V. The proposed converter had a high η_e over the entire input voltage and output current ranges.

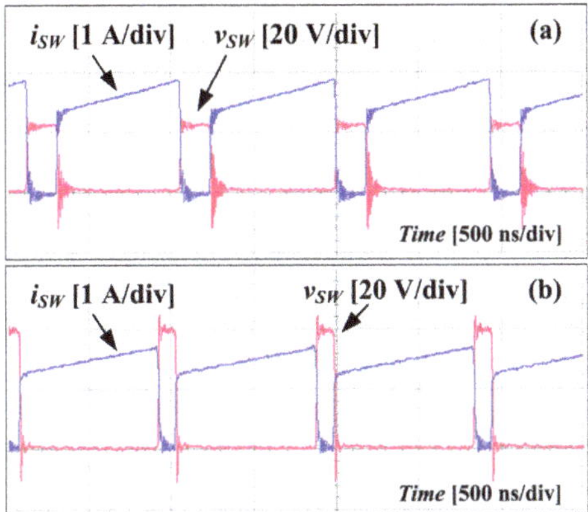

Figure 9. Voltage and current waveforms of switch measured at V_{IN} = 5 V, V_O = 40 V, I_O = 150 mA, and f_S = 1 MHz. (**a**) Proposed converter, (**b**) conventional boost converter.

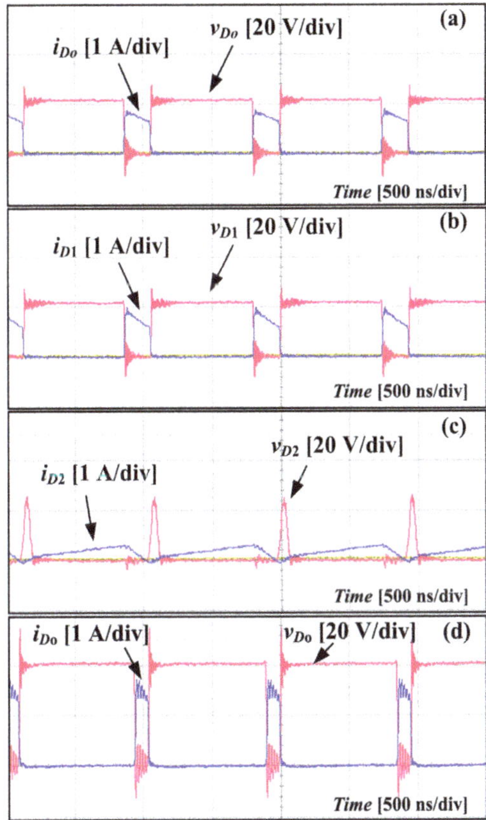

Figure 10. Diode waveforms of the proposed converter. (**a**) D_0, (**b**) D_1, (**c**) D_2, and diode waveform of the conventional boost converter. (**d**) D_o measured at V_{IN} = 5 V, V_O = 40 V, I_O = 150 mA and f_S = 1 MHz.

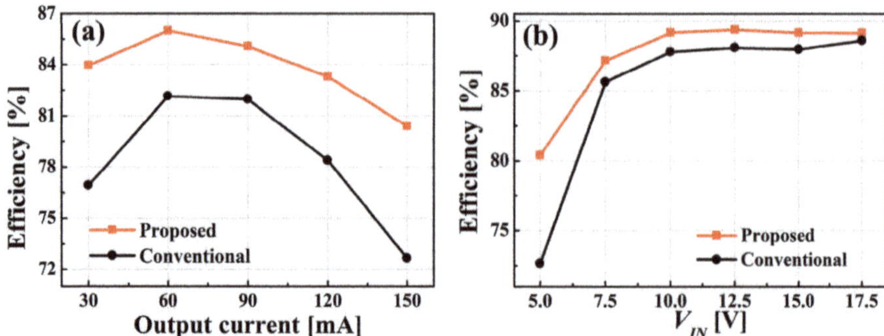

Figure 11. (**a**) η_e versus I_O at $V_{IN} = 5$ V, $V_O = 40$ V, $f_S = 1$ MHz, and $30 \leq I_O \leq 150$ mA. (**b**) η_e versus V_{IN} at $I_O = 150$ mA, $V_O = 40$ V, $f_S = 1$ MHz, and $5 \leq V_{IN} \leq 17.5$ V.

The temperatures of SW and inductors (Figure 12) were measured for 30 min using a digital temperature recorder (GL-220, GRAPHTEC), while the converters were operated at $V_{IN} = 5$ V, $V_O = 40$ V, $f_S = 1$ MHz, and $I_O = 150$ mA. In the conventional boost converter the switch stabilized at 86.6 °C and the inductor stabilized at 91.7 °C, but in the proposed converter, the switch stabilized at 49.5 °C and the inductor stabilized at 62.9 °C. These data indicate that the proposed converter generated less heat loss than the conventional boost converter.

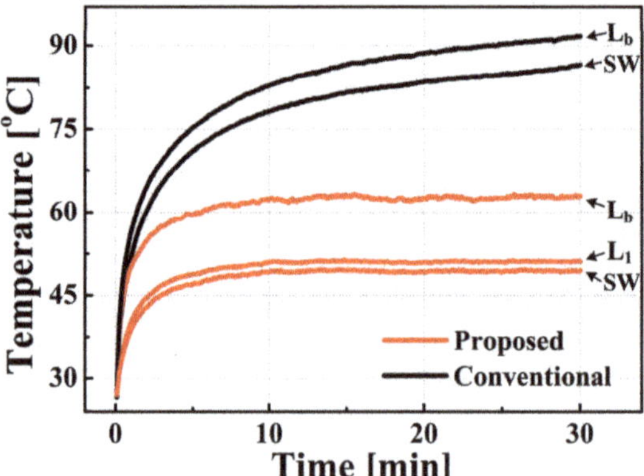

Figure 12. Temperature of switch and inductors versus operation time for the experimental converters, measured at $V_{IN} = 5$ V, $V_O = 40$ V, $I_O = 150$ mA, and $f_S = 1$ MHz.

The circuit losses were calculated using a circuit simulator at $V_O = 40$ V, $I_O = 150$ mA, $f_S = 1$ MHz, and $V_{IN} = 5$ V (Figure 13a) and 15 V (Figure 13b). The total power losses were 1.59 W (proposed) and 2.31 W (conventional) at $V_{IN} = 5$ V, and they were 0.68 W (proposed) and 0.78 W (conventional) at $V_{IN} = 15$ V. These results show that the proposed converter had lower power loss than the conventional boost converter at both $V_{IN} = 5$ V and 15 V. The losses in the switch were 0.65 W (proposed) and 1.41 W (conventional) at $V_{IN} = 5$ V, and they were 0.38 W (proposed) and 0.63 W (conventional) at $V_{IN} = 15$ V. The proposed converter reduced the switching loss by decreasing V_{SW}. Additionally, the snubber loss of the proposed converter was 0.16 W at $V_{IN} = 5$ V, and it was 0.15 W at $V_{IN} = 15$ V. The proposed converter reduced the snubber losses by placing L_1 at the output stage and eliminating the resonant current in the snubber circuit.

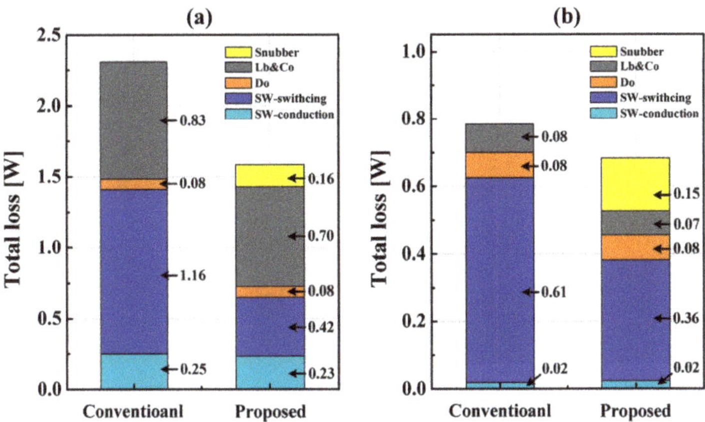

Figure 13. Circuit losses for V_O = 40 V, I_O = 150 mA and f_S = 1 MHz, calculated at (**a**) V_{IN} = 5 V and (**b**) V_{IN} = 15 V.

The voltage and current stress (Table 2) of the proposed and conventional boost converter were measured at V_{IN} = 5 V, V_O = 40 V, I_O = 150 mA, and f_S = 1 MHz. Without parasitic components, the voltage stress of SW and diodes were as follows: (1) in the conventional boost converter, $V_{SW} = V_O$ because D_O was turned ON when SW was turned OFF, and $V_{Do} = V_O$ because SW was turned ON when D_O was turned OFF; (2) in the proposed converter, $V_{SW} = V_O - V_{C1}$ because D_O, D_1, and D_2 were turned ON when SW was turned OFF, $V_{Do} = V_{D1} = V_O - V_{C1}$ because SW was turned ON when D_O and D_1 are turned OFF, and $V_{D2} = V_O - V_{C2}$ because D_1 and D_O were turned ON, and SW was turned OFF when D_2 was turned OFF. The proposed converter had lower voltage stress of SW and diodes than the conventional boost converter because $V_O = V_{C1} + V_{C2}$. The current stress of SW was similar in the conventional boost converter and the proposed converter. The current stresses of diodes were lower by half in the proposed converter than in the conventional boost converter.

Table 2. Voltage and current stresses of components in the experimental boost converters.

	Components	Conventional Boost Converter	Proposed Converter
	SW	45.1	24.9
Peak voltage stress (V)	D_O	53.5	27.7
	D_1	-	27.3
	D_2	-	26.0
	SW	1.72	1.87
Peak voltage stress (A)	D_O	1.78	0.90
	D_1	-	0.96
	D_2	-	0.32

The voltage stress, current stress, voltage gain, and efficiency of the existing boost converter were calculated using a circuit simulator at V_{IN} = 5 V, V_O = 40 V, I_O = 150 mA, and f_s = 1 MHz (Table 3). The efficiency η_e was 72.6% for the conventional boost converter, 80.4% for the proposed converter, 75.1% for the converter of Reference [12], and 76.6% for the converter of Reference [21]. The proposed converter had the highest η_e for given input and output conditions. The converters of References [12,21] had a lower efficiency than that of the proposed converter due to a high current stress and the loss of auxiliary switch, respectively. The voltage stress of the proposed converter was lower than that of the other converters because the voltage stress is V_O-V_{C1} in the proposed converter but V_O in the other converters. The current stresses of the proposed converter and the conventional boost converter were similar, but the converter of Reference [12] had a high current stress because it used resonance to reduce the switching loss at high frequency. At given V_{IN} and V_O, the proposed converter had the smallest duty, so the proposed converter had the highest voltage gain.

Table 3. Comparison with existing boost converters.

		Conventional	Proposed	Converter of Reference [12]	Converter of Reference [21]
Number of Components	SW	1	1	1	2
	Inductor	1	2	2	2
	Capacitor	1	3	3	2
	Diode	1	3	3	3
Switch Stress	Voltage stress (V)	45.1	24.9	45.2	SW$_1$ 42.2 / SW$_2$ 42.8
	Current stress (A)	1.72	1.87	2.92	SW$_1$ 2.42 / SW$_2$ 2.64
Duty		0.875	0.775	0.831	0.82 (main SW)
Efficiency (%)		72.6	80.4	75.1	76.6

5. Conclusions

A miniature DC-DC boost converter for driving a display panel of a notebook computer was proposed. This converter operates at a switching frequency of 1 MHz to miniaturize the circuit. The switching loss is reduced by using a passive snubber that lowers the voltage stresses of the switch. The conduction losses of the snubber and switch is also minimized by preventing high peak current due to the resonance at high f_S. The experimental converter was fabricated in 2.5 cm × 1 cm size and tested at $5 \leq V_{IN} \leq 17.5$ V, $30 \leq I_O \leq 150$ mA. Compared to previous boost converters, the proposed converter had a higher voltage conversion ratio, ~7.8% higher power conversion efficiency over the entire range of V_{IN} and I_O, ~1/2 as much voltage stress of the switch and diodes, and a much lower switch temperature. The results indicate that the proposed converter is a strong candidate for driving the display panel of a notebook computer.

Author Contributions: S.-H.H. conceived the main idea for the proposed converter and performed overall analysis and experiment. H.-J.C. led the project and gave technical advice.

Acknowledgments: This research was supported by the Tongmyong University Research Grants 2019(2019A003-1).

Conflicts of Interest: The authors have no conflict of interest.

References

1. Chen, M.; Rincón-Mora, G.A. Accurate, compact, and power-efficient Li-ion battery charger circuit. *IEEE Trans. Circuits Syst. II Express Briefs* **2006**, *53*, 1180–1184. [CrossRef]
2. Zhou, S.; Rincon-Mora, G.A. A high efficiency, soft switching DC-DC converter with adaptive current-ripple control for portable applications. *IEEE Trans. Circuits Syst. II Express Briefs* **2006**, *53*, 319–323. [CrossRef]
3. Salinas, F.; Krüger, L.; Neupert, S.; Kowal, J. A second life for li-ion cells rescued from notebook batteries. *J. Energy Storage* **2019**, *24*, 100747. [CrossRef]
4. Nordmann, H.; Frisch, M.; Sauer, D.U. Thermal Fault-Detection method and analysis of peripheral systems for large battery packs. *Measurement* **2018**, *114*, 484–491. [CrossRef]
5. Sun, J.; Xu, M.; Ying, Y.; Lee, F.C. High Power Density, High Efficiency System Two-Stage Power Architecture for laptop Computers. In Proceedings of the 37th IEEE Power Electronics Specialists Conference, Jeju, Korea, 18–22 June 2006; pp. 1–7.
6. Issa, J. Display power analysis and design guidelines to reduce power consumption. *J. Inf. Disp.* **2012**, *13*, 167–177. [CrossRef]

7. Tsai, Y.Y.; Tsai, Y.S.; Tsai, C.W.; Tsai, C.H. Digital noninverting-Buck–boost converter with enhanced duty-cycle-overlap control. *IEEE Trans. Circuits Syst. II Express Briefs* **2016**, *64*, 41–45. [CrossRef]

8. Datasheet: STLD41 White LED Driver for Mid-Size LCD Display Backlight. Available online: http://www.st.com/resource/en/datasheet/stld41.pdf (accessed on 1 October 2018).

9. Lee, S.W.; Choe, H.J.; Yun, J.J. Performance improvement of a boost LED driver with high voltage gain for edge-lit LED backlights. *IEEE Trans. Circuits Syst. II Express Briefs* **2017**, *65*, 481–485. [CrossRef]

10. Wu, C.H.; Chen, C.L. High-efficiency current-regulated charge pump for a white LED driver. *IEEE Trans. Circuits Syst. II Express Briefs* **2009**, *56*, 763–767.

11. Hsieh, Y.T.; Liu, B.D.; Wu, J.F.; Fang, C.L.; Tsai, H.H.; Juang, Y.Z. A high current accuracy boost white LED driver based on offset calibration technique. *IEEE Trans. Circuits Syst. II Express Briefs* **2011**, *58*, 244–248. [CrossRef]

12. Choe, H.J.; Chung, Y.C.; Sung, C.H.; Yun, J.J.; Kang, B. Passive snubber for reducing switching-power losses of an IGBT in a DC-DC boost converter. *IEEE Trans. Power Electron.* **2014**, *29*, 6332–6341. [CrossRef]

13. Bamgboje, D.O.; Harmon, W.; Tahan, M.; Hu, T. Low Cost High Performance LED Driver Based on a Self-Oscillating Boost Converter. *IEEE Trans. Power Electron.* **2019**, *34*, 10021–10034. [CrossRef]

14. Gao, S.; Wang, Y.; Guan, Y.; Xu, D. A High Frequency High Voltage Gain DCM Coupled-Inductor Boost LED Driver based on Planar Component. *IEEE Trans. Ind. Appl.* **2019**, in press. [CrossRef]

15. Hariharan, K.; Kapat, S. Near Optimal Controller Tuning in a Current-Mode DPWM Boost Converter in CCM and Application to a Dimmable LED Array Driving. *IEEE J. Emerg. Sel. Top. Power Electron.* **2018**, *7*, 1031–1043. [CrossRef]

16. Liu, X.; Li, X.; Zhou, Q.; Xu, J. Flicker-Free Single-Switch Quadratic Boost LED Driver Compatible with Electronic Transformers. *IEEE Trans. Ind. Electron.* **2018**, *66*, 3458–3467. [CrossRef]

17. Sathyan, S.; Suryawanshi, H.M.; Singh, B.; Chakraborty, C.; Verma, V.; Ballal, M.S. ZVS–ZCS high voltage gain integrated boost converter for DC microgrid. *IEEE Trans. Ind. Electron.* **2016**, *63*, 6898–6908. [CrossRef]

18. Choi, H.S.; Cho, B.H. Novel zero-current-switching (ZCS) PWM switch cell minimizing additional conduction loss. *IEEE Trans. Ind. Electron.* **2002**, *49*, 165–172. [CrossRef]

19. Barreto, L.H.S.C.; Coelho, E.A.A.; Farias, V.J.; de Oliveira, J.C.; de Freitas, L.C.; Vieira, J.J.B. A quasi-resonant quadratic boost converter using a single resonant network. *IEEE Trans. Ind. Electron.* **2005**, *52*, 552–557. [CrossRef]

20. Elasser, A.; Torrey, D.A. Soft switching active snubbers for dc/dc converters. *IEEE Trans. Power Electron.* **1996**, *11*, 710–722. [CrossRef]

21. Zhu, J.Y.; Ding, D. Zero-voltage-and zero-current-switched PWM DC-DC converters using active snubber. *IEEE Trans. Ind. Appl.* **1999**, *35*, 1406–1412. [CrossRef]

Article

Analytical and Simulation Fair Comparison of Three Level Si IGBT Based NPC Topologies and Two Level SiC MOSFET Based Topology for High Speed Drives

Jelena Loncarski [1,2,*,†,‡], Vito Giuseppe Monopoli [1,‡], Riccardo Leuzzi [1,‡], Leposava Ristic [3,‡] and Francesco Cupertino [1,‡]

1 Department of Electrical and Information Engineering, Politecnico di Bari, 70125 Bari, Italy;
   vitogiuseppe.monopoli@poliba.it (V.G.M.); riccardo.leuzzi@poliba.it (R.L.);
   francesco.cupertino@poliba.it (F.C.)
2 Department of Engineering Sciences, Division of Electricity, Uppsala University, S-751 21 Uppsala, Sweden
3 Department of Electrical Engineering, University of Belgrade, 11120 Belgrade, Serbia; lela@etf.bg.ac.rs
* Correspondence: jelena.loncarski@poliba.it
† Current address: Department of Electrical and Information Engineering, Politecnico di Bari, Via E. Orabona, 4, 70125 Bari, Italy.
‡ These authors contributed equally to this work.

Received: 29 October 2019; Accepted: 23 November 2019; Published: 30 November 2019

Abstract: Wide bandgap (WBG) power devices such as silicon carbide (SiC) can viably supply high speed electrical drives, due to their capability to increase efficiency and reduce the size of the power converters. On the other hand, high frequency operation of the SiC devices emphasizes the effect of parasitics, which generates reflected wave transient overvoltage on motor terminals, reducing the life time and the reliability of electric drives. In this paper, a SiC metal-oxide-semiconductor field-effect transistor (MOSFET) based two level (2L) inverter is systematically studied and compared to the performance of Si insulated-gate bipolar transistor (IGBT) based three level (3L) neutral point clamped (NPC) inverter topologies, for high speed AC motor loads, in terms of efficiency, overvoltages, heat sink design, and cost. A fair comparison was introduced for the first time, having the same output voltage capabilities, output current total harmonic distortion (THD), and overvoltages for the three systems. The analysis indicated the convenience of using the SiC MOSFET based 2L inverter for lower output power. In the case of the maximum output power, the heat sink volume was found to be 20% higher for the 2L SiC based inverter when compared to 3L NPC topologies. Simulations were carried out by realistic dynamic models of power switch modules obtained from the manufacturer's experimental tests and verified both in the LTspice and PLECS simulation packages.

Keywords: SiC devices; Si devices; three level NPC inverter; three level T-NPC inverter; two level SiC MOSFET inverter; overvoltages; heat sink volume

---

1. Introduction

Wide bandgap semiconductor devices, such as silicon carbide (SiC), offer many benefits due to their superior material properties, among which are increased junction operating temperature, low specific on resistance, high switching speed capability, low switching losses, etc. Consequently, motor drives supplied by a SiC based voltage source inverter (VSI) can provide lower losses, higher efficiency, and a smaller size when compared with their silicon (Si) counterparts [1,2]. All these features contribute to these devices generating interest in applications for electric traction systems, where a long life time is important, and by reducing the losses in such applications, the life time can be extended [3,4].

Even though the modern switching devices offer many benefits, they still experience many problems mainly connected to the fast switching such as high *dv/dt*-rates, also in combination with impedance mismatch (machine against cabling and surge), high di/dt, crosstalk [5], etc. Moreover, the fast switching leads to transient overvoltage which increases the strain for the machine's insulation and accelerated aging [4,6]. This, together with the fact that often, the power cable between the inverter and the motor is long, leads to significant voltage overshoot due to reflected wave phenomenon. All of these problems resulting from the high *dv/dt* spikes could degrade the reliability and efficiency of motor drive systems [7–9]. In some cases, motor transient peak voltages can be even up to 3–4-times the DC bus voltage [10]. Moreover, besides the effects of the overvoltages on a whole winding, also the stress on the inter-turn insulation of the motor windings must be considered, especially as the rise time becomes shorter. This can result in relevant voltage drops across one turn, going beyond the inter-turn insulation design limit considered under sinusoidal operation [6,7].

Currently, insulated-gate bipolar transistor (IGBT) based two level (2L) VSI, and three level (3L) neutral point clamped (NPC) and T-type neutral point clamped (T-NPC) inverters are commercially available and widely used by industry. However, in some cases, the application of 2L inverters is limited due to several drawbacks. They are usually connected with increased losses (in the case of higher fundamental frequencies (1 kHz); at the same time, they typically commutate large motor currents with a high switching frequency) and the generation of the high frequency ripple currents at the input DC-bus due to device switching (leading to the usage of oversized capacitors) [11]. Alternatively, 3L NPC and T-NPC inverters feature the following benefits: (1) lower switching losses; voltage across the device is half of the DC-link voltage; and (2) lower output current distortion. Furthermore, the T-NPC inverters incorporate additional advantages such as lower conduction losses when compared with the NPC inverter, being the better choice for low voltage applications [12,13].

Most of the state-of-the-art comparisons between the SiC metal-oxide-semiconductor field-effect transistor (MOSFET) and Si IGBT based inverters are done on the device level or in 2L VSIs [14,15] or hybrid topologies such as the H8 inverter [16]. Recently, also some multilevel Si IGBT based, SiC MOSFET based, or gallium nitride (GaN) based topologies have been compared, such as the single phase T-type inverter [17,18], advanced inverter topologies [19], or the single phase two stage decoupled active neutral point clamped (NPC) converter [20]. Generally, inverters with GaN and SiC MOSFET devices have shown benefits when compared to their Si counterparts such as lower losses, high efficiency at high frequency applications, reduced volume of the heat sink and output filter, etc.

A similar comparison was introduced in [21], where the 2L SiC inverter was compared with the 3L NPC inverter in terms of overvoltages and power losses, having the same output current total harmonic distortion (THD) and the same output voltage capabilities. This paper, following a similar principle, gives the comparative analysis of the 2L and 3L topologies mainly used by industry, introducing a fair comparison.

In this paper, a comparison between the 2L SiC MOSFET based VSI and 3L IGBT based NPC and T-NPC VSI is made, being widely used in industrial applications and offering additional advantages [22]. The three systems consider low voltage high speed electric drive applications with long power cables. The main contribution is the conducted fair comparison that has not been published yet in the literature, and that is based on the fact that the output voltages of the inverters are set in such a way to produce along the cable and at the motor terminals the same overvoltage for the three systems. The overvoltages are known to be the main causes of the partial discharge occurring in the stator winding, greatly influencing its life time and the reliability of the electric drive. In particular, in order to have a fair comparison, the three inverters have the same output voltage capabilities and the same output current THD. Moreover, the overvoltage *dv/dt* is set the same for the three inverters, by changing the gate resistance. The main idea is to analyze the three inverters in the same working condition. It is crucial in order to understand which topology is better and in which conditions. Even though the 2L SiC based inverters might seem a better solution for an electric drive application, having generally less losses and high switching operation, they still experience problems connected to the high switching speed, i.e., high overvoltages. On the other hand, 3L NPC inverters are widely used by industry, are considered reliable enough, and can be a good competitor of the 2L SiC MOSFET based solution. The high frequency equivalent circuit of the inverter-cable-motor system is introduced for the simulation modeling and transient analysis. The comparison includes the power loss difference of the three systems, as well as the heat sink design and total cost. The analysis is done on the power modules' real dynamic models obtained from the manufacturer's experimental test in the LTspice simulation tool (for the transient analysis of the overvoltages on the high frequency equivalent circuit, but also switching loss analysis with double pulse tests) and the PLECS simulation package for the loss and heat sink comparison, the access to the steady state being facilitated.

2. Inverter Topologies

Generally speaking, when considering the comparison of 2L and 3L inverter topologies, the 2L inverter features advantages such as the simple configuration, reliability, easy control, and lower cost. However, when it comes to the increase in the switching frequency, it is limited due to the switching losses. Moreover, the magnitude of the switching voltage over each device in the 2L inverter is the same as the entire DC-link voltage, bringing large harmonics in the inverter output. On the other hand, the 3L inverters offer several advantages such as: half the DC-link voltage across each device, lower harmonics, lower switching losses and electromagnetic interference (EMI).

In Figure 1, the three topologies considered for the comparison in this paper are represented. Figure 1a shows the 2L inverter with SiC MOSFET power switches, while Figure 1b,c show the 3L NPC and 3L T-NPC topology with IGBT devices, respectively. When compared to the 2L inverter, the 3L NPC has a more complex structure and control due to the voltage imbalance problem between the two DC-link capacitors caused by the parameter mismatch of non-ideal components. Currently, this problem is solved by the use of more advanced modulation schemes [23]. Historically, the 3L NPC topology was developed for medium voltage applications, because the available devices had limited voltage blocking capability, and there was a need to connect devices in series. The eventual increase in the conduction losses has been outperformed by the gain in voltage handling capability. In the case of low voltage applications, this was not necessary since the available devices have sufficient voltage ratings and fast switching speeds. For this reason, the T-type topology is the better choice for low voltage applications, since there is no series connection of devices that has to block the whole DC-link voltage [12,24].

The T-type inverter on Figure 1c, a member of the NPC inverter topologies, offers three output voltage levels. Switches that are forming one phase leg (S_1 and S_4) in Figure 1c are rated at V_{DC} and bidirectional switches S_2 and S_3 are rated at $V_{DC}/2$. For this reason, the two middle switches (S_2 and S_3) have very low switching and conduction losses. When compared to the 3L NPC topology, the T-type inverter has several benefits. Due to the fact that there is no series connection of the devices that has to block the whole DC-link voltage, the uneven share of the voltage to be blocked in the case when IGBTs in series turn off at the same time cannot occur [12]. Another benefit can be seen in the usage of a single device to block the full DC-link voltage instead of two devices in series, leading to reduced conduction losses, if bipolar devices are considered.

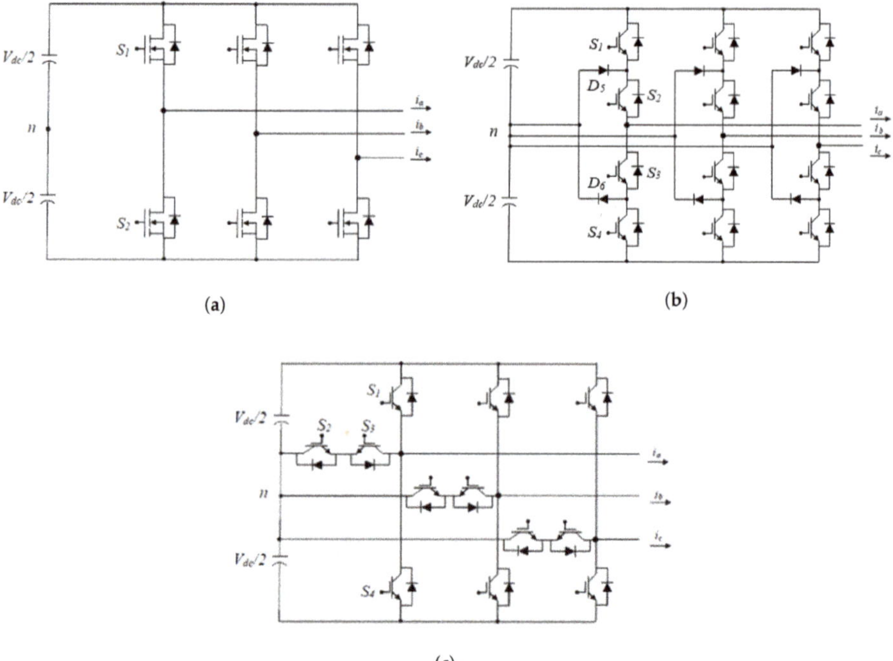

(a)   (b)

(c)

Figure 1. Inverter topologies in a motor drive application: (**a**) 2L SiC MOSFET, (**b**) 3L IGBT NPC, and (**c**) 3L IGBT T-NPC.

3. System Description and Overvoltage Comparison

For the overvoltage comparison, the three systems were analyzed using the simulation package LTspice. To have the same output voltage capabilities, the three inverters were supplied by total DC voltage V_{DC} = 600 V. The power modules' real models from the main producers were used for this analysis, i.e., the 1200 V SiC module SCT3080KLHR Rohm with the SiC Schottky anti-parallel diode SCS220KG Rohm for the 2L SiC based inverter, the 600 V IGBT module RGCL80TK60D Rohm for the 3L NPC IGBT based inverter, and the 1200 V RGS50TSX2DHR (leg switches) and 600 V RGCL80TK60D (middle switches) Rohm for the 3L T-NPC IGBT based inverter, with the main characteristics listed in Table 1. The reference was made to carrier based (CB) pulse width modulation (PWM) for the 2L inverter and phase disposition (PD) CB-PWM for the 3L inverter. The sizing of the inverters was done based on high speed drive requirements, considering the future application the high speed AC motor of 5 kW rated power and 48,000 rpm rated speed.

To have a fair comparison, the THD of the output current was set at the same value for the three systems, having in this way different switching frequencies, 60 kHz for the SiC based 2L inverter and 30 kHz for the 3L NPC and T-NPC inverter, while the fundamental frequency was set to 1.6 kHz. It was verified from the manufacturer's tests that two IGBTs can switch at a 30 kHz switching frequency for the maximum considered collector current.

For the overvoltage analysis, the transient analysis was introduced in LTspice, having the three inverter configurations supplying the high frequency (HF) motor model [25], as shown in Figure 2. The parameters of the model used for the simulation shown in Table 1 (motor and cable parameters) were obtained from the genetic algorithm, as explained in [25], and measurements on the real experimental setup, containing the SiC MOSFET based inverter, a 30 meter long cable, and a 0.37 kW induction motor, as explained in [7]. This system was used only to extract the high frequency motor circuit model parameters, which can approximately represent also other loads.

Table 1. Simulation parameters.

Device Parameters			
Parameters	**Rohm SiC MOSFET SCT3080KLHR**	**Rohm Si IGBT RGCL80TK60D**	**Rohm Si IGBT RGS50TSX2DHR**
V_{ds}	1200 V	600 V	1200 V
I_{ds} (25 °C)	31 A	35 A	50 A
I_{ds} (100 °C)	22 A	21 A	25 A
$R_{DS(on)}$ (25 °C)	80 mΩ	N/A	N/A
V_{CE-sat} (25 °C)	N/A	1.4 V	1.7 V
Q_g	60nC@18 V	98nC@15 V	67nC@15 V
V_{th}	2.7 V	5.5 V	6 V
V_{gs}	−4 to +22 V	±30 V	±30 V
T_j	175 °C	175 °C	175 °C
P_{diss} (25 °C)	165 W	57 W	395 W
r_{jc}	0.7 °C/W	2.62 °C/W	0.38 °C/W

Motor parameters	
Parameter (unit)	Value
R_{p1} (kΩ)	7.13
C_1 (pF)	761.02
R_{c1} (Ω)	63.47
L_1 (mH)	3.33
R_{l1} (Ω)	679.43
R_{p2} (MΩ)	55.55
C_2 (nF)	5.31
R_{c2} (kΩ)	21.04
L_2 (mH)	16.80
R_{l2} (Ω)	718.36
C_g (pF)	231.51
R_g (Ω)	2.56

Cable parameters	
Parameter (unit)	Value
R_{cable} (mΩ/m)	195.87
L_{cable} (µH/m)	0.63
C_{cable} (pF/m)	63.33

In particular, the three phase equivalent circuit in Figure 2 was used to model the cable lumped impedance (R_{cable}, L_{cable}, and C_{cable}) and the motor impedance. Each phase contained three resonators: the two external ones (accounting for the terminal part of the windings) and the central resonator (for the active part of the winding). Each resonator was made by the parallel connection of three branches: two resistances R_{p1} and R_{p2} took into account the eddy current HF path; the capacitive branch with C_1 or C_2 modeled the HF current path due to the turn-to-turn parasitic capacitance; and the inductive branch with L_1 or L_2 described the low frequency machine behavior. Moreover, two shunt capacitive branches were connected between the resonators and the ground, taking into account the parasitic coupling with the motor case.

Figures 3–5 show the terminal voltage of the AC motor (V_{mot_ab}, with the peak of the total voltage rise ΔV_{mot}) of the 2L SiC MOSFET based and 3L NPC and T-NPC IGBT based inverters, respectively. The transient analysis with the inverters supplying a high frequency motor load circuit was considered. The figures include a comparison when different power cable lengths are used, i.e., 3 m, 5 m, and 10 m (Figures 3a, 4a and 5a), and the case when gate resistance R_g is changed (Figures 3b, 4b and 5b). As can be seen from the figure, even in the case of a sufficiently short cable length of 3 m, the high dv/dt was induced at the motor terminals due to the fast rise time and the voltage reflection phenomenon. The 3L NPC and T-NPC inverter generally provided lower dv/dt than the 2L inverter. This was due to the fact that lower voltages were applied across each device in 3L inverters. In the case when the cable length was 10 m, a large high frequency transient could be noted, which did not expire, especially in the case of the 2L inverter. Overall, almost all dv/dt values were higher than the ones permitted by the standard on voltage stress in motors and drives.

Figures 3b, 4b and 5b show the change in the overvoltages for the fixed cable length (L_{cable}) and different values of the gate resistance. The external gate resistance (R_g) was varied as follows: R_g = 10 Ω, 20 Ω, 30 Ω. For the higher values of R_g, the decrease in the overvoltages can be seen, since both the rise time was increased and the peak was decreased. In order to have a fair comparison for the three systems, the same overvoltages were considered, i.e., the value of R_g for the 2L SiC MOSFET based inverter was increased from 12 Ω to 25 Ω (resulting in the decrease of dv/dt from 11.68 kV/μs to 8.6 kV/μs), while in the case of 3L inverters, R_g was kept the same, i.e., 10 Ω. In this way, the same dv/dt of 8.6 kV/μs for the three systems was considered, and the power loss comparison (using the steady state analysis) is given for that specific case in the following section. In addition, the THD of the output currents and the output voltage capabilities were the same.

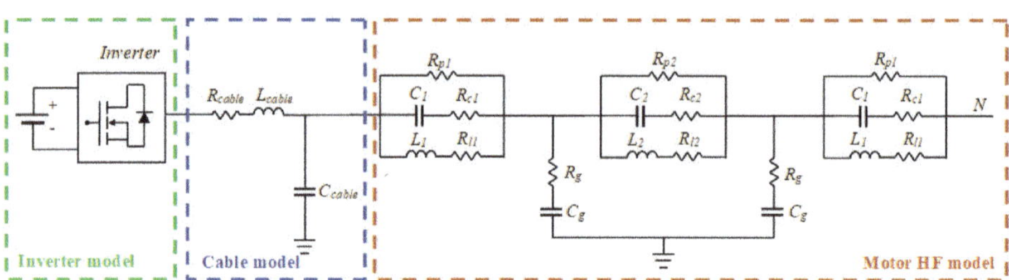

Figure 2. High frequency (HF) equivalent circuit of the inverter-cable-motor system.

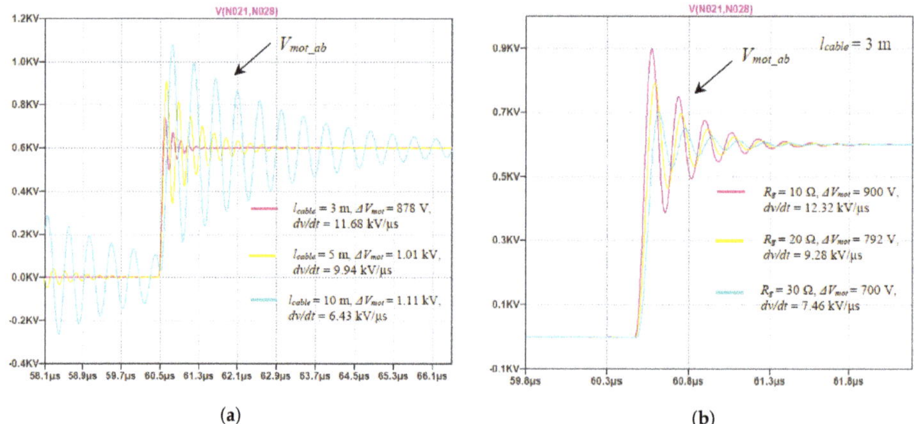

(**a**)

(**b**)

Figure 3. Simulation results of the motor line-to-line voltages for the 2L SiC MOSFET inverter in the case of: (**a**) different lengths of the power cable and (**b**) different values of gate resistance.

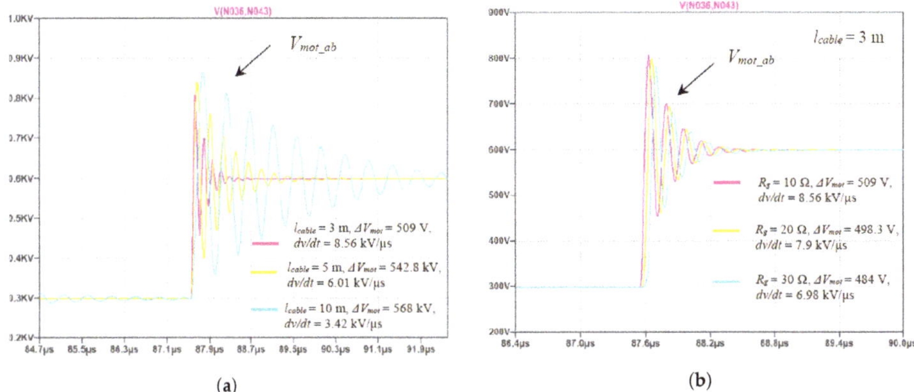

(a)   (b)

Figure 4. Simulation results of the motor line-to-line voltages for the 3L IGBT NPC inverter in the case of: (**a**) different lengths of the power cable and (**b**) different values of gate resistance.

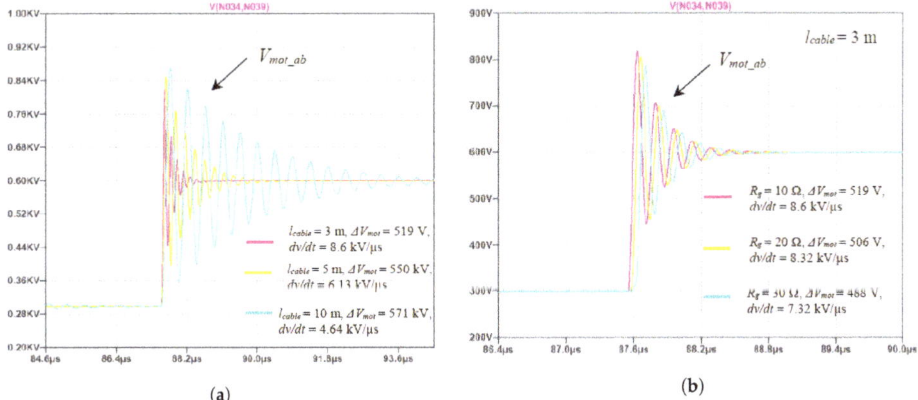

(a)   (b)

Figure 5. Simulation results of the motor line-to-line voltages for the 3L IGBT T-NPC inverter in the case of: (**a**) different lengths of the power cable and (**b**) different values of gate resistance.

4. Power Loss Analysis

Device losses were mainly resulting from two causes: the conduction losses and switching losses. The IGBT conduction loss model was obtained by using its dynamic on resistance $R_{on,IGBT}$ and zero on-state voltage V_0, i.e.,

$$P_{con,IGBT} = V_0 I_{av} + R_{on,IGBT} I_{rms}^2 \tag{1}$$

where I_{av} and I_{rms} are the average and rms currents through the device. For the SiC MOSFETs, the on resistance $R_{DS(on)}$ was needed to determine conduction losses, i.e.,

$$P_{con,MOSFET} = R_{DS(on)} I_{rms}^2 \tag{2}$$

The conduction losses for the diodes were based on their threshold voltage V_T and dynamic on resistance $R_{on,diode}$, i.e.,

$$P_{con,diode} = V_T I_{av} + R_{on,diode} I_{rms}^2 \tag{3}$$

The switching losses had a linear relationship to the switched current. The overall averaged switching losses can be expressed as:

$$P_{sw,device} = f_{sw}\frac{1}{T}\int_{0+\phi}^{\frac{T}{2}} (E_{on,device} + E_{off,device})dt \tag{4}$$

where $E_{on,device}$ and $E_{off,device}$ are the dissipated energy of the device during turning on and off and ϕ is the phase angle.

These equations are valid for the general case and are not connected to the specific application. In the next subsections are given the losses analysis connected to the specific applications, i.e., the 2L SiC MOSFET based inverter and NPC and T-NPC inverter.

4.1. 2L SiC MOSFET Based Inverter

When considering the specific applications, i.e., the 2L inverter adopting SiC MOSFET and the SiC Schottky diode, the losses can be calculated according to the following formulae [26]:

$$P_{con,MOSFET} = R_{DS(on)}I_{rms}^2 = R_{DS(on)}\hat{I}^2\left(\frac{1}{8} + \frac{m\cos\phi}{3\pi}\right) \tag{5}$$

where $\hat{I}$ is the peak value of the output current, m is the modulation index, $m = 2V^*/V_{DC}$, and V^* is the amplitude of the reference output voltage. Diode conduction losses can be expressed as:

$$P_{con,diode} = V_T I_{av} + R_{on,diode}I_{rms}^2 = V_T\hat{I}\left(\frac{1}{2\pi} - \frac{m\cos\phi}{8}\right) + R_{on,diode}\hat{I}^2\left(\frac{1}{8} - \frac{m\cos\phi}{3\pi}\right) \tag{6}$$

For the switching losses, we can write:

$$P_{sw,device} = f_{sw}\frac{1}{\pi}\frac{\hat{I}}{I_{ref}}E_{sw,device} \tag{7}$$

where I_{ref} is the reference current value of the switching loss measurement, available in the component's datasheet.

4.2. NPC and T-NPC Inverter

In the case of the NPC and T-NPC inverter adopting the Si IGBT and Si diode, the conduction and switching losses can be expressed as follows [18,27]:

- NPC inverter:

$$P_{con,S_1\&S_4} = V_0 I_{av} + R_{on,IGBT}I_{rms}^2 = \frac{m\hat{I}}{12\pi}\{3V_0[(\pi - \phi)\cos\phi + \sin\phi] + 2R_{on,IGBT}\hat{I}[1 + \cos\phi]^2\} \tag{8}$$

m being the modulation index, $m = 2V^*/\sqrt{3}\,V_{DC}$.

$$P_{sw,S_1\&S_4} = f_{sw}E_{sw,device}\left(\frac{\hat{I}}{I_{ref}}\right)^{K_I}\left(\frac{V_{cc}}{V_{ref}}\right)^{K_V}\left\{\frac{1}{2\pi}[1 + \cos\phi]\right\}G_I \tag{9}$$

where V_{cc} is the collector-emitter supply voltage and I_{ref} and V_{ref} are the reference current and voltage of the switching loss measurement available in the component's datasheets, respectively. The coefficients K_I, K_V, and G_I [27] are defined as:

- $K_I = 1$ for IGBT and $K_I = 0.6$ for diode,
- $K_V = 1.4$ for IGBT and $K_V = 0.6$ for diode,
- $G_I = 1$ for IGBT and $G_I = 1.15$ for diode

For diodes, we can write:

$$P_{con,D_1\&D_4} = V_T I_{av} + R_{on,diode} I_{rms}^2 = \frac{m\hat{I}}{12\pi}\{3V_T[-\phi\cos\phi + \sin\phi] + 2R_{on,diode}\hat{I}[1 - \cos\phi]^2\} \tag{10}$$

$$P_{sw,D_1\&D_4} = f_{sw}E_{sw,device}\left(\frac{\hat{I}}{I_{ref}}\right)^{K_I}\left(\frac{V_{cc}}{V_{ref}}\right)^{K_V}\left\{\frac{1}{2\pi}[1 - \cos\phi]\right\}G_I \tag{11}$$

For the remaining two switches, S_2 and S_3 in Figure 1b, we can write:

$$P_{con,S_2\&S_3} = \frac{\hat{I}}{12\pi}\{V_0[12 + 3m(\phi\cos\phi - \sin\phi)] + R_{on,IGBT}\hat{I}[3\pi - 2m(1 - \cos\phi)^2]\} \tag{12}$$

$$P_{sw,S_2\&S_3} = f_{sw}E_{sw,device}\left(\frac{\hat{I}}{I_{ref}}\right)^{K_I}\left(\frac{V_{cc}}{V_{ref}}\right)^{K_V}\left\{\frac{1}{2\pi}[1 - \cos\phi]\right\}G_I \tag{13}$$

$$P_{con,D_2\&D_3} = \frac{m\hat{I}}{12\pi}\{3V_T[-\phi\cos\phi + \sin\phi] + 2R_{on,diode}\hat{I}[1 - \cos\phi]^2\} \tag{14}$$

$$P_{sw,D_2\&D_3} = 0 \tag{15}$$

For diodes D_5 and D_6, we can write:

$$P_{con,D_5\&D_6} = \frac{\hat{I}}{12\pi}\{V_T[12 + 3m[(2\phi - \pi)\cos\phi - 2\sin\phi)]] + R_{on,diode}\hat{I}[3\pi - 4m(1 + \cos^2\phi)]\} \tag{16}$$

$$P_{sw,D_5\&D_6} = f_{sw}E_{sw,device}\left(\frac{\hat{I}}{I_{ref}}\right)^{K_I}\left(\frac{V_{cc}}{V_{ref}}\right)^{K_V}\left\{\frac{1}{2\pi}[1 + \cos\phi]\right\}G_I \tag{17}$$

Similarly, for the T-NPC inverter, we can write the following equations:

- T-NPC inverter:

$$P_{con,S_1\&S_4} = \frac{m\hat{I}}{12\pi}\{3V_0[(\pi - \phi)\cos\phi + \sin\phi] + 2R_{on,IGBT}\hat{I}[1 + \cos\phi]^2\} \tag{18}$$

$$P_{sw,S_1\&S_4} = f_{sw}E_{sw,device}\left(\frac{\hat{I}}{I_{ref}}\right)^{K_I}\left(\frac{V_{cc}}{V_{ref}}\right)^{K_V}\left\{\frac{1}{2\pi}[1 + \cos\phi]\right\}G_I \tag{19}$$

For diodes, we can write:

$$P_{con,D_1\&D_4} = \frac{m\hat{I}}{12\pi}\{3V_T[-\phi\cos\phi + \sin\phi] + 2R_{on,diode}\hat{I}[1 - \cos\phi]^2\} \tag{20}$$

$$P_{sw,D_1\&D_4} = f_{sw}E_{sw,device}\left(\frac{\hat{I}}{I_{ref}}\right)^{K_I}\left(\frac{V_{cc}}{V_{ref}}\right)^{K_V}\left\{\frac{1}{2\pi}[1 - \cos\phi]\right\}G_I \tag{21}$$

For the switches S_2 and S_3 in Figure 1c, we can write:

$$P_{con,S_2\&S_3} = \frac{\hat{I}}{12\pi}\{V_0[12 + 6m(\phi\cos\phi - \sin\phi) - 3m\pi\cos\phi] + R_{on,IGBT}\hat{I}[3\pi - 4m(1 + \cos^2\phi)]\} \tag{22}$$

$$P_{sw,S_2\&S_3} = f_{sw}E_{sw,device}\left(\frac{\hat{I}}{I_{ref}}\right)^{K_I}\left(\frac{V_{cc}}{V_{ref}}\right)^{K_V}\left\{\frac{1}{2\pi}[1 - \cos\phi]\right\}G_I \tag{23}$$

$$P_{con,D_2\&D_3} = \frac{\hat{I}}{12\pi}\{V_T[12 + 6m(\phi\cos\phi - \sin\phi) - 3m\pi\cos\phi] + R_{on,diode}\hat{I}[3\pi - 4m(1 + \cos^2\phi)]\} \tag{24}$$

$$P_{sw,D_2\&D_3} = f_{sw}E_{sw,device}\left(\frac{\hat{I}}{I_{ref}}\right)^{K_I}\left(\frac{V_{cc}}{V_{ref}}\right)^{K_V}\left\{\frac{1}{2\pi}[1+cos\phi]\right\}G_I \qquad (25)$$

4.3. Double Pulse Test and PLECS Analysis

The simulation package PLECS enables the user to merge the thermal design with the electrical design, providing in this way the requirements for the cooling solution suitable for the particular application. The switches are ideal, and the switching and conduction losses are inserted in terms of 3D look-up tables for the specific semiconductor's operating condition (forward current, blocking voltage, junction temperature) before and after each switch operation. In this way, depending also on the application, the simulation speed is not necessarily affected. Moreover, there is the possibility to use steady state analysis, which skips the long transient in the thermal analysis and displays the stead state. The thermal description of the device is done by a thermal network, either Cauer or Foster.

To determine the device switching and conduction losses, the standard double pulse test was used for each device in LTspice, as presented in Figure 6a in the case of IGBT. In Figure 6b are presented the turn on losses in the case of the 600 V IGBT RGCL80TK60D Rohm. In the double pulse test, the first gate pulse was used to charge the inductive load L to a desired current level, and the freewheeling diode (FWD) was used to keep the current level when the device under test (DUT) was off. A second short pulse was used for IGBT switching transient characterization. It is to be noted that when IGBT was used with a fast recovery diode, the diode losses were not added to the IGBT's turn on losses, but presented with turn off losses as the negative part. Similarly, also the losses for the SiC MOSFET can be introduced.

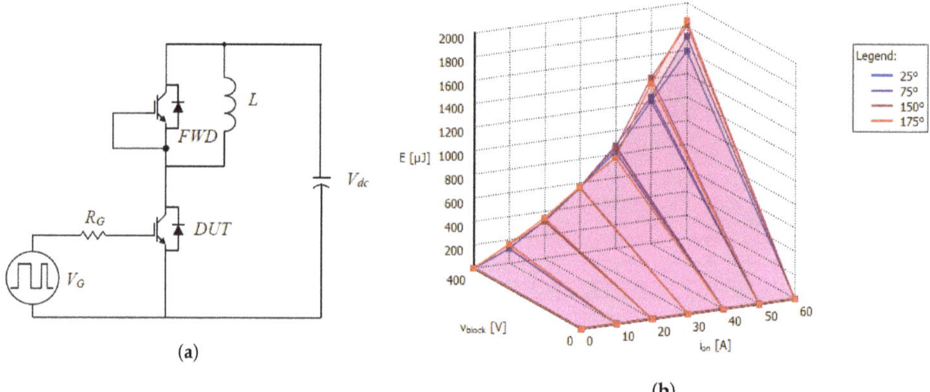

Figure 6. Double pulse test: (**a**) schematic of the double pulse test and (**b**) example of the IGBT's switching on losses for different temperatures.

After the losses' description was added in PLECS, it was possible to place the heat sink and obtain the device junction and case temperatures, as well as the respective losses for a specific application.

4.4. Power Loss Comparison

Switching and conduction losses were analyzed using the power electronics simulation package LTspice and PLECS, as explained in Section 4.3. Namely, from the double pulse test on real device models in LTspice, the conduction and switching losses were obtained. These losses were introduced in the look-up tables that the PLECS simulation package was using for the characterization of each device. The complete inverter-passive RL load systems were then developed in PLECS, having the three inverter configurations. Moreover, for a fair comparison, the same overvoltages were set for the three systems, having a 25 Ω gate resistance value in the case of the 2L SiC based inverter, while for the NPC and T-NPC inverters, the values from the datasheet were kept (R_g = 10 Ω). Consequently, also the double pulse tests and analytical considerations were done considering these values. The fact that the gate resistance for the 2L SiC inverter was much higher than the value from the datasheet, i.e., 12 Ω, can be considered a somewhat unrealistic condition for the 2L SiC inverter, but it is useful for the comparison; in this way, the three systems could be compared for the same value of *dv/dt*.

The same conditions were kept as in Section 3, i.e., the same output voltage capabilities and the same output current THD, just in this case, the inverters were supplying passive RL load. The PLECS simulation package used for the steady state analysis enabled a fast way to obtain the steady state and also provided the cooling solution. Figure 7 shows the comparison of the analytical (as presented in Section 4.2) and simulation losses in the case of the T-NPC inverter for the maximum output power, i.e., 11.4 kW. The good agreement between the analytical results and the results obtained with the PLECS simulation package can be noted from the figure.

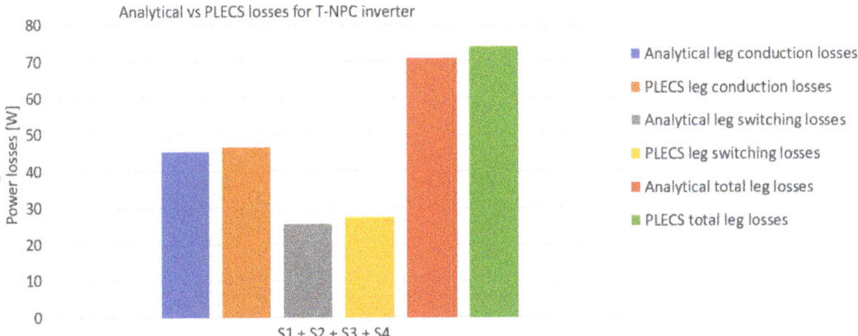

Figure 7. Analytical and simulation switching, conduction, and total losses of one inverter leg with maximum output power for the 3L T-NPC IGBT based inverter

Figure 8 summarizes the analytical and simulation total losses of one inverter leg in the case of the inverter output power having different values. Total losses of the three configurations in the case of the fair comparison, i.e., same voltage capabilities, same output current THD, and same overvoltage, were almost the same for low output powers, i.e., 25% of the maximum output power. For higher values of the output power, the 2L SiC inverter had a greater increase in the total losses when compared to the NPC and T-NPC inverter. In the case of 50% of the max power, the 2L SiC inverter had 14% more losses than the NPC inverter and 30% more losses than the T-NPC inverter. For the maximum output power, the 2L SiC inverter had 42% more losses when compared to the lowest, which in this case was the T-NPC inverter. The T-NPC inverter had also slightly lower losses than the NPC inverter, as expected.

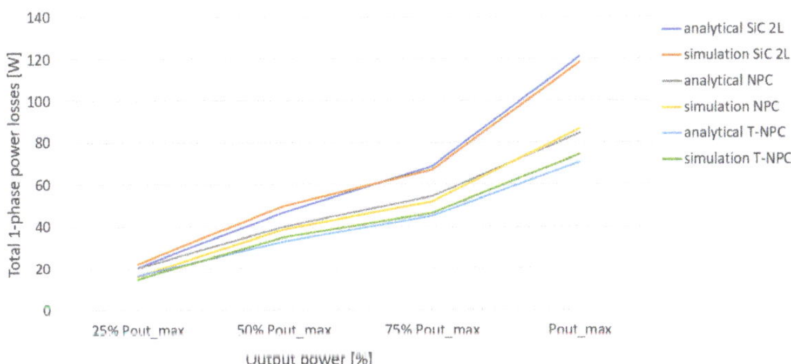

Figure 8. Analytical and simulation total one phase power device losses of the inverters with different values of the output power.

4.5. Heat Sink Volume

The thermal network of switching devices is presented in Figure 9a. N is the number of devices; T_j is the junction temperature; r_{jc} is the junction to case thermal resistance; r_{ch} is the case to heat sink thermal resistance; r_h is the heat sink thermal resistance; T_c is the case temperature; and T_a is the ambient temperature.

The heat sink volume analysis was based on the power losses of three systems at maximum output power and considering heat sink temperatures between 75 °C and 125 °C. According to Figure 9a, we can write the following expressions:

$$T_{j1...n} = P_{loss1...n} \frac{r_{jc1...n} + r_{ch1...n}}{2} + T_h \tag{26}$$

$$r_h = \frac{T_h - T_a}{P_{tloss}} \tag{27}$$

where P_{tloss} is the total inverter loss.

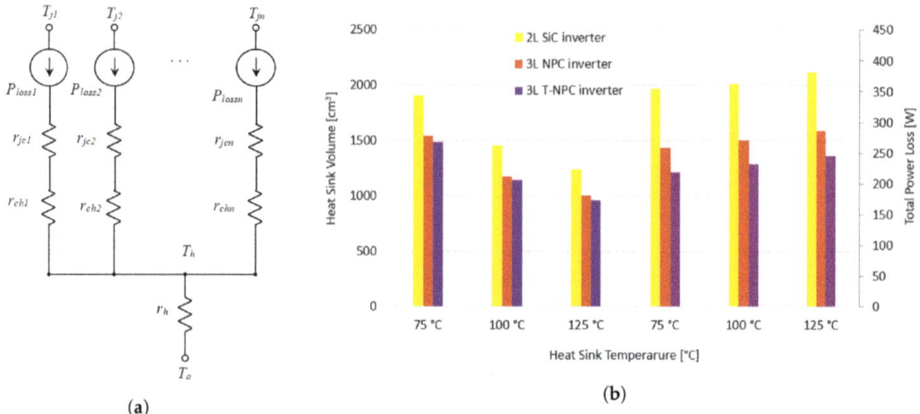

(a)

(b)

Figure 9. Heat sink design: (**a**) general thermal network; (**b**) heat sink volume and total inverter power loss for three inverters and different heat sink temperatures.

The calculated r_h can be used to calculate the heat sink volume based on natural air convection. The calculation of the fitting function applied to the minimum heat sink volume available at a given r_h value can be found in [18]:

$$Vol_{heatsink} = 3263e^{-13.09r_h} + 1756e^{-1.698r_h} \tag{28}$$

Equation (28) [18] is obtained from the curve fitting based on the volume of various extruded naturally cooled heat sinks against heat sink thermal resistance.

Heat sink volume calculations for three different inverters with respect to heat sink temperature are presented in Figure 9b. A room temperature of 25 °C was chosen for the ambient temperature. The results showed that the 2L SiC MOSFET based converter had an increase of around 20% in the heat sink volume when compared to the 3L NPC and T-NPC inverter at a 75 °C heat sink temperature. The same trend can be seen also for the other temperature values. Additionally, the volume of the heat sink can be reduced by a factor of 1.5 for the SiC based inverter by increasing the temperature from 75 °C to 125 °C, which on the other hand would bring 7% more losses. The 3L NPC and T-NPC inverters had a similar heat sink volume, with T-NPC being slightly lower (4% in the case of a 75 °C heat sink temperature).

5. System Cost Analysis

In this section are given the costs of the inverter-motor system, taking into account only the switching devices together with gate drivers and the heat sink. Since the fair comparison considered the same dv/dt for the three systems, the output filters were considered of the same volume and were not taken into account for this analysis. The same price was considered for the other components as the price difference was none or negligible, and therefore was not included in this comparison. In particular, two cases were considered, when inverters were working with their maximum rated capacity and when they were working with 50% of the maximum rated capacity. Note that the prices in Table 2 are given in per unit, referring to the sample prices of the main parts' manufacturers in Europe.

Table 2. System cost comparison.

Component	NPC	T-NPC	2L SiC
Switching device/price trend (p.u.)	24.85/stable	27.43/stable	55.71/strong reduction
Heat sink/P_{out_max} (p.u.)	20.74	17.68	27.23
Heat sink/50%P_{out_max} (p.u.)	12.55	10.8	14.32
Total cost/P_{out_max} (p.u.)	45.6	45.1	82.94 ↓
Total cost/50%P_{out_max} (p.u.)	37.4	38.23	70.03 ↓

For all the considered cases, the 3L NPC IGBT based inverter showed the lowest total cost and the 2L SiC MOSFET based inverter the highest. When considering only the device cost, the 2L SiC MOSFET based inverter had more then two-times the cost of the NPC and T-NPC inverter. In the case of P_{out_max}, the total cost of the 2L SiC inverter was around 1.8-times the cost of the NPC inverter and T-NPC inverter. It was similar also in the case of 50%P_{out_max}. It is interesting to notice that the total cost of the T-NPC inverter was similar to the NPC inverter, but offering a 16% reduction of the losses when considering the case when inverters were working with maximum output power. The price of the 2L SiC inverter was dominant in all the considered cases, also when taking into account the passive components. However, a strong cost reduction in upcoming years for SiC devices is expected, so that fact is likely to change.

In Table 3 are summarized all the features analyzed in this paper for the three inverters, in terms of transient analysis where the dv/dt are given and steady state analysis where the losses, heat sink volume, and costs are given for the fair comparison.

Table 3. Summary.

Descriptors	NPC	T-NPC	2L SiC
Transient analysis/LTspice			
dv/dt for l_{cable}=3 m (kV/µs)	8.56	8.6	11.68
dv/dt for l_{cable}=5 m (kV/µs)	6.01	6.13	9.94
dv/dt for l_{cable}=10 m (kV/µs)	3.42	4.64	6.43
Steady-state analysis/PLECS			
Total losses @75 °C/50% P_{out_max} (W)	120	98.34	139.4
Total losses @75 °C/P_{out_max} (W)	258.9	218.6	355
Heat sink volume/50% P_{out_max} (cm^3)	878.9	756	1002.6
Heat sink volume/P_{out_max} (cm^3)	1452.9	1237.8	1906.5
Total cost/50% P_{out_max} (p.u.)	37.4	38.23	70.03 ↓
Total cost/P_{out_max} (p.u.)	45.6	45.1	82.94 ↓

By looking at the transient analysis, it is clear that the 2L SiC inverter has a higher dv/dt, in some cases almost double. When it comes to the steady state analysis, for 50% of the output power and less, we could say that the 2L SiC inverter could be a better choice for a high speed motor drive system. This is, however, premature to say since the cost of the 2L SiC inverter and SiC devices in general is still rather high. With the reduction in the SiC devices' price, the need to mitigate the overvoltage problem in SiC based inverters is essential and inevitable, as they offer several benefits such as high frequency operation and reduction in heat sink volume and output inductor volume when operated in realistic conditions, i.e., with lower values of gate resistance.

6. Conclusions

This paper discussed the comparison of the 2L SiC MOSFET based and 3L NPC and T-NPC Si IGBT based inverters in terms of the efficiency, overvoltages on motor terminals, heat sink design, and cost of the inverter-motor load system. The three systems considered a low voltage high speed electric drive application with a long power cable. A fair comparison was introduced for the first time and was based on the fact that the output voltages of the inverters were set in such a way as to produce along the cable and at motor terminals the same overvoltage for the three systems. The overvoltages are known to be the main cause of the partial discharge occurring in the stator winding, greatly influencing the life time and the reliability of an electric drive. To have a fair comparison, inverters had the same output voltage capabilities, the same output current THD, and the same overvoltages on motor terminals. The analysis was conducted on power modules' real dynamic models obtained from the manufacturer's experimental tests in the LTspice simulation tool (for the overvoltage and double pulse tests analysis) and the PLECS simulation package for the power losses and heat sink comparison.

The overvoltages were compared by using a high frequency motor-load circuit and considering three lengths of the power cable (3, 5, and 10 m). In most of the cases, high dv/dt could be noted, not complying with the standards on voltage stress in motors and drives.

Power loss comparison and steady state analysis where the inverters were supplying passive RL load were done in the PLECS simulation package. Switching and conduction losses were obtained by performing double pulse tests in LTspice on the devices' realistic dynamic models. In this case, a fixed length of the power cable (3 m) was considered, and the gate resistance was set in order to have the same overvoltages for the three systems. A power loss comparison was conducted for that specific case in PLECS, showing similar switching losses for the 2L SiC based VSI and 3L VSIs in the case of lower output power (less than 50% of the maximum output power). Moreover, the power loss analysis showed good agreement between the analytical and simulation results.

The heat sink volume was compared for the maximum inverter output power for the three systems. The SiC MOSFET based inverter, in this specific case of the fair comparison, had a 20% increase in the volume when compared to 3L Si IGBT based inverters for all the considered heat sink temperatures. It could, however, be reduced by the factor of 1.5 by increasing the heat sink temperature from 75 °C to 125 °C, which on the other hand would bring 7% more losses. The 3L T-NPC inverter had the lowest heat sink volume and lowest losses for all the considered temperatures.

For the inverter-motor cost analysis, two cases were considered, when inverters were working with their maximum rated capacity and when they were working with 50% of the maximum rated capacity. For all the considered cases, the 3L NPC IGBT based inverter had the lowest cost and 2L SiC MOSFET based inverter the highest, with 1.8-times the total cost of the NPC inverter and T-NPC inverter. With the price reduction in the next few years for SiC devices, this is likely to change, and 2L SiC based inverters can become competitive for high speed motor drives in the future.

Speaking for the present, the 2L SiC MOSFET based VSIs can be seen as a better solution when compared to the 3L NPC inverter in the case when the inverter is working at low output power. When considering higher output power and the upcoming price reduction of the SiC devices, the need to mitigate the overvoltage problem in SiC based inverters is inevitable, as they can offer several benefits such as high frequency operation, reduction in heat sink volume, etc. Moreover, the 2L inverter configuration is more reliable when compared to 3L configuration and could benefit from the usage of SiC devices.

Future work will include more detailed comparisons, including the experimental verification, but also other hybrid topologies. The investigation for the potential solutions for switching performance improvements and mitigation of reflected waves, as well as the future study of electric aging phenomena can then be performed.

Author Contributions: J.L.: conceptualization, data curation, formal analysis, investigation, methodology, software, writing, original draft. V.G.M.: formal analysis, investigation, writing, review and editing. R.L.: formal analysis, investigation, writing, review and editing. L.R.: formal analysis, writing, review and editing. F.C.: formal analysis, writing, review and editing, funding acquisition.

Funding: This research was supported by the project "FURTHER - Future Rivoluzionarie Tecnologie per velivoli più Elettrici" Code ARS01_01283, funded by the Italian Ministry of University and Research within Programma Operativo Nazionale (PON) "Ricerca e Innovazione" 2014–2020.

Acknowledgments: The authors would like to thank ROHM Semiconductor GmbH for providing technical support.

Conflicts of Interest: The authors declare no conflict of interest. The funders had no role in the design of the study; in the collection, analyzes, or interpretation of data; in the writing of the manuscript; nor in the decision to publish the results.

References

1. Zhang, Z.; Wang, F.; Tolbert, L.M.; Blalock, B.J.; Costinett, D.J. Evaluation of switching performance of SiC devices in PWM inverter-fed induction motor drives. *IEEE Trans. Power Electron.* **2015**, *30*, 5701–5711. [CrossRef]
2. Biela, J.; Schweizer, M.; Waffler, S.; Kolar, J.W. Evaluation of potentials for performance improvement of inverter and DC–DC converter systems by SiC power semiconductors. *IEEE Trans. Ind. Electron.* **2011**, *58*, 2872–2882. [CrossRef]
3. Ding, X.; Cheng, J.; Chen, F. Impact of Silicon Carbide Devices on the Powertrain Systems in Electric Vehicles. *Energies* **2017**, *10*, 533. [CrossRef]
4. Zoeller, C.; Vogelsberger, M.A.; Wolbank, T.M.; Ertl, H. Impact of SiC semiconductors switching transition speed on insulation health state monitoring of traction machines. *IET Power Electron.* **2016**, *9*, 2769–2775. [CrossRef]
5. Zaman, H.; Wu, X.; Zheng, X.; Khan, S.; Ali, H. Suppression of Switching Crosstalk and Voltage Oscillations in a SiC MOSFET Based Half-Bridge Converter. *Energies* **2018**, *11*, 3111. [CrossRef]
6. Haq, S.U.; Jayaram, S.H.; Cherney, E.A. Insulation problems in medium-voltage stator coils under fast repetitive voltage pulses. *IEEE Trans. Ind. Appl.* **2008**, *44*, 1004–1012. [CrossRef]
7. Leuzzi, R; Monopoli, V.G.; Cupertino, F.; Zanchetta, P. Active Ageing Control of Winding Insulation in High Frequency Electric Drives. In Proceedings of the 2018 IEEE Energy Conversion Congress and Exposition (ECCE), Portland, OR, USA, 23–27 September 2018; pp. 1–7.
8. Swamy, M.M.; Kang, J.; Shirabe, K. Power loss, system efficiency, and leakage current comparison between Si IGBT VFD and SiC FET VFD with various filtering options. *IEEE Trans. Ind. Appl.* **2015**, *51*, 3858–3866. [CrossRef]
9. De Caro, S.; Foti, S.; Scimone, T.; Testa, A.; Scelba, G.; Pulvirenti, M.; Russo, S. Over-voltage mitigation on SiC based motor drives through an open end winding configuration. In Proceedings of the 2017 IEEE Energy Conversion Congress and Exposition (ECCE), Cincinnati, OH, USA, 1–5 October 2017; pp. 4332–4337.
10. Yi, P.; Murthy, P.K.S.; Wei, L. Performance evaluation of SiC MOSFETs with long power cable and induction motor. In Proceedings of the 2016 IEEE Energy Conversion Congress and Exposition (ECCE), Milwaukee, WI, USA, 18–22 September 2016; pp. 1–7.
11. Bhattacharya, S.; Mascarella, D.; Joós, G.; Cyr, J.; Xu, J. A Dual Three-Level T-NPC Inverter for High-Power Traction Applications. *IEEE J. Emerg. Sel. Top. Power Electron.* **2016**, *4*, 668–678. [CrossRef]
12. Schweizer, M.; Kolar, J.W. Design and Implementation of a Highly Efficient Three-Level T-Type Converter for Low-Voltage Applications. *IEEE Trans. Power Electron.* **2013**, *28*, 899–907. [CrossRef]
13. Lee, K.; Shin, H.; Choi, J. Comparative analysis of power losses for 3-Level NPC and T-type inverter modules. In Proceedings of the 2015 IEEE International Telecommunications Energy Conference (INTELEC), Osaka, Japan, 18–22 October 2015; pp. 1–6.
14. Zhang, L.; Yuan, X.; Wu, X.; Shi, C.; Zhang, J.; Zhang, Y. Performance Evaluation of High-Power SiC MOSFET Modules in Comparison to Si IGBT Modules. *IEEE Trans. Power Electron.* **2019**, *34*, 1181–1196. [CrossRef]
15. Vechalapu, K.; Bhattacharya, S.; Van Brunt, E.; Ryu, S.; Grider, D.; Palmour, J.W. Comparative Evaluation of 15-kV SiC MOSFET and 15- kV SiC IGBT for Medium-Voltage Converter Under the Same dv/dt Conditions. *IEEE J. Emerg. Sel. Top. Power Electronics* **2017**, *5*, 469–489. [CrossRef]

16. Concari, L.; Barater, D.; Toscani, A.; Concari, C.; Franceschini, G.; Buticchi, G.; Liserre, M.; Zhang, H. Assessment of Efficiency and Reliability of Wide Band-Gap Based H8 Inverter in Electric Vehicle Applications. *Energies* **2019**, *12*, 1992. [CrossRef]
17. Anthon, A.; Zhang, Z.; Andersen, M.A.E.; Holmes, D.G.; McGrath, B.; Teixeira, C.A. The Benefits of SiC mosfets in a T-Type Inverter for Grid-Tie Applications. *IEEE Trans. Power Electron.* **2017**, *32*, 2808–2821. [CrossRef]
18. Gurpinar, E.; Castellazzi, A. Single-Phase T-Type Inverter Performance Benchmark Using Si IGBTs, SiC MOSFETs, and GaN HEMTs. *IEEE Trans. Power Electron.* **2016**, *31*, 7148–7160. [CrossRef]
19. Anthon, A.; Zhang, Z.; Andersen, M.A.E.; Holmes, D.G.; McGrath, B.; Teixeira, C.A. Comparative Evaluation of the Loss and Thermal Performance of Advanced Three-Level Inverter Topologies. *IEEE Trans. Ind. Appl.* **2017**, *53*, 1381–1389. [CrossRef]
20. Zhang, D.; He, J.; Madhusoodhanan, S. Three-Level Two-Stage Decoupled Active NPC Converter With Si IGBT and SiC MOSFET. *IEEE Trans. Ind. Appl.* **2018**, *54*, 6169–6178. [CrossRef]
21. Loncarski, J.; Monopoli, V.G.; Leuzzi, R.; Cupertino, F. Operation analysis and comparison of Multilevel Si IGBT and 2 level SiC MOSFET inverter based high speed drives with long power cable. In Proceedings of the 2019 IEEE International Conference on Clean Electrical Power (ICCEP), Otranto, Italy, 2–4 July 2019.
22. Schweizer, M.; Friedli, T.; Kolar, J.W. Comparative evaluation of advanced three-phase three level inverter/converter topologies against two level systems. *IEEE Trans. Ind. Electron.* **2013**, *60*, 5515–5527. [CrossRef]
23. Rodriguez, J.; Bernet, S.; Steimer, P.K.; Lizama, I.E. A Survey on Neutral-Point-Clamped Inverters. *IEEE Trans. Ind. Electron.* **2010**, *57*, 2219–2230. [CrossRef]
24. Kaminski, B.; Koczara, W.; Al-Khayat, N. A three level inverter concept for low voltage applications. In Proceedings of the 2007 European Conference on Power Electronics and Applications, Aalborg, Denmark, 2–5 September 2007.
25. Degano, M.; Zanchetta, P.; Empringham, L.; Lavopa, E.; Clare, J. HF induction motor modeling using automated experimental impedance measurement matching. *IEEE Trans. Ind. Electron.* **2012**, *59*, 3789–3796. [CrossRef]
26. Kolar, J.W.; Ertl, H.; Zach, F.C. Influence of the modulation method on the conduction and switching losses of a PWM converter system. *IEEE Trans. Ind. Appl.* **1991**, *27*, 1063–1075. [CrossRef]
27. Semikron. 2015. 3L NPC & TNPC Topology (Application Note No. AN-11001). Available online: https://www.semikron.com/service-support/downloads/ (accessed on 29 November 2019).

Article

Solid-State Transformers in Locomotives Fed through AC Lines: A Review and Future Developments

Stefano Farnesi, Mario Marchesoni *, Massimiliano Passalacqua and Luis Vaccaro

Department of Electrical, Electronic, Tlc Engineering and Naval Architecture (DITEN), University of Genova, via all'Opera Pia 11a, 16145 Genova, Italy; massimiliano.passalacqua@edu.unige.it (M.P.); luis.vaccaro@unige.it (L.V.)
* Correspondence: marchesoni@unige.it

Received: 31 October 2019; Accepted: 6 December 2019; Published: 10 December 2019

Abstract: One of the most important innovation expectation in railway electrical equipment is the replacement of the on-board transformer with a high power converter. Since the transformer operates at line-frequency (i.e., 50 Hz or 16 2/3 Hz), it represents a critical component from weight point of view and, moreover, it is characterized by quite poor efficiency. High power converters for this application are characterized by a medium frequency inductive coupling and are commonly referred as Power Electronic Transformers (PET), Medium Frequency Topologies or Solid-State Transformers (SST). Many studies were carried out and various prototypes were realized until now, however, the realization of such a system has some difficulties, mainly related to the high input voltage (i.e., 25 kV for 50 Hz lines and 15 kV for 16 2/3 Hz lines) and the limited performance of available power electronic switches. The aim of this study is to present a survey on the main solutions proposed in the technical literature and, analyzing pros and cons of these studies, to introduce new possible circuit topologies for this application.

Keywords: medium frequency transformer; power electronic transformer; Solid State Transformer (SST); railway electric traction; Modular Multilevel Converter (MMC); soft-switching

---

1. Introduction

In last decades, many changes appeared in the railway sector thanks to the development of power electronics; as a matter of facts, nowadays railway vehicles are characterized by high specific power, high speed and high tractive effect. Thanks to these characteristics, they are able to manage both freight and passenger tasks with a single gear ratio.

The main problem in locomotives fed by AC lines is that an on-board transformer is needed, which main aims are:

- To supply the traction drive with a voltage furtherly lower than the voltage line (usually lower than 2 kV starting from 15 kV or 25 kV).
- To supply auxiliary services.
- To supply the conditioning system, which is fed at 1500 V if the line frequency is 50 Hz and at 1000 V if the line frequency is 16 2/3 Hz.
- To ensure galvanic insulation between the overhead line and vehicle equipment, mainly for safety reasons. Moreover, the transformer reduces also the effects of transient overvoltage, very frequent events in railway lines.

The transformer has to be sized according to line frequency, which can be 50 Hz or 16 2/3 Hz, and therefore is a critical component from the weight point of view, which has a drawback on rolling stock total weight; moreover, on-board transformers are characterized by quite low efficiency, which affects the traction chain overall efficiency. Many studies were carried out in order to reduce transformer weight, or to replace it with a power electronics converter. These studies, which were performed especially for high speed railways, where high AC voltage lines are used (i.e., 15 kV or 25 kV), aimed also to increase the traction chain efficiency. This last improvement was possible thanks to the lower losses obtainable with power electronics in comparison to traditional on-board transformers. Nevertheless, despite theoretical studies and prototypes, none of the proposed solutions seems to be capable of replacing the line-frequency transformer.

On-board transformers are single-phase, column-type and oil-immersed, designed for this specific application and therefore with very different characteristics in comparison to distribution transformers. Robustness is a required feature as well as high short-circuit impedance, which is a requirement to supply the four-quadrant converters connected to the transformer (Figure 1). Indeed, it has several low voltage secondaries which supply the converters. Other secondaries are dedicated to auxiliary services, heating, conditioning and, sometimes, to supply the L-C filter for harmonic suppression. One can observe that such type of machine is rather complex and bulky, therefore it is designed to reduce as much as possible size and weight, at the cost of low efficiency. In order to reduce transformer sizing, its rated power is generally lower that the traction power, since traction power is not constant and the transformer thermal constant is quite high. Winding current density is much higher than that in distribution machines and this aspect implies drawbacks in terms of losses. In addition, high losses imply the design of an effective cooling system, with forced-oil circulation and oil-air heat exchanger, is needed. Analogously, the core is designed for induction maximum exploitation, in order to minimize iron weight. However, transformer sizing is obviously related to the machine rated power and to the overhead line voltage frequency; for this reason, a 16 2/3 Hz transformer is heavier than a 50 Hz one, for same rated power. Considering a 6 MVA transformer, which is the typical transformer sizing of a modern locomotive, power density is about 0.55 kVA/kg for a 16 2/3 Hz machine and about 0.63 kVA/kg for a 50 Hz machine. Transformer weight is about 30% of locomotive weight and, since it is designed to minimize volume and mass, its efficiency is around 90% for 16 2/3 Hz machines [1–4]. For all these considerations, one can understand how transformer sizing is a critical task, especially for high speed trains and light locomotives, where the electrical equipment is generally installed in the underframe and the available space is limited, especially regarding components height, since the aim is to maximize the space available for passengers. In this scenario, it can be necessary to size the transformer taking into account the instantaneous required power, evaluated thanks to simulations which consider route characteristic (i.e., slopes and vehicle speed).

Starting from the end of '70s, when the first three-phase fed locomotives came out, many studies were carried out to reduce transformer weight, volume and to increase its efficiency. Many solutions were proposed: transformer with high-temperature superconductive windings [5,6], transformer-less traction drives, with the traction chain directly fed by the overhead HV-line [7–11] and, the most promising, different solutions based on one or more medium frequency transformers with the opportune power electronic converters, from which the name Power Electronic Transformer (PET) [12–36] or other similar denominations. The interest in PET started in the early 2000s [12,13], continued in years 2007–2009 [14–16] and recorded a significant increase in last ten years. Indeed, from 2010 to 2016, a large number of technical papers have been published on this topic [17–27] and in last three years the interest has further increased [28–36]. PET solutions for railway rolling stock are conceptually analogous to Solid State Transformer (SST) used in grid applications [37–43]. These solutions should reduce transformer sizing preserving the galvanic insulation and the voltage matching function. The PET solutions aim to replicate, for high power and high voltage levels, architectures and structures which were already exploited and used for different low and medium power applications. In addition, the electronic transformer is not only a topic of great interest for railway traction, but also for electricity distribution, renewables, industrial applications and automotive sector [44]. In this applications the denomination Solid State Transformer (SST) is usually used. However, in this article, only railway application is taken into account, analyzing the features and constraints needed in this sector. The objectives of the

PET should be reaching 1 kVA/kg specific power, in order to allow PET installation in the underframe or on the roof of high speed trains; moreover, an efficiency increase of about 2–5% is desirable. Indeed, even if the CO_2 emissions associated to railways in the world are just 4.2% of the emissions associated to transports, the global high-speed train lines increase every year of about 8%. Therefore, the train efficiency issue will become of prior importance in next decades [45]. Annual CO_2 world emissions are about 336 million tonnes, so an increase of just 1% in rolling stock efficiency would lead to a reduction of about 3.4 million tonnes of CO_2 per year. In addition to the efficiency challenge, the potential reduction of weight, achievable with the use of PET technology, would allow, for the same rolling stock, to transport more passengers or freight.

It is interesting to observe that transformer weight problem is much critical in multiple unit train, generally used for high speed applications. As a matter of facts, in a traditional train, the weight of the locomotive has to be significantly higher than the weight of carriages, in order to increase the tractive force. In such scenario, the on-board transformer contributes to locomotive weight and therefore the reduction of weight is not the key factor for the introduction of PET technology.

Despite the above mentioned expected benefits, PET has some limits and problems which are hard to get around. Primarily, reliability, availability and maintenance costs play a very significant role in establishing the commercial success in the field of transport vehicles field [46]; this is particularly significant for PET, since it involves a large number of components: power switches, gate drivers, passive components, electronic components for control and monitoring, insulation shields and an adequate cooling system.

The aim of this article is to provide a survey on the main conversion architectures proposed in the last years, particularly focusing on the solution tested with large scale prototypes. After this first review, a possible future solution, exploiting Modular Multilevel Converter (MMC) is analyzed. In particular, simulation results, demonstrating the necessity of soft-switching modules for MMC application in railway rolling stock, are shown.

2. Power Electronic Transformer (PET)

The powertrain principle scheme of a typical modern railway vehicle is reported in Figure 1 and it is composed by the following components:

- Line-frequency transformer.
- AC/DC bidirectional single-phase rectifier (Four Quadrant Converter, 4QC).
- L-C resonant filter.
- DC-link.
- Inverter.
- Motor (which can be either an Induction Motor, IM, or a Permanent Magnet Synchronous Motor, PMSM).

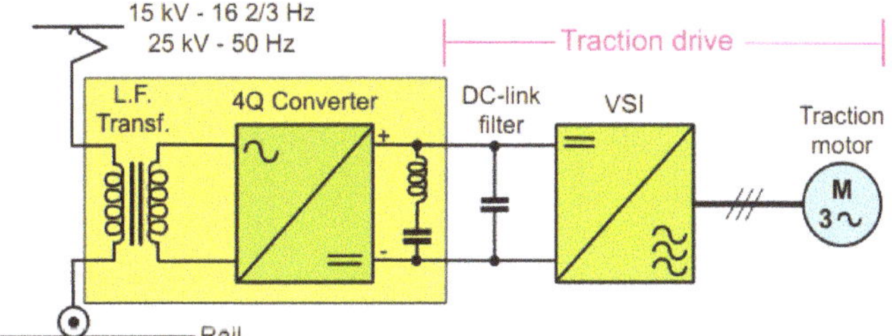

Figure 1. Principle scheme of an electrical powertrain for modern AC fed railway rolling stock.

The primary winding of the transformer is connected to the catenary voltage and usually there are several secondary windings. For example, in a modern Bo'Bo' locomotive (4-axles, 2-bogies, 4-motors), four secondary windings are used to supply the 4QCs, as reported in Figure 2. Secondary output voltage is about 1–1.6 kV and therefore the DC-link has a voltage of about 2.4–2.8 kV. The DC-link resonant filter is tuned at twice the line frequency. One of the main functions of the 4QC is to make the rolling stock working at unit power factor, with obvious advantages on fixed installation sizing.

The drive concept is to be modular, for redundancy reason. In the Bo'Bo' locomotive in Figure 2, for example, two identical semi-drives are used, one for each bogie. Minor variants can regard the number of secondary windings, DC-links, inverters, motors of Active Front Ends (AFEs) and AFEs mutual connections. The number of components depends on vehicle power, number of motors, desired degree of redundancy or on other constrains. Railway operators nowadays prefer multisystem traction vehicles, i.e., vehicles that can operate with different line voltages. This feature is important in order to use trains in international traffic and because there are different electrification systems, sometimes also in the same country (e.g., different voltages for traditional trains and high-speed trains).

As a matter of fact, in Europe there are four prevalent feeding systems:

- AC, 25 kV, 50 Hz.
- AC, 15 kV, 16 2/3 Hz.
- DC, 1.5 Kv.
- DC, 3 kV.

In multisystem vehicles, electrical equipment is easy to reconfigure. AFEs for AC feeding can work as choppers for DC supply; the transformer can be designed and sized to work both at 25 kV–50 Hz and 15 kV–16 2/3 Hz varying winding turns with taps; for DC feeding the secondary windings can be series connected and used as inductance input filter. Therefore, interoperability is a key feature that must be taken into account while considering PET technology.

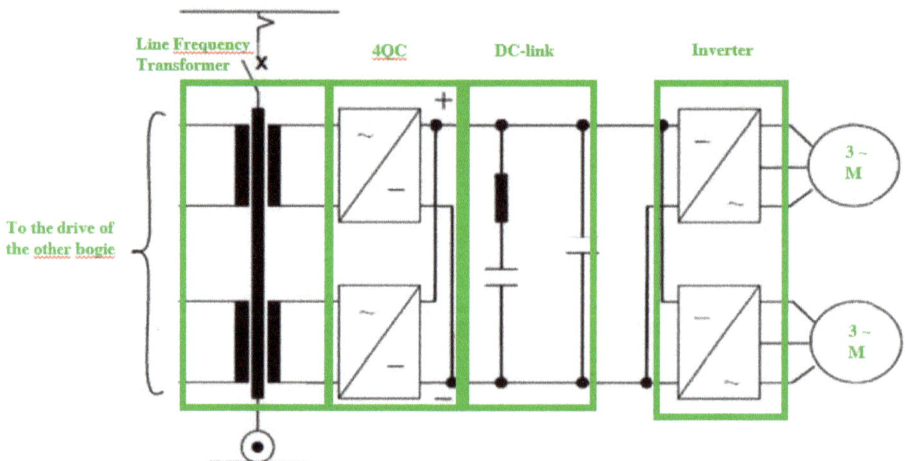

Figure 2. Classical drive for a modern Bo' Bo' high power AC-fed locomotive (one bogie).

A PET is basically an AC/DC conversion system with a medium frequency transformer, where the AC input is the line voltage and the DC output is a voltage of about 1–3 kV. Since the PET includes an AC/DC conversion stage, PET features have to be compared with on-board transformer plus 4QCs. The various topologies proposed until know can be divided in two main categories, depending on the input conversion stage:

- AC/AC input converter.
- AC/DC input converter.

The principle scheme of the first type of conversion chain in shown in Figure 3. This structure is assembled with the following components:

- Input inductance (between the overhead line and the 1st stage).
- AC/AC (1st stage).
- Medium frequency transformer.
- AC/DC (2nd stage).

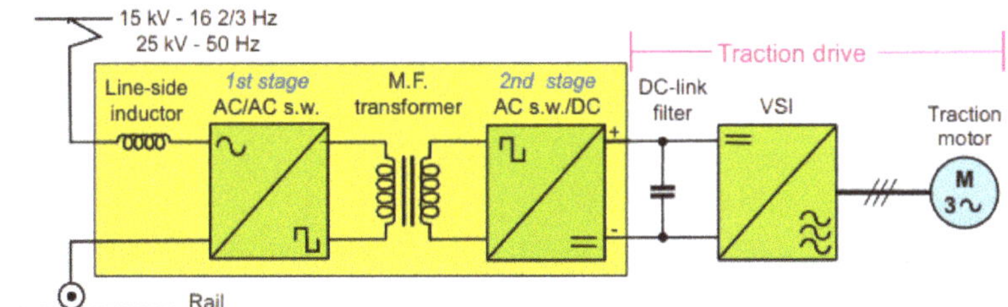

Figure 3. Conversion scheme for a two stages PET.

After the above mentioned components, there is the traditional traction drive: DC-link filter, inverter and motor. The 1st stage aim is to increase the frequency from the line frequency to a medium frequency (several hundreds or several thousands of Hz); the medium frequency output is a square wave voltage (s.w.), in order to minimize transformer sizing and therefore the 2nd stage input is a square wave too.

The principle scheme of the second type of conversion chain in shown in Figure 4. This structure is made with the following components:

- Input inductance (between the overhead line and the 1st stage).
- AC/DC (1st stage).
- DC filter.
- DC/AC (2nd stage).
- Medium frequency transformer.
- AC/DC (3rd stage).

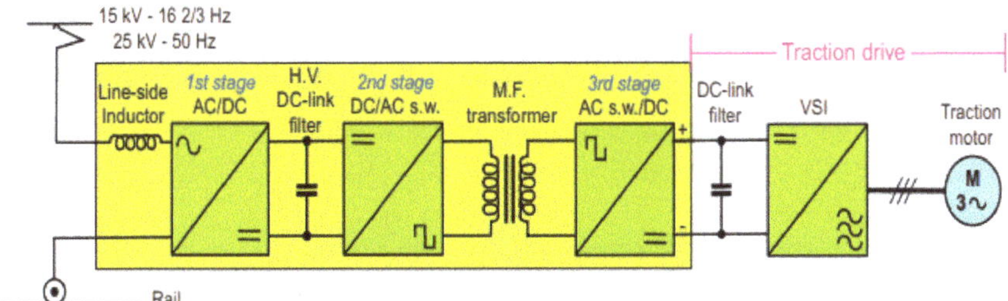

Figure 4. Conversion scheme for a three stages PET.

Compared to the first type of PET, which involves two conversion stages, in the second type, three conversion stages are involved, plus, obviously, the traction drive, which is the same in both cases. For this reason, authors will refer to the first type as 2-Conversion Stages PET (2CS PET) and to the second type as 3-Conversion Stages PET (3CS PET).

In both cases a multilevel converter is needed for the 1st stage, since the input voltage (15 kV or/and 25 kV) is too high to be tolerated by available power switches in conventional two-level converters.

Since the transformer has to work at sufficiently high frequency, in order to obtain a significant weight reduction, soft-switching is mandatory for the converter that supplies the transformer. Whatever the soft-switching circuit is, an extremely high number of semiconductor devices and gate-drivers is needed, with obvious problems in terms of costs, reliability, efficiency, cooling and insulation. This is the main reason why, despite the efforts in studies and prototypes, PET has not found a practical implementation yet. It is quite possible that PET will find an easy success with the availability of switches with two-three times greater rating voltage than that of semiconductor devices available today (6.5 kV) and able to operate at frequencies of some tens of kHz.

At this point, it is necessary to point out the main targets that have to be achieved with the PET technology:

- Weight reduction (at least 50%, or, analogously, increasing the specific power at least to 1 kVA/kg).
- Volume reduction (at least from 30% to 50% less than line frequency transformer).
- Efficiency increase (at least 2–5%).
- Capability of supplying traction loads, auxiliary services, air conditioning, heating, etc.
- Capability of reconfiguration for multisystem power supply.
- Capability of delivering a stabilized DC-link voltage, independently from line voltage variations.
- Reversibility of the system to allow electrical braking with recovery on the supply line.
- High reliability.
- Simple cooling of semiconductors and medium frequency transformer.
- Capability of reconfiguration in case of failure.
- Easily maintainable and repairable.
- Easily controllable.
- Absence of noise and vibrations.
- Robustness.
- Capability of working in a temperature wide range.
- Low cost.

Regarding efficiency increase, the PET includes two or three conversion stages, so the efficiency of the all chain should be greater than the efficiency of the low frequency transformer plus the 4QC. For this reason, the conversion stages and the medium-frequency transformer have to be particularly efficient, which in obviously difficult with such a complex structure and such a high number of components. The high number of components is of course in contrast also with the requirement of high reliability, even if this aspect can be mitigated thanks to a modular structure, which can be reconfigured in case of failure of a module.

The study of PET involves a widespread range of topics that require accurate investigations:

- Multilevel conversion structures and related control strategies [47–53].
- Soft-switching conversion structures to reduce switching losses and therefore increase switching frequency [29,42,54–59].
- Power electronics devices insulation (materials, shields, covers).
- Medium frequency, high power transformer.
- Development of power semiconductors specifically designed for soft-switching operation, with switching losses minimization at the expense of conduction losses, obtainable with particular treatments such as doping and irradiation [20].
- Gate driver powering when installed on system operating at medium voltage.
- Modular design of compact and lightweight building modules, Power Electronics Building Blocks (PEBB).
- Reliability of system with such a high number of semiconductor devices and gate drivers
- Strategies and hardware for control and diagnostics.

- Transient behavior in case of pantograph bouncing from the line (frequent event and lasting < 15 ms).
- Study of failure modes, fault detection and fault tolerance with bypass of the faulted module [43,60–68].

3. Review of Main Conversion Architectures

With the availability of 3.3 kV high-power IGBTs (Insular Gate Bipolar Transistor), many studies were carried out on railway traction at the end of the '90s [10,11], proposing innovative architectures, modular and based on the use of high performance IGBTs. These studies consisted of a four quadrant multilevel converter in a cascade connection of H-bridges, as 1st stage to interface with the 15 kV, 16 2/3 Hz overhead line voltage. As 2nd and 3rd stage, a reversible DC/DC converter composed of a DC/AC with H-bridge, a medium frequency transformer and a AC/DC with H-bridge was proposed. It was hypothesized the use of a series-resonant circuit in the AC-link, in order to reach sufficiently high frequency for transformer operation. Further studies were carried out by the same researchers regarding an architecture which could work both at 15 kV, 16 2/3 Hz and 3 kV DC [69].

Regarding IGBT peak voltage and the required number of stages, it is necessary to observe that the maximum admitted voltage is 29 kV RMS for 25 kV/50 Hz systems and 18 kV RMS for 15 kV/16 2/3 Hz systems. Therefore, the required number of stages is reported in Table 1.

Table 1. Required number of stages as a function of IGBT peak voltage and line voltage.

Switch Collector-Emitter Peak Voltage V_{CEPK}	DC-BUS E/V_{CEPK} = 0.55	Number of Stages Network: 25 kV/50 Hz 29 kV_{RMS}/50Hz *	Number of Stages Network: 15 kV/16 2/3 Hz 18 kV_{RMS}/16 2/3 Hz *
3.3 kV	1.8 kV	23	14
4.5 kV	2.5 kV	17	10
6.5 kV	3.6 kV	12	7
10 kV	5.5 kV	8	5

 * RMS Catenary voltage in the worst case.

3.1. Alstom-SMA "eTransformer"

In 2003 Alstom LHB GmbH and SMA Regelsysteme GmbH presented a prototype called eTransformer [12]. The prototype is a three-stage converter (3SC). Its main feature are reported in Table 2, whereas its scheme is shown in Figure 5.

Table 2. eTransformer features.

Feature	Value
Input voltage	15 kV
Input frequency	16.7 Hz
Output voltage	1.65 kV DC
Rating power	1500 kVA
Maximum power	2250 kVA (30 s)
Efficiency	94%
Transformer frequency	5 kHz
Transformer + electronic weight	2830 kg
Output LC filter weight	385 kg
Heat exchanger weight	255 kg
Overall weight	<3600 kg
Power density	0.42 kVA/kg
Total number of IGBTs	52
Cooling system	Forced oil circulation

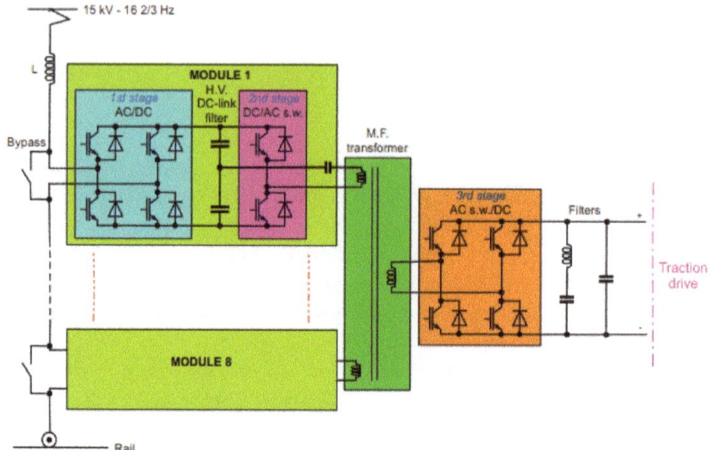

Figure 5. Alstom SMA eTransformer.

The HV input (1st and 2nd stages) consists of an input reactor and eight AC/AC s.w. (i.e., converter from AC to square wave, s.w.) conversion modules in cascade connection (multilevel concept). The 1st stage is a full H-bridge AC/DC whereas the 2nd stage is an half H-bridge DC/AC s.w. The primary of the transformer is fed with a resonant circuit (capacitor and leakage inductance of the transformer), which is tuned at 5 kHz. While there are eight modules for 1st and 2nd stage, there is only one H-bridge for the 3rd stage, which feeds the low voltage DC-link, equipped also with an LC filter. The 1st stage converter works in hard-switching, since its switching frequency can be relatively low, and it is equipped with 6.5 kV–400 A IGBTs. The 2nd stage converter works in soft switching, with a quasi-zero current switching (quasi-ZCS), and it is equipped with 3.3 kV IGBTs.

In case of fault of one of eight modules, the failed one can be bypassed and the operation can continue without any limitation on the power. Semiconductors and transformer are cooled with forced oil. This solution uses 52 IGBTs with the related gate-drivers, plus additional passive components and auxiliary items. Despite the number of semiconductors is about three times greater in comparison to a traditional line-frequency transformer solution, the eTransformer has the benefit of using a relatively low number of devices with respects to other PET solutions.

3.2. Bombardier Transportation "Medium Frequency Topology"

In 2007 Bombardier Transportation presented a laboratory prototype called "Medium Frequency Topology" [14], which exploits the principle shown in [10,11] and which is designed for 15 kV, 16 2/3 Hz. PET main features are reported in Table 3, whereas PET scheme is shown in Figure 6. Analogously to the eTransformer, it is a 3SC and it is made of eight identical modules cascade connected. Differently from eTransformer, where only one medium frequency transformer is used, in Medium Frequency Topology a transformer is used for each module and the outputs of the modules (LV DC-link) are parallel connected. The 1st stage is a AC/DC full H-bridge, the 2nd stage a full H-bridge DC/AC s.w. (differently from eTransformer that has an half-bridge) and the 3rd stage is again a full H-bridge AC s.w./DC. If one of the modules fails, the Medium Frequency Topology can work with a reduction of power of 1/8; indeed it uses 6.5 kV IGBTs for the 1st stage and therefore 7 modules are sufficient for 15 kV line voltage. The transformers are just 18 kg each and they have 1:1 winding ratio. The main disadvantage of this PET is that a great number of IGBTs, and of course of auxiliary devices, is needed. A reduction of number of components can be performed using half H-bridges, as shown in next solutions.

Table 3. Medium frequency topology features.

Feature	Value
Input voltage	15 kV
Input frequency	16.7 Hz
Output voltage	3.6 kV DC
Rating power	3000 kVA
Transformer frequency	8 kHz
Transformers weight [1]	18 kg
Total number of IGBTs	96
Cooling system	Deionized water

Notes: Weight for each transformer.

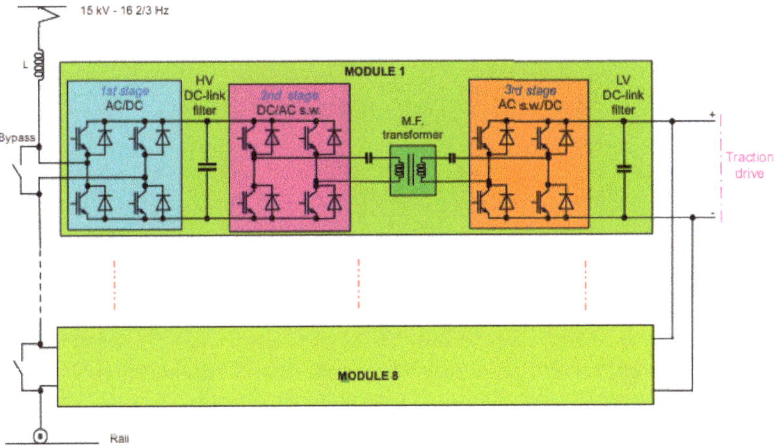

Figure 6. Bombardier "Medium Frequency Topology".

3.3. ABB First Converter Prototype

In 2007 ABB presented its first laboratory prototype [15], whose main features are reported in Table 4 and whose scheme is shown in Figure 7. Differently from the abovementioned PETs, this prototype is a 2SC. It uses all 3.3 kV, 400 A IGBTs in 16 modules, therefore double number of modules, compared to previous PETs, is needed since it uses 3.3 kV IGBTs. A medium frequency transformer with 1:1 ratio is used for each module. The transformer frequency is quite low (400 Hz), therefore it is not possible to exploit a significant transformer size reduction. For this reason, dimensions related to this solution are greater than those of a traditional architecture (line frequency transformer + 4QCs). Regarding the efficiency, even if at medium and high loads the efficiency of the prototype is 3% higher than that of a traditional transformer, at low loads the efficiency is even lower. In additional to these disadvantages, this prototype involves the use of 192 IGBTs. For these reasons it is outdated.

Table 4. ABB first converter prototype features.

Feature	Value
Input voltage	15 kV
Input frequency	16.7 Hz
Output voltage	1.8 kV DC
Rating power	1200 kVA
Transformer frequency	400 Hz
Total number of IGBTs	192
Cooling system	Forced oil circulation

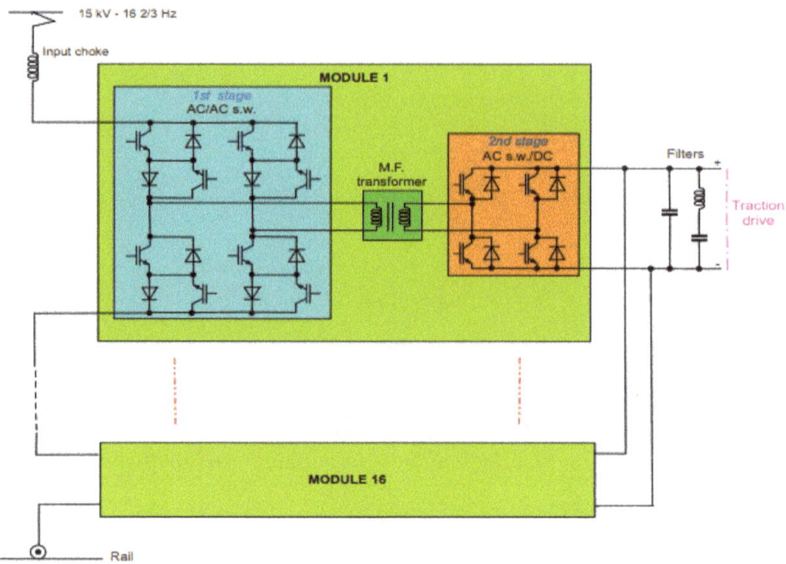

Figure 7. ABB first converter prototype.

3.4. ABB Power Electronic Traction Transformer (PETT)

In 2012 ABB presented probably the most advanced prototype of PET, which is called Power Electronic Traction Transformer (PETT) [19–24], whose main features are reported in Table 5 and whose scheme is shown in Figure 8. The PETT is a 3SC, includes nine cascade connected modules, one for redundancy, and has a medium frequency transformer for each module. The 1st stage is a full H-bridge AC/DC which uses 6.5 kV–400 A IGBTs and which supplies the 3.6 kV DC-link; the 2nd stage is a half H-bridge DC/AC s.w. which uses again 6.5 kV–400 A IGBTs. Finally, the 3rd stage is a half H-bridge AC s.w./DC which uses 3.3 kV–800 A IGBTs. In the LLC resonant circuit, both the leakage inductance and magnetizing inductance participate at the resonance, whereas, in order to minimize weight and dimension, a ripple in the output voltage is accepted, so the LC filter is not necessary.

The specific power, related to rating and maximum power, is only 0.266 kVA/kg and 0.4 kVA/kg respectively, which is quite low. Even if the power density should be increased at least to 0.5–0.75 kVA/kg, the PETT is the first answer related to this technology and, moreover, the efficiency of 96% is significantly higher than a traditional line frequency transformer with 4QCs.

Table 5. ABB first converter prototype features.

Feature	Value
Input voltage	15 kV
Input frequency	16.7 Hz
Output voltage	1.5 kV DC
Rating power	1200 kVA
Maximum power	1800 kVA
Efficiency	96%
Total number of IGBTs	72
Total weight	4500 kg
Power density	0.266 kVA/kg

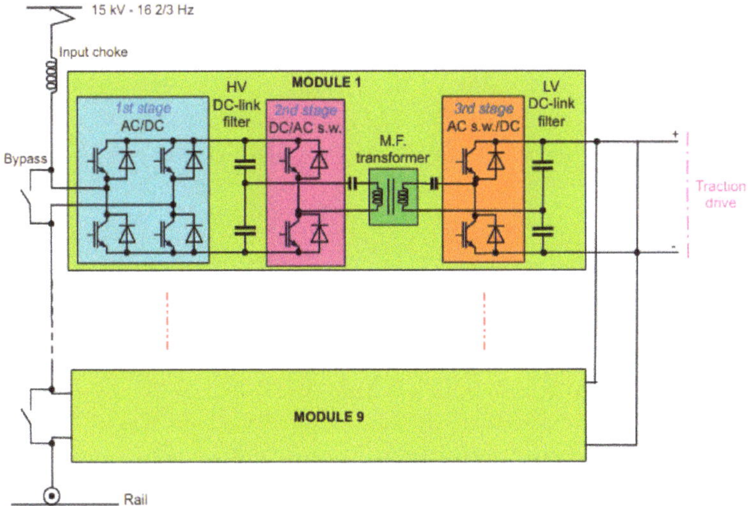

Figure 8. ABB Power Electronic Traction Transformer (PETT).

4. Bombardier and ABB AC-Link

In this section, Bombardier and ABB DC/DC converters (i.e., 2nd stage, medium frequency AC link and 3rd stage of Figures 6 and 8) are analyzed in detail. Both converter, using full H-bridge or half H-bridge, are fully reversible, so they can perform regenerative braking. Bombardier uses a full H-bridge solution both for 2nd and 3rd stages, whereas ABB PETT uses a half H-bridge solution for both stages. ABB solution needs half semiconductor devices (4 IGBTs per module instead of 8) and half of related auxiliaries (gate-drivers, etc.). However, for the floating DC-links, four capacitors are necessary instead of the two capacitors of Bombardier solution, in order to create the central taps. Both solutions have pros and cons. From reliability point of view, the failure rate of an IGBT with related components is higher than the failure rate of a capacitor; however, DC-link capacitors are usually bulky and heavy, therefore ABB solution is not optimal from sizing point of view.

Bombardier uses a classical LC resonant circuit, where the inductance is the transformer leakage inductance; transformer magnetizing inductance is quite high and it is not exploited in this solution. The resonant frequency is:

$$f_0 = \frac{1}{2\pi \sqrt{L_d C_r}} \tag{1}$$

where L_d is the transformer leakage inductance and C_r the capacitance of the AC link. The operating conditions depend on the ratio between the resonant frequency and the switching frequency [70,71].

ABB solution uses a more performant LLC resonant circuit, which exploit both the transformer leakage inductance L_d and the magnetizing inductance L_m. For this reason, the circuit has two different resonance frequencies:

$$f_0 = \frac{1}{2\pi \sqrt{L_d C_r}} \tag{2}$$

$$f_p = \frac{1}{2\pi \sqrt{(L_d + L_m) C_r}} \tag{3}$$

Being, usually, $L_m \gg L_d$ it follows that $f_0 \gg f_p$ LLC resonant circuit has many advantages compared to the LC circuit, which make this configuration particularly suitable for this application:

- Higher range of output voltage.
- Primary winding switching losses reduction in a wide range of operation, thanks to Zero-Voltage Switching (ZVS).
- Low circulating energy and Zero-Current Switching (ZCS) in secondary side rectifier diodes.

After this brief introduction on converter main features, the two structures are analyzed in detail in the following subsections.

4.1. Bombardier Converter

The LC full H-bridge used by Bombardier (Figure 6) allows, from a theoretical point of view, ZVS and ZCS of all active components, which make possible to reach relatively high switching frequencies, up to 10 kHz. Both for the input bridge and the output bridge, 6.5 kV IGBTs are used, even if experimental tests on prototypes were performed also with 3.3 kV and 4.5 kV IGBTs. The medium frequency transformer has a 1:1 ratio and therefore same size of IGBTs can be used both for primary and secondary size. The transformer has concentric windings, that guarantee optimal electric and magnetic field symmetry (which allows symmetric current in the devices), optimal insulation exploitation and low flux leakage. As further advantage, switching losses are minimized also in non-ZCS working operation. Switching frequency is chosen close to resonance frequency. Current and voltage sensors are not necessary for close-loop control, indeed open-loop control is used. IGBT commands of input and output bridge are synchronized, with a 50% duty cycle and additional blanking time. Therefore, on both windings there are square wave synchronized voltages. It is interesting to note that it could be sufficient to control the IGBTs of the input converter (2nd stage), since the output converter (3rd stage) works as a rectifier and the current flows in the diodes. However also 3rd stage IGBTs are commanded, in order to prevent malfunctioning and delays in case of regenerative braking; indeed, in this working condition, the 3rd stage IGBTs have to be commanded.

Even if the converter theoretically guarantees soft-switching (i.e., ZCS and/or ZVS), which allows negligible switching losses, one has to take into account the real behavior of the switches. Indeed high power IGBTs are optimized to work in hard-switching, therefore the physical structure is optimized to minimize conduction losses, rather than switching losses. This aspect implies that, in practical operation, soft-switching condition is partially respected: switching losses are produces and they depend on the behavior of the components during the transients. Converter optimization, and therefore efficiency optimization (which is evaluated around 97% on prototypes), can be therefore obtained with an accurate tuning between converter parameters and semiconductor features.

4.2. ABB Power Electronic Traction Transformer

ABB PETT (Figure 8) uses two half H-bridges for 2nd and 3rd stage and the LLC resonant circuit allows turn-on ZVS for 2nd stage IGBTs and ZCS turn-on for 3rd stage diodes; 2nd stage turn-off is quasi-ZCS and can be obtained with an opportune choice of component parameters and imposing a very low current during the turn-off (I_{turn_off}). This current is related to transformer magnetizing current and depends on transformer magnetizing inductance, on switching frequency and on DC-link voltage, but is independent from converter load. Therefore, this resonant scheme considers by default switching losses, caused by I_{turn_off}, which is different from zero (despite it is low and independent from the load) on the input half-bridge IGBTs. The turn-off losses are therefore constant and load-independent, but predominant in comparison to conduction losses; for this reason it is necessary to minimize them. Even if I_{turn_off} causes turn-off switching losses, it allows IGBT ZVS turn-on, therefore an excessive reduction of I_{turn_off} can cause undesired turn-on switching losses. Moreover, IGBT turn-off losses are related not only to I_{turn_off}, but they strongly depend also on IGBT dynamic characteristics. Moreover, turn-on ZVS condition is strongly related to dead time, which can be quite high, tens of microseconds, for high power IGBTs. From these considerations, one can notice that switching losses maximum reduction can be achieved with an accurate choice of converter parameters and through the optimization of semiconductor dynamic behavior, as observed before also for Bombardier converter.

In [20], an accurate description of an experimental test regarding the interaction between semiconductor properties and circuit properties is reported. The optimization of IGBT physical structure is investigated, with the redesigning of anodic structure, which has an effect on charge carriers in silicon thickness and can therefore modify the behavior during the switching, potentially improving switching phenomenon and reducing the losses. The aim of the study is to find a structure that could reduce switching losses at high switching frequency (higher than 1 kHz) but that could, at the same time, avoid conduction losses increase. Indeed, previous studies using electron irradiation show a reduction of switching losses but at the cost of conduction losses increase.

ABB studies have been performed on a 225 kW converter. The DC/DC converter has 3.6 kV input voltage and 1.5 kV output voltage; the 225 kW transformer has 2.4:1 winding ratio. On the input half-bridge 6.5 kV–400 A IGBTs are used, whereas 3.3 kV–800 A IGBTs are used for the output half-bridge. Switching frequency is 1.8 kHz, which is just under the resonance frequency of 2 kHz. With this choice of f_s and f_0, the relation between input and output voltage depends only on transformer winding ratio. For this reason, also this converter works in open loop.

5. A New Conversion Architecture for PET Application: Modular Multilevel Converter (MMC)

From the analysis of architectures and prototypes shown in the previous section, some considerations can be pinpointed at this stage:

- The research focuses mainly on new architectures. Other aspects like converter lay-out, converter on-board placement, cooling, converter management, failure modes etc., are neglected (except in ABB PETT, since a demonstrator was realized).
- The high number of requirements that a traction transformer should ensure is not always taken into account in the new conversion systems.
- The proposed topologies are inspired to solutions used in low power and low voltage applications.
- The medium frequency for the transformer is quite low, despite significantly higher than line frequency.
- All the solutions require a high number of semiconductor devices and of passive components. From this point of view, there are significant differences between the architectures but the high number of devices is a common feature.
- In some solutions, the total power is subdivided among various medium frequency transformers. This could be an advantage from a reliability point of view, but it implies problems related to dimensions, weight and layout (large number of electrical connections, of sensors, cooling connectors, etc.).
- All the solutions are extremely modular, in order to guarantee high reliability levels, even if with complex circuits and high number of devices.
- It is shown that soft-switching is necessary to increase the semiconductor maximum operating frequency, which is quite low. Indeed the optimal switching frequency for hard-switching operation of IGBTs used in railway rolling stock is about 600 Hz [72], which is too low for medium frequency transformer exploitation.
- In some studies, in particular in ABB PETT, the possibility of using IGBTs optimized for soft-switching operation is taken into account. The performance increase is obtained intervening in the device physical layout or with opportune treatments in the production process. The aim of switching losses reduction is indeed to increase the medium frequency value of the transformer.

None of the proposed architecture is really satisfying; the ABB PETT, despite an accurate study, an excellent engineering of the converter and the use of advanced techniques such as devices optimized for soft-switching working condition, is far from meeting volume and weight target.

The main limits and problems to resolve are:

- The high number of devices.
- The low frequency of the transformer.
- The use of modular and redundant structures (such as the use of various number of transformers and the use of three conversion stages). The complexity of modular and redundant structures is in contrast with the necessity of volumes reduction, since in addition to electrical connection, one has to take into account the insulation distance (since the line voltage is relatively high), the hydraulic connections for the cooling system, etc. The use of three stages instead of two stages is conceptually disadvantageous also from an efficiency point of view.
- The adaptation of schemes and circuits used in the low/medium power and low voltage may not be the optimal solution. It could be more opportune to base on schemes designed for high power—high voltage.
- The complexity of the converter involves high cost during the life cycle.

5.1. MMC Concept

A solution that appeared some years ago and that is opportune to analyze, integrates in the conversion chain a Modular Multilevel Converter (MMC). The MMC (also referred as M2C, M^2C or M^2LC) was presented for the first time in 2003 and is characterized by very high quality of modularity, flexibility, voltage and current waveforms [1,13,73–76]. More than a conversion scheme, the MMC is a concept of a family of converters, which can generate different types of conversion: AC/AC or DC/AC, single-phase or multiphase. The circuit structure is divided into phase modules (two or more), each of which is made of two arms, an upper one and a lower one, as it can be noticed from Figure 9. Conceptually, the scheme in Figure 9a is analogous to a two-level inverter, in which the MMC arms are the inverter switches plus the free-wheeling diodes. In the MMC, instead, each arm is made of a connection of submodules. Each submodule can be a half H-bridge or a full H-bridge, with a DC-link capacitor. The submodules are half-bridges for DC/AC conversion and full-bridges for the AC/AC conversion. The operating principle of a MMC is presented in [73].

Each arm of N submodules, if properly controlled, behaves like a variable voltage generator, in which the source is made of the series connection of the DC-link capacitors of the arm. Connecting in a proper way two or more phase modules, and realizing a suitable control, it is possible to obtain whatever type of conversion with whatever number of phases. The MMC control algorithm manages the power flow between input and output sources controlling at each switching period the stored energy in the DC-link capacitors.

It is necessary that at a certain instant, the stored energy is higher than a certain threshold and, moreover, there should be an equalization of the voltage between the different capacitors. Therefore, there should be an accurate control of the voltage on each capacitor and this implies the use of complex algorithms.

The general scheme of a railway rolling stock drive with the use of a MMC converter is shown in Figure 10 [13,74–76]. This structure uses a single medium frequency transformer with primary voltage of 13.5 kV and one or more secondary windings at 2.7 kV, each of which supplies a conventional traction drive (4QC-DC, filter, inverter and motor). The use of a single transformer is evaluated suitable since the construction of a single machine, sized for the locomotive full power and insulated for relatively high voltage (thus, with a bulky insulation), is simpler than that of various smaller transformers. Indeed, it implies a reduction in materials (especially the iron of the magnetic core) and therefore a reduction of dimensions, weight and costs.

In [13] a possible multisystem application is presented, highlighting how simply varying the number of active submodules for each arm and acting on the control, it is possible to easily pass from 15 kV–16.7 Hz to 25 kV–50 Hz. However, no experimental results related to such a type of converter are known in the technical literature.

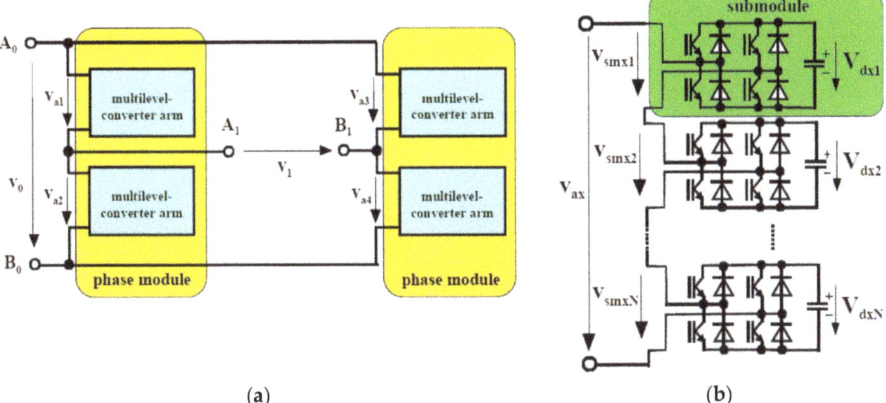

Figure 9. (a) MMC structure: v_0 is the input voltage and v_1 the output voltage (b) Submodule structure: N identical submodules in series connection, each of which consist of an H-bridge with a DC-link capacitor.

On the contrary, the MMC concept attracts the attention in high-power and high-voltage applications [40,62,63,77–79]: electrical drives [80], HVDC [81–85], photovoltaic [86] and in stationary installations that supply the primary lines of German railways (i.e., from 110 kV–50 Hz three-phase to 110 kV–16.7 Hz single-phase).

The MMC has many interesting advantages compared to other structures:

- Conversion topology of proven validity, able to work at very high power and voltage (hundreds of kV and MW in HVDC applications).
- Highly modular, with identical and interchangeable submodules. The submodule includes 4 switches and a DC-link capacitor, which can assembled in a unique rack (Power Electronic Building Block—PEBB) that includes also the heat sink and the sensors. More submodules can be put side by side or on top of each, with a great exploitation of the available space.
- No need of low inductance connection between submodules.
- Only two stages, with benefits on the efficiency.
- Single medium frequency transformer.
- Flexible and intrinsically multisystem, thanks to the possibility of realizing different type of conversion just modifying the control.
- Perfectly reversible.
- Line current with low harmonic distortion.
- MMC allows different possibility of optimization, all of which obtainable just modifying the control but without intervention on hardware: switching losses reduction, line harmonics reduction, etc.
- Excellent static and dynamic behavior.
- No need of filter tuned at double line frequency: the converter can compensate the pulsation of the line power thanks to the energy stored in the submodules.
- Intrinsically fault tolerant.
- Simple interface with control electronic: gate-drivers can be supplied from the DC-link capacitor and the command signals are sent through fiber optic.

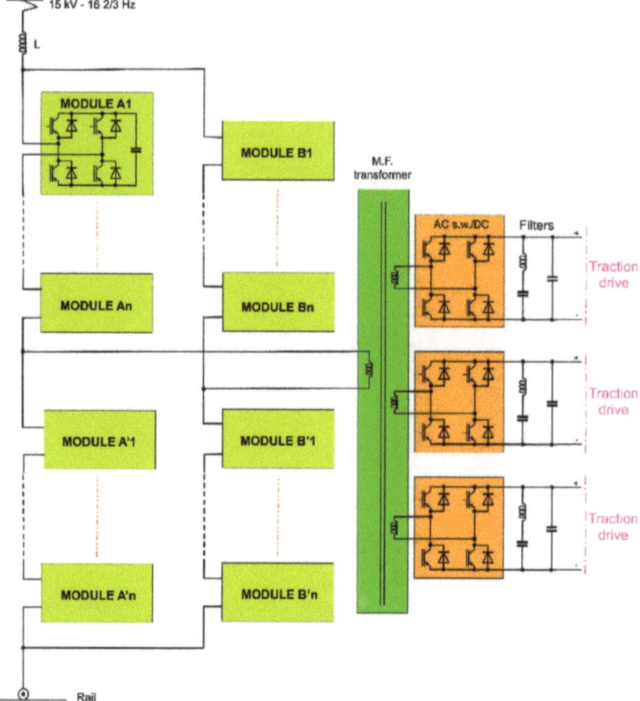

Figure 10. MMC application in a railway rolling stock working at 15 kV–16 2/3 Hz or 25 kV–50 Hz.

Besides these advantages, there are some critical aspects in the use of MMC in railway rolling stock:

- High number of components. Using 3.3 kV switches and 940 μF–1900 V capacitors, with 12 submodules per each of the four arms, 192 switches (IGBT plus diode) and 48 capacitors are needed [75,76]. Using 6.5 kV IGBTs, 7 submodules can be sufficient thus 112 switches and 28 capacitors can be used, which is still a high number of devices. To this number of devices, components of 4QC should be added, thus the MMC solution has more components than the other solutions.
- Since the MMC works in hard-switching, the medium frequency is quite low (from hundreds of Hz to about 2 kHz).
- The control logic is quite complex.
- From an efficiency point of view, the comparison should be between the MMC + medium frequency transformer + 4QC and line frequency transformer + 4QC. In order to be more efficient, medium frequency transformer and MMC efficiency should be in the range of 98–99%. There is no experience with such a high power (about 5 MW) medium frequency transformer to determine the efficiency and, even if the MMC is very efficient, the target could be difficult to reach.

Even if there are these critical aspects, the MMC concept is interesting to be investigated in railway application. This paper studies and analyzes the main problems, in order to find possible solutions.

5.2. Soft-Switching MMC: Description and Waveform Simulation

One of the main limitations in the use of MMC for railway application is connected to its switching frequency. In order to reduce transformer volume and weight, it should be necessary to have at least a frequency of 7–8 kHz and significant advantages will be obtained with a frequency of 10–20 kHz.

Even if the MMC is quite refined and exploit optimally the switching frequency, the features of high power switches (some kV and hundreds of A) limits the MMC frequency at 1–2 kHz. In the first studies [13,76] the switching frequency was 4 kHz for the two-voltage version (18 submodules per arm), with a submodule medium frequency of 350 Hz and an output voltage switching at 2 kHz; 4 kHz switching frequency for the one-voltage version (12 submodules per arm), with an output voltage switching at 1 kHz.

The MMC is intrinsically hard-switching but, being composed by identical submodules, there could be the possibility of increasing the switching frequency using a soft-switching topology [42,54] in the H-bridge; the aim of the study is to investigate this aspect. One has to note that if the switching frequency of the single submodule increases, once the desired output frequency is fixed, the number of submodules can decrease, and therefore the number of devices.

Among the various soft-switching solutions, it is necessary to identify the most suitable architecture for the application in the MMC concept and, in particular, to the use in railway rolling stock. Typically, resonant circuits introduce some problems that have to be taken into account:

- Increase in voltage/current stresses on the main converter power components, which in some cases can be quite high, more than 2 p.u.
- Not safe soft-switching operation, i.e., soft-switching depending on load current and other constraints.
- Impact of the soft-switching circuit with the overall operation of the converter.
- Inadequacy or only partial adequacy to the PWM modulation.

The soft-switching solutions can be classified in three main families:

- Resonant switch: the resonant circuit is applied to each single switch.
- Resonant pole: the resonant circuit is applied to a branch of the module.
- Resonant link: the resonant circuit is applied to the DC-link of the module.

Since the MMC is itself a complex system, with a large number of components, the additional components used to obtain the soft-switching condition, should be as less as possible. For this reason only the resonant pole and the resonant link families are taken into account. Several topologies were examined in this study:

- Resonant DC-Link Converter (RDCL) [8–11,44,69,87].
- Active Clamped Resonant DC-link Converter (ACRDCL) [88–94].
- Notch Commutated PWM Inverter [8,12].
- Quasi-Resonant PWM Inverter [8,14,95–100].
- Zero Switching Loss PWM Converter with Resonant Circuit [8,18,101,102].
- Active Resonant Commutated Pole Converter (ARCP) [8,73,103–109].
- Auxiliary Quasi Resonant DC-link Inverter (AQRDCL) [13,110–115].

All the above mentioned soft-switching topologies are designed for three-phase inverters, however, the adaptation to a single phase H-bridge of the MMC is still quite easy. Among the above mentioned topologies, the most interesting to realize soft-switching submodules for the MMC converter seem to be the ARCP (resonant pole family) and the AQRDCL (resonant link family), which are reported in Figure 11. The other topologies seem to be unsuitable for the MMC application due to the high stress on devices, the dependence of soft-switching condition from the load or the impossibility of operating in PWM.

The ARCP seems to be particularly suitable for this application and has some interesting features:

- ZVS both in turn on and turn off of the main devices (T_1, T_2, T_3, T_4) and ZCS or at least quasi-ZCS of the auxiliary devices of the resonant circuit.
- Oversizing of main devices is not necessary since there are not overvoltages and overcurrents.

- The output voltage and current waveforms are about the same as in hard-switching under same operating conditions. The only different is that the slope of the output voltage steps are less steep.
- dv/dt and di/dt on the main devices are lower than in hard-switching.
- Possibility of operating at high switching frequencies, from 10 kHz to 30 kHz with low losses [13].
- The resonant inductor may be designed with a small size, since it is subjected to a limited current stress (indeed it conducts only during the switching transient).
- Auxiliary switches are involved by same inductor current and they work in ZCS.
- Low losses on resonant circuit passive components.
- Fully controllable resonant soft switching process.
- High overall efficiency.
- Possible operation in full PWM.

The AQRDCL is derived from the ARCP, but unlike the previous type, which is a resonant-pole, the AQRDCL is a resonant-link. While the ARCP needs two resonant circuit for each H-bridge, the AQRDCL uses a single circuit acting on the DC-link. Therefore, the AQRDCL needs less auxiliary components, but at the price of a more complex control and some more limitations. Moreover, it needs a clamp switch that has the task of disconnecting the DC-link from the source during the commutation intervals and which commutates at the switching frequency, therefore its losses could be not negligible. The AQRDCL control is more complex and it requires faster devices, indeed the timing sequence in device command requires a very short time, and it is therefore necessary to use a proper control to handle situations where a resonant cycle is triggered before the previous one is finished. For this reason, after a preliminary analysis, it seems that the AQRDCL is more suitable for lower power than ARCP, about hundreds of kW, but it is anyway a topology to be taken into account too.

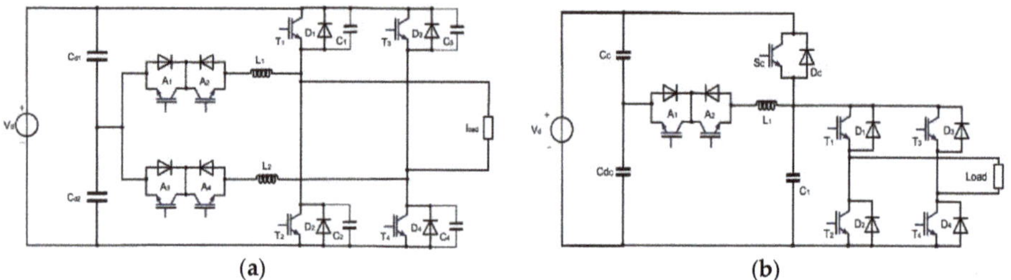

(a)	(b)

Figure 11. (**a**) ARCP full H-bridge (single submodule or cell). Each of the two half bridges needs its own resonant circuit. (**b**) AQRDCL full H-bridge. The resonant circuit is single and requires less components since the resonant capacity is unique; however it is necessary to add an active clamping device.

In the ARCP the capacitive snubbers guarantee turn-off ZVS, whereas the resonant circuit (in which snubber capacitors act as resonant capacities) allows turn-on ZVS. The auxiliary switches work in quasi-ZCS both in turn-on and turn-off. Unlike other topologies, the resonant circuit is parallel connected to the main switch only around the switching transient; outside this short time, the resonant circuit is not involved by the load current. This is an advantage for the sizing of the resonant inductor, which should not sustain continuously the load current (therefore also its losses are low). Another advantage is that the behavior of the H-bridge is very similar to the one in hard-switching, since the resonant circuit is disconnected for the major time of the switching period. In other simplest topologies, resonant inductor is connected in series to the DC-link and it must therefore sustain the load current continuously.

Some simulation results, using the control strategy presented in [116], are shown in Figure 12. One can notice that devices are less subjected to dv/dt stress, but apart from this, output voltage and current waveform is quite similar to PMW modulation waveform.

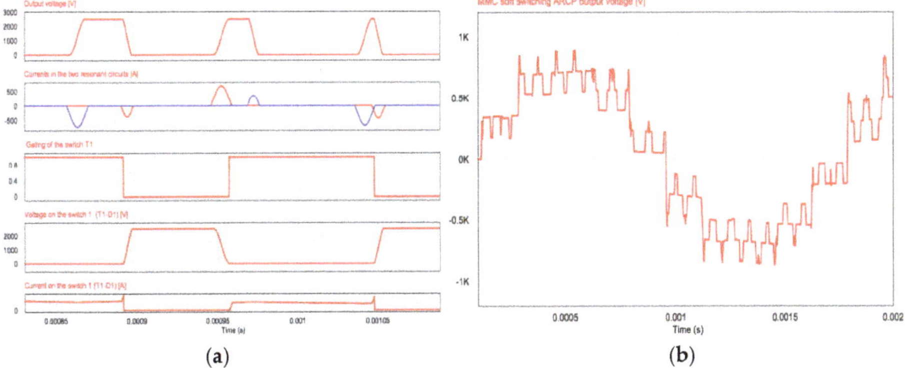

(a)   (b)

Figure 12. (**a**) Full H-bridge ARCP single submodule simulation diagram: output voltage, currents in the resonant circuits, gating of switch T_1, voltage on the switch 1, current on the switch 1. (**b**) Output voltage of a full ARCP soft-switching MMC (4-submodules for branch, all ARCP). Note: current and voltage values are scaled, for testing only.

In the AQRDCL, differently from the ARCP, there are not snubber capacitors, but a clamp switch (Sc-Dc) is necessary. During the switching, the commutation of the main switches, the turn-off of the clamp switch and the turn-on of the auxiliary switches, properly coordinated, occur. The consequence is the zeroing of the DC-link voltage for a short time, so that the switching of the main switches occur in ZVS both for turn-on and turn-off. At the end of the switching process, the resonant circuit is deactivated and the DC-link voltage grows at its value, since the clamp switch starts to conduce. Like in the ARCP, the auxiliary switches work in ZCS. Some simulation diagram of a single submodule, together with the output voltage of a 4-submodules for branch MMC are reported in Figure 13. Also in this case, one can notice that voltage waveform in quite similar to a MMC working in hard-switching.

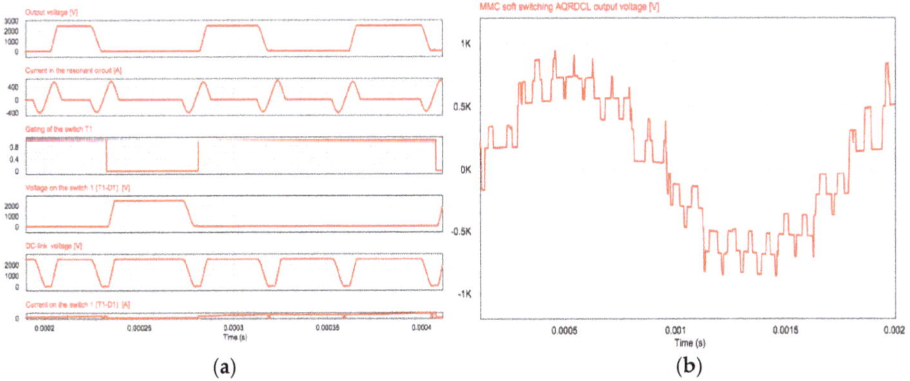

(a)   (b)

Figure 13. (**a**) Full H-bridge AQRDCL single submodule simulation diagrams: output voltage, current in the resonant circuit, gating of the switch T_1, voltage on the switch T_1, DC-link voltage, current on $(T_1–D_1)$ (**b**) phase output voltage of a full AQRDCL soft-switching MMC converter (4 submodules for branch, all AQRDCL). Note: current and voltage values are scaled, for testing only.

5.3. Soft-Switching MMC: Losses Evaluation

In order to evaluate the benefits of soft-switching in the MMC, a model with two full H-bridges (one working in hard-switching and one working in soft-switching) with same rating and in same working points, is created in PSIM software (8.0.7, Powersim, Rockville, MD, USA) equipped with Thermal Module.

The DC-link capacitor voltage is set to 2500 V, the switching frequency to 6 kHz, the modulation index is set respectively to 0.25, 0.5 and 0.75, and the converter works on a R-L load with $\cos\varphi = 0.8$, so the load currents are 77 A, 155 A and 230 A RMS, respectively, for the three modulation indexes. The output frequency is 450 Hz and the resonant frequency 70 kHz. IGBTs ABB 5SNA650J450300, 4500 V–650 A are used. Simulation losses in hard-switching configuration are reported in Table 6.

Table 6. Submodule losses in hard-switching.

Type of Loss	77 A	155 A	230 A
Overall conduction losses	235 W	556 W	962 W
Overall switching losses	10.8 kW	15.6 kW	20.3 kW
Total losses (efficiency)	**11 kW (69%)**	**16.1 kW (86%)**	**21.3 kW (91%)**

Of course the switching frequency of 6 kHz is too high for hard-switching working condition and the losses are obviously unsustainable. Losses in the same working conditions with the ARCP soft-switching submodule are reported in Table 7. Same ABB 5SNA 650J450300, 4500 V–650 A are used both for the main switches and for the auxiliary switches.

Table 7. Submodule losses in ARCP with auxiliary IGBTs.

Type of Loss	77 A	155 A	230 A
Overall conduction losses on the main devices	211 W	468 W	794 W
Overall conduction losses on the auxiliary devices	328 W	348 W	387 W
Overall switching losses on the main switches	93 W	101 W	156 W
Overall switching losses on the auxiliary devices	5.52 kW	5.53 kW	5.61 kW
Total losses (efficiency)	**6.2 kW (80%)**	**6.4 kW (94%)**	**6.9 kW (97%)**

The results in Table 7 show that the switching losses on the main devices are nearly zero, whereas the switching losses on the auxiliary devices are quite low but not negligible, since the turn-on and the turn-off do not completely occur in ZCS. In addition, the effect of the recovery losses is not negligible, especially in the auxiliary diodes, caused by relatively high steep of the resonant current pulses. Moreover, considering the series connection of the two auxiliary IGBTs and diodes, the connection is not beneficial from conduction losses point of view. It could be therefore advantageous to use two antiparallel fast switching thyristors; this solution could also reduce circuit complexity, since the two conduction paths include only one component. A possible solution for the thyristors could be the Proton-Electrotex TFI253-800-22, 2200 V–800 A. Results for the ARCP using ABB 5SNA 650J450300, 4500 V–650 A for main switches and Proton-Electrotex TFI253-800-22, 2200 V–800 A for auxiliary switches are reported in Table 8, where a great reduction of switching losses on the auxiliary devices is shown.

Simulation results for the AQRDCL submodule equipped with ABB 5SNA 650J450300, 4500 V–650 A for main switches and Proton-Electrotex TFI253-800-22, 2200 V–800 A for auxiliary switches are reported in Table 9.

The ARCP with thyristors for auxiliary circuit is the most efficient topology and, moreover, it has a simpler control and less limitation; therefore, this could be the most suitable configuration for a MMC medium frequency topology for railway applications.

Table 8. Submodule losses in ARCP with auxiliary thyristors.

Type of Loss	77 A	155 A	230 A
Overall conduction losses on the main devices	211 W	468 W	792 W
Overall conduction losses on the auxiliary devices	152 W	158 W	169 W
Overall switching losses on the main switches	85.5 W	89 W	136 W
Overall switching losses on the auxiliary devices	2.86 kW	2.86 kW	2.87 kW
Total losses (efficiency)	**3.3 kW (88%)**	**3.6 kW (96%)**	**4.0 kW (98.2%)**

Table 9. Submodule losses in AQRDCL.

Type of Loss	77 A	155 A	230 A
Overall conduction losses on the main devices	258 W	402 W	698 W
Overall conduction losses on the auxiliary devices	173 W	168 W	169 W
Overall conduction losses on the clamp device	19 W	44 W	138 W
Overall switching losses on the main switches	502 W	439 W	640 W
Overall switching losses on the auxiliary devices	3.33 kW	3.89 kW	4.17 kW
Overall switching losses on the clamp device	991 W	1.06 kW	1.11 kW
Total losses (efficiency)	**5.3 kW (82%)**	**6.0 kW (94%)**	**6.9 kW (96.9%)**

It is necessary to note that the devices available on the market are built as a compromise between conduction losses and switching losses, according to a switching frequency limited to about 1 kHz maximum. Special devices for fast applications can be obtained by electronic irradiation process, or with a reconfiguration of the anodic structure of the silicon chip, in order to modify the carrier distribution inside the silicon width. A comparison between hard and soft switching submodules, both equipped with the same devices, which are optimized for hard switching operation, it is therefore partially appropriate, since the full exploitation of soft switching technique cannot overlook the use of specifically designed devices. However, these results show that soft switching can potentially introduce significant benefits in the MMC in terms of efficiency and allows the single sub-module to operate at high frequency, which is unsustainable with common devices working in hard-switching. The increase of switching frequency is indeed a fundamental condition to use MMC in PET applications.

6. Conclusions

In this paper, a review on PET state of the art for railway applications was carried out. The aim of this technology is to substitute the on-board transformer, working at line frequency, with an electronic converter, which supplies a medium frequency transformer. The review focused on the solutions which were tested with experimental tests and prototypes. At first, a detailed literature review on technical papers published on this topic in last years was performed. After, authors showed the two PET principle schemes, i.e., two stage PET and three stage PET, together with main targets and topics which should be investigated. The study focused on the main prototypes that were conceived in the last decades. In 2003 Alstom LHB GmbH and SMA Regelsysteme GmbH presented a prototype called eTransformer, which was a 3SC using 6.5 kV–400 V IGBTs. In 2007 Bombardier Transportation presented a laboratory prototype called "Medium Frequency Topology", which was again a 3SC using 6.5 kV IGBTs. In 2007 ABB presented its first laboratory prototype, a 3SC with 3.3 kV–400 A IGBTs, which nevertheless has some problems in terms of efficiency and number of IGBTs. However, in 2012, ABB presented probably the most advanced prototype of PET, called Power Electronic Traction Transformer (PETT), which was a 3SC with 6.5 kV–400A IGBTs for the first and second stages. Although the power density of PETT was not very high, the efficiency of 96% was significantly higher than a traditional line frequency transformer with 4QCs. After the comparison between the main PET prototypes, Bombardier and ABB AC-links were deeply analyzed.

However, since none of the proposed PET prototypes seemed to have the correct features to be used in railway application, authors investigated a possible new configuration exploiting MMC, which has some interesting characteristics for the purpose. However, in order to achieve high switching frequency, and therefore to reduce significantly the weight and volume of the medium frequency transformer, a soft-switching topology should be used. Among the configurations proposed in the technical literature, authors identified the ARCP and the AQRDCL as the most suitable solution for this application. For each of the two soft-switching architectures, a thermal model was created in PSIM environment. Losses and efficiency of a single submodule were evaluated for both configurations with the same ABB 5SNA 650J450300, 4500 V–650 A IGBTs. The ARCP configuration, using Proton-Electrotex TFI253-800-22, 2200 V–800 A thyristors for the soft-switching circuit, was the configuration with the highest efficiency (98.2% for 230 A working operation on a RL load). Moreover, this architecture is characterized by a simpler control and less limitation and therefore it was judged as the most suitable configuration.

Although 98.2% is already a satisfying result, one has to note that it was obtained using components designed for hard-switching application. An additional reduction in switching losses could be achieved using components specifically designed for soft-switching or using innovative technologies such as SiC MOSFET.

Author Contributions: Conceptualization, M.M. and L.V.; Data curation, M.P.; Funding acquisition, M.M.; Investigation, S.F.; Methodology, S.F.; Software, S.F. and L.V.; Supervision, M.M. and L.V.; Visualization, M.P.; Writing—original draft, M.P.; Writing—review & editing, M.M.

Funding: This research received no external funding.

Conflicts of Interest: The authors declare no conflict of interest.

Nomenclature

PET	Power Electronic Transformer
SST	Solid State Transformer
MMC	Modular Multilevel Converter
IM	Induction Motor
PMSM	Permanent Magnet Synchronous Motor
4QC	Four Quadrant Converter
VSI	Voltage Source Inverter
LF	Line Frequency
MF	Medium Frequency
Bo'Bo'	4-axles, 2-bogies, 4-motors Locomotive
2CS PET	Two Conversion Stage PET
3CS PET	Three Conversion Stage PET
ZVS	Zero Voltage Switching
ZCS	Zero Current Switching
$I_{\text{turn-OFF}}$	Current during turn-off in ABB PETT
PEBB	Power Electronics Building Blocks
f_0	Resonant frequency associated to L_d
f_p	Resonant frequency associated to L_m
L_d	Transformer leakage inductance
L_m	Transformer magnetizing inductance
C_r	Capacitance of the AC link
SiC	Silicon Carbide

References

1. Steimel, A. *Electric Traction: Motive Power and Energy Supply*; Division Deutscher Industrie Verlag: Munchen, Germany, 2014.
2. Allenbach, J.-M.; Chapas, P.; Comte, M.; Kaller, R. *Traction Électrique*; PPUR Presses Polytechniques: Lausanne, Switzerland, 2008; Volume 1.
3. Mermet-Guyennet, M. New power technologies for traction drives. In Proceedings of the SPEEDAM 2010, Pisa, Italy, 14–16 June 2010; pp. 719–723.
4. Henning, U.; Schlosser, R.; Meinert, M. Supraleitende Transformatoren für elektrische Triebfahrzeuge. *Zev Det Glas. Ann. Die Eisenb.* **2002**, *126*, 86–92.
5. Schlosser, R.; Schmidt, H.; Leghissa, M.; Meinert, M. Development of high-temperature superconducting transformers for railway applications. *IEEE Trans. Appl. Supercond.* **2003**, *13*, 2325–2330. [CrossRef]
6. Meinert, M.; Leghissa, M.; Schlosser, R.; Schmidt, H. System test of a 1-MVA-HTS-transformer connected to a converter-fed drive for rail vehicles. *IEEE Trans. Appl. Supercond.* **2003**, *13*, 2348–2351. [CrossRef]
7. Dieckerhoff, S.; Schäfer, U. Transformerless drive system for main line rail vehicle propulsion. In Proceedings of the 9th European Conference on Power Electronics and Applications (EPE'01), Graz, Austria, 27–29 August 2001.
8. Dieckerhoff, S.; Bernet, S.; Krug, D. Evaluation of IGBT multilevel converters for transformerless traction applications. In Proceedings of the IEEE 34th Annual Conference on Power Electronics Specialist, PESC'03, Acapulco, Mexico, 15–19 June 2003; pp. 1757–1763.
9. Steiner, M.; Deplazes, R.; Stemmler, H. A new transformerless topology for AC-fed traction vehicles using multi-star induction motors. *EPE J.* **2000**, *10*, 45–53. [CrossRef]
10. Rufer, A.; Schibli, N.; Briguet, C. A direct coupled 4-quadrant multilevel converter for 16 2/3 Hz traction systems. In Proceedings of the 1996 Sixth International Conference on Power Electronics and Variable Speed Drives (Conf. Publ. No. 429), Nottingham, UK, 23–25 September 1996; pp. 448–453.
11. Schibli, N.; Rufer, A. Single-and Three Phased Multilevel Converters for Traction Systems 50Hz/16 2/3 Hz. In Proceedings of the European Conference on Power Electronics and Applications, Trondheim, Norway, 8–10 September 1997.
12. Engel, B.; Victor, M.; Bachmann, G.; Falk, A. 15 kV/16.7 Hz energy supply system with medium frequency transformer and 6.5 kV IGBTs in resonant operation. In Proceedings of the 10th European Conference on Power Electronics and Applications, EPE 2003, Toulouse, France, 2-4 September 2003; pp. 2–4.
13. Glinka, M.; Marquardt, R. A new single phase AC/AC-multilevel converter for traction vehicles operating on AC line voltage. *EPE J.* **2004**, *14*, 7–12. [CrossRef]
14. Steiner, M.; Reinold, H. Medium frequency topology in railway applications. In Proceedings of the 2007 European Conference on Power Electronics and Applications, Aalborg, Denmark, 2–5 September 2007; pp. 1–10.
15. Hugo, N.; Stefanutti, P.; Pellerin, M.; Akdag, A. Power electronics traction transformer. In Proceedings of the 2007 European Conference on Power Electronics and Applications, Aalborg, Denmark, 2–5 September 2007; pp. 1–10.
16. Weigel, J.; Ag, A.N.S.; Hoffmann, H. High voltage IGBTs in medium frequency traction power supply. In Proceedings of the 2009 13th European Conference on Power Electronics and Applications, Barcelona, Spain, 8–10 September 2009; pp. 1–10.
17. Casarin, J.; Ladoux, P.; Chauchat, B.; Dedecius, D.; Laugt, E. Evaluation of high voltage SiC diodes in a medium frequency AC/DC converter for railway traction. In Proceedings of the International Symposium on Power Electronics Power Electronics, Electrical Drives, Automation and Motion, Sorrento, Italy, 20–22 June 2012; pp. 1182–1186.
18. Das, M.K.; Capell, C.; Grider, D.E.; Leslie, S.; Ostop, J.; Raju, R.; Schutten, M.; Nasadoski, J.; Hefner, A. 10 kV, 120 A SiC half H-bridge power MOSFET modules suitable for high frequency, medium voltage applications. In Proceedings of the 2011 IEEE Energy Conversion Congress and Exposition, Phoenix, AZ, USA, 17–22 September 2011; pp. 2689–2692.
19. Zhao, C.; Lewdeni Schmid, S.; Steinke, J.K.; Weiss, M.; Chaudhuri, T.; Pellerin, M.; Duron, J.; Stefanutti, P. Design, implementation and performance of a modular power electronic transformer (PET) for railway application. In Proceedings of the 2011 14th European Conference on Power Electronics and Applications, Birmingham, UK, 30 August–1 September 2011; pp. 1–10.

20. Dujic, D.; Steinke, G.K.; Bellini, M.; Rahimo, M.; Storasta, L.; Steinke, J.K. Characterization of 6.5 kV IGBTs for high-power medium-frequency soft-switched applications. *IEEE Trans. Power Electron.* **2013**, *29*, 906–919. [CrossRef]

21. Behmann, U. Erster Mittelfrequenz-Traktionstransformator im Betriebs-Einsatz. *eb* **2012**, *110*, 408–413.

22. Dujic, D. Power electronic transformer for railway on-board applications—An overview. In Proceedings of the 7th International Power Electronics and Motion Control (PEMC), Harbin, China, 2–5 June 2012.

23. Dujic, D.; Mester, A.; Chaudhuri, T.; Coccia, A.; Canales, F.; Steinke, J.K. Laboratory scale prototype of a power electronic transformer for traction applications. In Proceedings of the 2011 14th European Conference on Power Electronics and Applications, Birmingham, UK, 30 August–1 September 2011; pp. 1–10.

24. Dujic, D.; Lewdeni-Schmid, S.; Mester, A.; Zhao, C.; Weiss, M.; Steinke, J.; Pellerin, M.; Chaudhuri, T. Experimental Characterization of LLC Resonant DC/DC Converter for Medium Voltage Applications. In Proceedings of the 2015 International Conference on Electrical Systems for Aircraft, Railway, Ship Propulsion and Road Vehicles (ESARS), Aachen, Germany, 3–5 March 2015; pp. 1–6.

25. Casarin, J.; Ladoux, P.; Lasserre, P. 10kV SiC MOSFETs versus 6.5kV Si-IGBTs for medium frequency transformer application in railway traction. In Proceedings of the 2015 International Conference on Electrical Systems for Aircraft, Railway, Ship Propulsion and Road Vehicles (ESARS), Aachen, Germany, 3–5 March 2015; pp. 1–6.

26. Wang, G. Design, Development and Control of >13 kV Silicon-Carbide MOSFET Based Solid State Transformer (SST). Ph.D. Thesis, North Carolina State University, Raleigh, NC, USA, 20 August 2013.

27. Kadavelugu, A.; Bhattacharya, S.; Ryu, S.-H.; Van Brunt, E.; Grider, D.; Leslie, S. Experimental switching frequency limits of 15 kV SiC N-IGBT module. In Proceedings of the 2014 International Power Electronics Conference (IPEC-Hiroshima 2014-ECCE ASIA), Hiroshima, Japan, 18–21 May 2014; pp. 3726–3733.

28. John, N.; James, J.; Koshy, D.E. The Concept of Power Electronic Traction Transformer For Indian Railway. In Proceedings of the 2018 International Conference on Control, Power, Communication and Computing Technologies (ICCPCCT), Kannur, India, 23–24 March 2018; pp. 340–346.

29. Vladimir, B.; Iurie, E.; Sergiu, I. SST Medium Voltage Transformer with Bidirectional Power Transmission for Electric Railway Transport. In Proceedings of the 2018 International Conference on Applied and Theoretical Electricity (ICATE), Craiova, Romania, 4–6 October 2018; pp. 1–6.

30. Oliveira, D.S.; Honório, D.d.A.; Barreto, L.H.S.; Praça, P.P.; Kunzea, A.; Carvalho, S. A two-stage AC/DC SST based on modular multilevel converterfeasible to AC railway systems. In Proceedings of the 2014 IEEE Applied Power Electronics Conference and Exposition-APEC 2014, Fort Worth, TX, USA, 16–20 March 2014; pp. 1894–1901.

31. Stackler, C.; Morel, F.; Ladoux, P.; Fouineau, A.; Wallart, F.; Evans, N. Optimal sizing of a power electronic traction transformer for railway applications. In Proceedings of the IECON 2018—44th Annual Conference of the IEEE Industrial Electronics Society, Washington, DC, USA, 21–23 October 2018; pp. 1380–1387.

32. Niyitegeka, G.; Park, J.; Cho, G.; Park, G.; Choi, J. Control Algorithm Design and Implementation of Solid State Transformer. In Proceedings of the 2019 10th International Conference on Power Electronics and ECCE Asia (ICPE 2019-ECCE Asia), Busan, Korea, 27–30 May 2019; pp. 2052–2058.

33. Ronanki, D.; Williamson, S.S. Topological Overview on Solid-state Transformer Traction Technology in High-speed Trains. In Proceedings of the 2018 IEEE Transportation Electrification Conference and Expo (ITEC), Long Beach, CA, USA, 13–15 June 2018; pp. 32–37.

34. Feng, J.; Chu, W.; Zhang, Z.; Zhu, Z. Power electronic transformer-based railway traction systems: Challenges and opportunities. *IEEE J. Emerg. Sel. Top. Power Electron.* **2017**, *5*, 1237–1253. [CrossRef]

35. Yao, R.; Zheng, Z. Design and control of a SiC-based three-stage AC to DC power electronic transformer for electric traction applications. *J. Eng.* **2018**, *2019*, 2947–2952. [CrossRef]

36. Ronanki, D.; Williamson, S.S. Evolution of power converter topologies and technical considerations of power electronic transformer-based rolling stock architectures. *IEEE Trans. Transp. Electrif.* **2017**, *4*, 211–219. [CrossRef]

37. Yun, C.-g.; Cho, Y. Active Hybrid Solid State Transformer Based on Multi-Level Converter Using SiC MOSFET. *Energies* **2018**, *12*, 66. [CrossRef]

38. Yun, H.-J.; Kim, H.-S.; Kim, M.; Baek, J.-W.; Kim, H.-J. A DAB Converter with Common-Point-Connected Winding Transformers Suitable for a Single-Phase 5-Level SST System. *Energies* **2018**, *11*, 928. [CrossRef]
39. Wang, R.; Sun, Q.; Cheng, Q.; Ma, D. The Stability Analysis of a Multi-Port Single-Phase Solid-State Transformer in the Electromagnetic Timescale. *Energies* **2018**, *11*, 2250. [CrossRef]
40. Rodrigues, W.A.; Oliveira, T.R.; Morais, L.M.F.; Rosa, A.H.R. Voltage and Power Balance Strategy without Communication for a Modular Solid State Transformer Based on Adaptive Droop Control. *Energies* **2018**, *11*, 1802. [CrossRef]
41. Montoya, R.J.G.; Mallela, A.; Balda, J.C. An evaluation of selected solid-state transformer topologies for electric distribution systems. In Proceedings of the 2015 IEEE Applied Power Electronics Conference and Exposition (APEC), An evaluation of Selected Solid-State Transformer Topologies for Electric Distribution Systems, Charlotte, NC, USA, 15–19 March 2015; pp. 1022–1029.
42. Farnesi, S.; Marchesoni, M.; Passalacqua, M.; Vaccaro, L. Soft-Switching Power Converters for Efficient Grid Applications. In Proceedings of the 19th EEEIC International Conference on Environment and Electrical Engineering, Genova, Italy, 11–14 June 2019.
43. Zeng, Y.; Zou, G.; Wei, X.; Sun, C.; Jiang, L. A Novel Protection and Location Scheme for Pole-to-Pole Fault in MMC-MVDC Distribution Grid. *Energies* **2018**, *11*, 2076. [CrossRef]
44. Kolar, J.W.; Ortiz, G. Solid state transformer concepts in traction and smart grid applications. In Proceedings of the Tutorial 15th International Power Electronics and Motion Control Conference (ECCE Europe), Novi Sad, Serbia, 4–6 September 2012.
45. Railway Handbook 2017. Energy Consuption and CO_2 Emissions. International Energy Agency (IEA). Available online: https://uic.org/IMG/pdf/handbook_iea-uic_2017_web3.pdf (accessed on 3 December 2019).
46. Marchesoni, M.; Savio, S. Reliability analysis of a fuel cell electric city car. In Proceedings of the 2005 European Conference on Power Electronics and Applications, Toulouse, France, 11–14 September 2005; p. 10.
47. Marchesoni, M.; Vaccaro, L. Extending the operating range in diode-clamped multilevel inverters with active front ends. In Proceedings of the 2012 IEEE International Energy Conference and Exhibition (ENERGYCON), Florence, Italy, 9–12 September 2012; pp. 63–68.
48. Fazio, P.; Maragliano, G.; Marchesoni, M.; Vaccaro, L. A new capacitor balancing technique in Diode-Clamped Multilevel Converters with Active Front End for extended operation range. In Proceedings of the 2011 14th European Conference on Power Electronics and Applications, Birmingham, UK, 30 August–1 September 2011; pp. 1–10.
49. Marchesoni, M.; Vaccaro, L. Operating limits in multilevel MPC inverters with active front ends. In Proceedings of the SPEEDAM, Pisa, Italy, 14–19 June 2010; pp. 192–197.
50. Carpaneto, M.; Marchesoni, M.; Vaccaro, L. A new cascaded multilevel converter based on NPC cells. In Proceedings of the 2007 IEEE International Symposium on Industrial Electronics, Vigo, Spain, 4–7 June 2007; pp. 1033–1038.
51. Leskovar, S.; Marchesoni, M. Control techniques for dc-link voltage ripples minimization in cascaded multilevel converter structures. In Proceedings of the 11th European Conference on Power Electronics and Applications EPE 2005, Dresden, Germany, 11–14 September 2005. [CrossRef]
52. Carpaneto, M.; Maragliano, G.; Marchesoni, M.; Vaccaro, L.R. A Novel approach for DC-link voltage ripple reduction in cascaded multilevel converters. In Proceedings of the International Symposium on Power Electronics, Electrical Drives, Automation and Motion SPEEDAM 2006, Taormina, Italy, 23–26 May 2006; pp. 571–576.
53. Maragliano, G.; Carpita, M.; Kissling, S.; Marchesoni, M.; Passalacqua, M.; Pidancier, T.; Vaccaro, F.; Vaccaro, L. A New Common Mode Voltage Reduction Method In A Generalized Space Vector Modulator For Cascaded Multilevel Converters. In Proceedings of the 21st Conference and Exhibition on Power Electronics and Applications EPE'19, Genova, Italy, 3–5 September 2019.
54. Farnesi, S.; Marchesoni, M.; Passalacqua, M.; Vaccaro, L. Soft-switching cells for Modular Multilevel Converters for efficient grid integration of renewable sources. *AIMS Energy* **2019**, *7*, 246–263. [CrossRef]
55. Xing, Z.; Ruan, X.; You, H.; Yang, X.; Yao, D.; Yuan, C. Soft-Switching Operation of Isolated Modular DC/DC Converters for Application in HVDC Grids. *IEEE Trans. Power Electron.* **2016**, *31*, 2753–2766. [CrossRef]
56. Zhang, J.; Zheng, T.Q.; Yang, X.; Wang, M. Single resonant cell based multilevel soft-switching DC-DC converter for medium voltage conversion. In Proceedings of the 2016 IEEE Energy Conversion Congress and Exposition (ECCE), Milwaukee, WI, USA, 18–22 September 2016; pp. 1–5.

57. Moosavi, M.; Toliyat, H.A. A Multicell Cascaded High-Frequency Link Inverter With Soft Switching and Isolation. *IEEE Trans. Ind. Electron.* **2019**, *66*, 2518–2528. [CrossRef]

58. Matsumura, K.; Koizumi, H. Interleaved soft-switching multilevel boost converter. In Proceedings of the IECON 2013—39th Annual Conference of the IEEE Industrial Electronics Society, 10–13 November 2013; pp. 936–941.

59. Cui, S.; Soltau, N.; Doncker, R.W.D. A High Step-Up Ratio Soft-Switching DC–DC Converter for Interconnection of MVDC and HVDC Grids. *IEEE Trans. Power Electron.* **2018**, *33*, 2986–3001. [CrossRef]

60. Xue, S.; Lu, J.; Liu, C.; Sun, Y.; Liu, B.; Gu, C. A Novel Single-Terminal Fault Location Method for AC Transmission Lines in a MMC-HVDC-Based AC/DC Hybrid System. *Energies* **2018**, *11*, 2066. [CrossRef]

61. Correa, P.; Pacas, M.; Rodriguez, J. Modulation strategies for fault-tolerant operation of H-bridge multilevel inverters. In Proceedings of the 2006 IEEE International Symposium on Industrial Electronics, Montreal, QC, Canada, 9–13 July 2006; pp. 1589–1594.

62. Zhu, Q.; Dai, W.; Guan, L.; Tan, X.; Li, Z.; Xie, D. A Fault-Tolerant Control Strategy of Modular Multilevel Converter with Sub-Module Faults Based on Neutral Point Compound Shift. *Energies* **2019**, *12*, 876. [CrossRef]

63. Li, J.; Yin, J. Fault-Tolerant Control Strategies and Capability without Redundant Sub-Modules in Modular Multilevel Converters. *Energies* **2019**, *12*, 1726. [CrossRef]

64. Li, K.; Yuan, L.; Zhao, Z.; Lu, S.; Zhang, Y. Fault-Tolerant Control of MMC With Hot Reserved Submodules Based on Carrier Phase Shift Modulation. *IEEE Trans. Power Electron.* **2017**, *32*, 6778–6791. [CrossRef]

65. Farnesi, S.; Fazio, P.; Marchesoni, M. A new fault tolerant NPC converter system for high power induction motor drives. In Proceedings of the SDEMPED 2011—8th IEEE Symposium on Diagnostics for Electrical Machines, Power Electronics and Drives, Bologna, Italy, 5–8 September 2011; pp. 337–343. [CrossRef]

66. Fazio, P.; Maragliano, G.; Marchesoni, M.; Parodi, G. A new fault detection method for NPC converters. In Proceedings of the 2011 14th European Conference on Power Electronics and Applications, EPE 2011, Birmingham, UK, 30 August–1 September 2011; pp. 1–10.

67. Bordignon, P.; Carpaneto, M.; Marchesoni, M.; Tenca, P. Faults analysis and remedial strategies in high power neutral point clamped converters. In Proceedings of the PESC Record—IEEE Annual Power Electronics Specialists Conference, Rhodes, Greece, 15–19 June 2008; pp. 2778–2783. [CrossRef]

68. Fazio, P.; Marchesoni, M.; Parodi, G. Fault detection and reconfiguration strategy for ANPC converters. In Proceedings of the 15th International Power Electronics and Motion Control Conference and Exposition, EPE-PEMC 2012 ECCE Europe, Novi Sad, Serbia, 4–6 September 2012; pp. 171–175. [CrossRef]

69. Rufer, A.; Schibli, N.; Chabert, C.; Zimmermann, C. Configurable front-end converters for multicurrent locomotives operated on 16 2/3 Hz AC and 3 kV DC systems. *IEEE Trans. Power Electron.* **2003**, *18*, 1186–1193. [CrossRef]

70. Mohan, N.; Undeland, T.M. *Power Electronics: Converters, Applications, and Design*; John Wiley & Sons: Hoboken, NJ, USA, 2007.

71. Bühler, H. *Convertisseurs Statiques*; PPUR Presses Polytechniques: Lausanne, Switzerland, 1991.

72. Frugier, D.; Coquery, G.; Jeunesse, A. *Mise in Oeuvre des IGBT sur les Materiel Roulants Ferroviaires de la SNCF. Revue Générale des Chemins de Fer, Parte I –II*; Éditions RGRA: Paris, France, 2016.

73. Marquardt, R. A new modular voltage source inverter topology. In Proceedings of the 10th European Conference on Power Electronics and Applications, EPE 2003, Toulouse, France, 2–4 September 2003.

74. Glinka, M. Prototype of multiphase modular-multilevel-converter with 2 MW power rating and 17-level-output-voltage. In Proceedings of the 2004 IEEE 35th Annual Power Electronics Specialists Conference (IEEE Cat. No. 04CH37551), Aachen, Germany, 20–25 June 2004; pp. 2572–2576. [CrossRef]

75. Glinka, M.; Marquardt, R. A new AC/AC-multilevel converter family applied to a single-phase converter. In Proceedings of the Fifth International Conference on Power Electronics and Drive Systems, PEDS 2003, Singapore, 17–20 November 2003; pp. 16–23. [CrossRef]

76. Glinka, M.; Marquardt, R. A new AC/AC multilevel converter family. *IEEE Trans. Ind. Electron.* **2005**, *52*, 662–669. [CrossRef]

77. Ishfaq, M.; Uddin, W.; Zeb, K.; Khan, I.; Ul Islam, S.; Adil Khan, M.; Kim, H.J. A New Adaptive Approach to Control Circulating and Output Current of Modular Multilevel Converter. *Energies* **2019**, *12*, 1118. [CrossRef]

78. Martinez-Rodrigo, F.; Ramirez, D.; Rey-Boue, A.B.; De Pablo, S.; Herrero-de Lucas, L.C. Modular Multilevel Converters: Control and Applications. *Energies* **2017**, *10*, 1709. [CrossRef]

79. Wu, Z.; Chu, J.; Gu, W.; Huang, Q.; Chen, L.; Yuan, X. Hybrid Modulated Model Predictive Control in a Modular Multilevel Converter for Multi-Terminal Direct Current Systems. *Energies* **2018**, *11*, 1861. [CrossRef]

80. Liu, Z.; Li, K.; Sun, Y.; Wang, J.; Wang, Z.; Sun, K.; Wang, M. A Steady-State Analysis Method for Modular Multilevel Converters Connected to Permanent Magnet Synchronous Generator-Based Wind Energy Conversion Systems. *Energies* **2018**, *11*, 461. [CrossRef]

81. Marchesoni, M.; Vaccaro, L. Study of the MMC circulating current for optimal operation mode in HVDC applications. In Proceedings of the 2015 17th European Conference on Power Electronics and Applications (EPE'15 ECCE-Europe), Geneva, Switzerland, 8–10 September 2015; pp. 1–10.

82. Farnesi, S.; Marchesoni, M.; Vaccaro, L. Reliability improvement of Modular Multilevel Converter in HVDC systems. In Proceedings of the 2016 Power Systems Computation Conference (PSCC), Genoa, Italy, 20–24 June 2016; pp. 1–7.

83. Maragliano, G.; Marchesoni, M.; Vaccaro, L. Optimal operation mode for Modular Multilevel Converter based HVDC. In Proceedings of the 2014 International Symposium on Power Electronics, Electrical Drives, Automation and Motion, Ischia, Italy, 18–20 June 2014; pp. 784–789. [CrossRef]

84. Bordignon, P.; Marchesoni, M.; Parodi, G.; Vaccaro, L. Modular multilevel converter in HVDC systems under fault conditions. In Proceedings of the 2013 15th European Conference on Power Electronics and Applications (EPE), Lille, France, 2–6 September 2013; pp. 1–10.

85. Marchesoni, M.; Passalacqua, M.; Vaccaro, L.; Carpita, M.; Gavin, S.; Kissling, S. Capacitor Voltage Ripple Minimization in Voltage Source Converter for HVDC Applications. In Proceedings of the 2019 AEIT HVDC International Conference (AEIT HVDC), Florence, Italy, 9–10 May 2019; pp. 1–6. [CrossRef]

86. Acharya, A.B.; Ricco, M.; Sera, D.; Teodorescu, R.; Norum, L.E. Arm Power Control of the Modular Multilevel Converter in Photovoltaic Applications. *Energies* **2019**, *12*, 1620. [CrossRef]

87. McLyman, C.W.T. *Magnetic Core Selection for Transformers and Inductors: A User's Guide to Practice and Specifications*; CRC Press: Boca Raton, FL, USA, 2018.

88. Dujic, D.; Kieferndorf, F.; Canales, F. Power electronic transformer technology for traction applications—An overview. *Electronics* **2012**, *16*, 50–56. [CrossRef]

89. Narayanan, P.T.R.; Balamurugan, R. An efficient active clamp resonant DC link for BLDCM drive system. In Proceedings of the IEEE-International Conference On Advances In Proceedings of the Engineering, Science And Management (ICAESM-2012), Tamil Nadu, India, 30–31 March 2012; pp. 709–714.

90. Cuzner, R.M.; Abel, J.; Luckjiff, G.A.; Wallace, I. Evaluation of actively clamped resonant link inverter for low-output harmonic distortion high-power-density power converters. *IEEE Trans. Ind. Appl.* **2001**, *37*, 847–855. [CrossRef]

91. Nakaoka, M.; Yonemori, H.; Yurugi, K. Zero-voltage soft-switched PDM three phase AC-DC active power converter operating at unity power factor and sinewave line current. In Proceedings of the IEEE Power Electronics Specialist Conference—PESC '93, Seattle, WA, USA, 20–24 June 1993; pp. 787–794. [CrossRef]

92. Katsutoshi, Y.; Hideo, Y.; Mutsuo, N. Zero-voltage soft-switching discrete pulse modulated three-phase AC-DC power converter with an active filtering function and its extended application. In Proceedings of the 15th International Telecommunications Energy Conference, Paris, France, 27–30 September 1993; Volume 392, pp. 392–399. [CrossRef]

93. Miyagawa, K.; Nakaoka, M.; Ogino, Y.; Murakami, Y.; Hayashi, K. Space-vector controlled soft-switching three-phase PDM AC/DC converter with unity power factor and sinusoidal line current. In Proceedings of the 1992 International Conference on Industrial Electronics, Control, Instrumentation, and Automation, San Diego, CA, USA, 13 November 1992; Volume 201, pp. 209–216. [CrossRef]

94. Yurugi, K.; Muneto, K.; Yonemori, H.; Nakaoka, M. New space-vector controlled soft-switching three-phase PDM AC/DC converter with unity power factor and sinusoidal line current shaping functions. In Proceedings of the Fourteenth International Telecommunications Energy Conference—INTELEC '92, Washington, DC, USA, 4–8 October 1992; pp. 286–293. [CrossRef]

95. Divan, D.M.; Skibinski, G. Zero-switching-loss inverters for high-power applications. *IEEE Trans. Ind. Appl.* **1989**, *25*, 634–643. [CrossRef]

96. Malesani, L.; Tomasin, P.; Toigo, V. Space vector control and current harmonics in quasi-resonant soft-switching PWM conversion. *IEEE Trans. Ind. Appl.* **1996**, *32*, 269–278. [CrossRef]

97. Shukla, J.; Fernandes, B.G. Quasi-resonant dc-link soft-switching PWM inverter with active feedback clamp circuit for motor drive applications. *IEE Proc. Electr. Power Appl.* **2006**, *153*, 75–82. [CrossRef]

98. Shaotang, C.; Lipo, T.A. A novel soft-switched PWM inverter for AC motor drives. *IEEE Trans. Power Electron.* **1996**, *11*, 653–659. [CrossRef]

99. Jafar, J.J.; Fernandes, B.G. A new quasi-resonant DC-link PWM inverter using single switch for soft switching. *IEEE Trans. Power Electron.* **2002**, *17*, 1010–1016. [CrossRef]

100. Kedarisetti, J.; Mutschler, P. A Motor-Friendly Quasi-Resonant DC-Link Inverter With Lossless Variable Zero-Voltage Duration. *IEEE Trans. Power Electron.* **2012**, *27*, 2613–2622. [CrossRef]

101. Reinold, H. Characterization of semiconductor losses in series resonant DC-DC converters for high power applications using transformers with low leakage inductance. In Proceedings of the 8th European Power Electron and Appl. Conf. Epe'99, Lausanne, Switzerland, 7–9 September 1999; pp. 1–10.

102. Reinold, H.; Flerlage, H.; Hartung, G.; Magin, O. Medium Frequency Technology in Auxiliary Power Converter for Traction Applications. Available online: http://www.railway-research.org/IMG/pdf/224.pdf (accessed on 8 December 2019).

103. Ladoux, P.; Mermet, M.; Casarin, J.; Fabre, J. Outlook for SiC devices in traction converters. In Proceedings of the 2012 Electrical Systems for Aircraft, Railway and Ship Propulsion, Bologna, Italy, 16–18 Octorber 2012; pp. 1–6.

104. Pfisterer, H.; Spath, H. Switching behaviour of an auxiliary resonant commutated pole (ARCP) converter. In Proceedings of the 7th IEEE International Power Electronics Congress Technical Proceedings CIEP 2000 (Cat. No.00TH8529), Mexico, Mexico, 15–19 October 2000; pp. 359–364.

105. Karys, S. Selection of resonant circuit elements for the ARCP inverter. In Proceedings of the 2009 10th International Conference on Electrical Power Quality and Utilisation, Lodz, Poland, 15–17 September 2009; pp. 1–6. [CrossRef]

106. Nagafuchi, S.; Abe, T.; Higuchi, T. Loss evaluation for ARCP Matrix Converter. In Proceedings of the 2013 IEEE 10th International Conference on Power Electronics and Drive Systems (PEDS), Kitakyushu, Japan, 22–25 April 2013; pp. 608–612. [CrossRef]

107. Morisaki, S.; Abe, T.; Higuchi, T. Experimental loss estimate for matrix converter using ARCP soft switching technique. In Proceedings of the 2015 18th International Conference on Electrical Machines and Systems (ICEMS), Pattaya, Thailand, 25–28 October 2015; pp. 1898–1902. [CrossRef]

108. Morii, H.; Yamamoto, M.; Funabiki, S. Capacitor-Less Auxiliary Resonant Commutated Pole (ARCP) voltage source soft switching inverter suitable for EV. In Proceedings of the 2009 13th European Conference on Power Electronics and Applications, Barcelona, Spain, 8–10 September 2009; pp. 1–8.

109. Toda, H.; Yamamoto, M. 1/3 weight core of a capacitor-less ARCP method three-phase voltage source soft-switching inverter suitable for EV. In Proceedings of the 2011 IEEE Energy Conversion Congress and Exposition, Phoenix, AZ, USA, 17–22 September 2011; pp. 4101–4106.

110. Doncker, R.W.D.; Lyons, J.P. The auxiliary quasi-resonant DC link inverter. In Proceedings of the PESC '91 Record 22nd Annual IEEE Power Electronics Specialists Conference, Phoenix, AZ, USA, 24–27 June 1991; pp. 248–253. [CrossRef]

111. Abeyratne, S.G.; Aydemir, M.T.; Lipo, T.A.; Murai, Y.; Yoshida, M. Current clamped, PWM, quasi-resonant DC link series resonant converter. In Proceedings of the 1994 IEEE Industry Applications Society Annual Meeting, Denver, CO, USA, 2–6 October 1994; Volume 822, pp. 820–826. [CrossRef]

112. Ahmed, T.; Nagai, S.; Nakaoka, M.; Hyun Woo, L. Two-switch auxiliary quasi-resonant DC link snubber-assisted voltage source three-phase V-connection soft-switching sinewave inverter with bidirectional soft-switching chopper for solar PV power conditioner. In Proceedings of the 30th Annual Conference of IEEE Industrial Electronics Society, IECON 2004, Busan, Korea, 2–6 November 2004; Volume 3103, pp. 3106–3110. [CrossRef]

113. Attaianese, C.; Tomasso, G. A new quasi-resonant DC link topology for soft switching of voltage-source inverter. In Proceedings of the IEEE 34th Annual Conference on Power Electronics Specialist, PESC '03., Acapulco, Mexico, 15–19 June 2003; Volume 522, pp. 525–530. [CrossRef]

114. Nagai, S.; Sato, S.; Yamamoto, M.; Nakaoka, M. Two switch auxiliary quasi-resonant DC link three-phase PWM inverter and two switch auxiliary resonant commutated pole link three-phase PWM rectifier. In Proceedings of the The 25th International Telecommunications Energy Conference, INTELEC '03., Yokohama, Japan, 23–23 October 2003; pp. 657–663.
115. Khalilin, M.; Adib, E.; Karimian, M.R.; Farzanehfard, H. Analysis of a new quasi resonant DC link inverter. In Proceedings of the 2011 2nd Power Electronics, Drive Systems and Technologies Conference, Tehran, Iran, 16–17 February 2011; pp. 421–426. [CrossRef]
116. Glinka, M.; Mehrpunkt-Umrichtersystem, M. Ein Ne aurtiges Konzept am Beispiel der Elektrischen Traktion–. Ph.D. Thesis, Munich University, Munich, Germany, 2011.

Article

A Refined Loss Evaluation of a Three-Switch Double Input DC-DC Converter for Hybrid Vehicle Applications

Mario Marchesoni *, Massimiliano Passalacqua and Luis Vaccaro

Department of Electrical, Electronic, Tlc Engineering and Naval Architecture (DITEN), University of Genoa, via all'Opera Pia 11a, 16145 Genova, Italy; massimiliano.passalacqua@edu.unige.it (M.P.); luis.vaccaro@unige.it (L.V.)
* Correspondence: marchesoni@unige.it

Received: 20 November 2019; Accepted: 30 December 2019; Published: 1 January 2020

Abstract: In this paper, an accurate efficiency evaluation of an innovative three-switch double input DC–DC converter for hybrid vehicle applications was carried out. The converter was used to interface two storages, (e.g., supercapacitor and battery) to the DC link. A refined model was created in MATLAB/Simulink Plecs environment and it was used to compare the traditional four-switch converter (i.e., two DC–DC converters in parallel connection) with the innovative three-switch converter. Loss and efficiency contour maps were obtained for both converters and a comparison between them was performed. A prototype of the three-switch converter was realized and used to validate the simulation thermal model by comparing both efficiency and current waveforms obtained with simulations and experimental tests.

Keywords: DC–DC converter; multi-input converter; supercapacitor; battery; hybrid electric vehicle (HEV), efficiency

1. Introduction

The increasing spread of hybrid electric vehicles (HEV) has led to the study and development of new storage systems and storage architectures. As far as battery storages are concerned, lithium batteries have increased their performance, reaching a key role in hybrid medium size car applications [1–6]. On the other hand, many studies have been carried out considering supercapacitor storages on hybrid vehicle, both combining batteries and supercapacitors [7–16] or using only supercapacitors as the storage system [17–28]. Regarding the combined use of batteries and supercapacitors, this configuration matches the pros of both type of storages (i.e., battery high specific energy and supercapacitor high specific power). In order to interface battery and supercapacitor storages with the vehicle DC link (i.e., with motor drive input), a double bidirectional converter is necessary and four full-controlled switches are generally used (Figure 1a) [29–31]. An innovative converter architecture using only three switches is proposed in [32–35] (Figure 1b). Such a converter not only can interface two storage systems with the DC link using a reduced number of switches, but it allows also a higher conversion efficiency. Indeed, on the one hand, reducing the number of switches in multi-input or multi-output converters is a topic of great interest in the technical literature [36–39], on the other, DC–DC converter efficiency is a key point in hybrid vehicles. In [32], a conduction loss comparison between the traditional four-switch converter (4SC) and the innovative three-switch converter (3SC) is carried out; however, regarding switching losses, only a qualitative study was performed. Since the switching losses are generally higher than conduction losses for such types of converters (as is confirmed from the results in Section 4), their accurate evaluation is of primary importance. A study regarding the 3SC efficiency is still missing in the technical literature and is the aim of this paper.

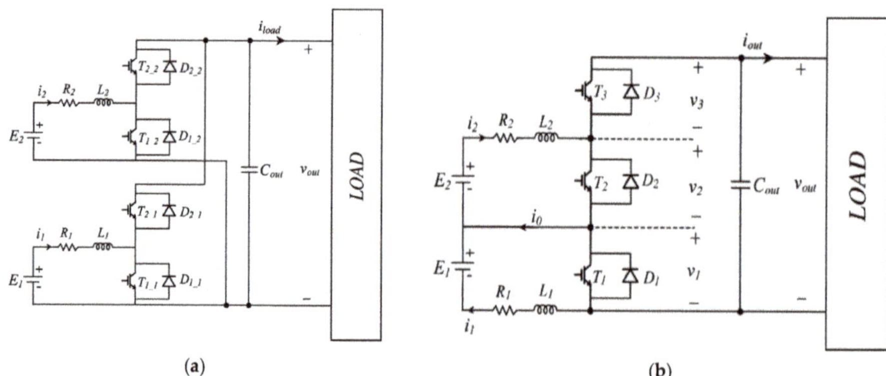

Figure 1. (**a**) 4SC architecture (**b**) 3SC architecture.

A refined model in MATLAB/Simulink Plecs environment was developed and used to perform simulations. Moreover, in order to verify the effectiveness of the model, a prototype of the 3SC was built and used to carry out experimental tests.

This paper is structured as follows. In Section 2, 3SC architecture and modulation are presented, together with the MATLAB/Simulink Plecs model. The simulation results regarding converter efficiency are shown in Section 3, and the experimental results are shown in Section 4. Finally, conclusions are carried out in Section 5.

2. Three-Switch Converter (3SC) Architecture and Modulation

The 3SC architecture is shown in Figure 1b. The DC-link voltage (i.e., V_{out} in Figure 1) has to be higher than the sum of E_1 and E_2, which are the storage voltages (i.e., battery voltage and supercapacitor voltage). The converter is bidirectional and can operate both in boost mode (i.e., power flow from storages to DC-link) or in buck mode (i.e., power flow from DC-link to storages). Battery and supercapacitor power flows in both directions are allowed and are independent of each other. For instance, the converter can work in boost mode for battery storage and in buck mode for supercapacitor storage simultaneously or vice versa.

2.1. Modulation Strategy

If one observes the converter structure, it is easy to notice that the condition in which all switches are closed (on) has to be avoided since it causes DC-link short circuit. Moreover, only one switch can be open at the same time [32]. Indeed, if two or three switches are open, voltages depend on current direction, since currents flow in the free-wheeling diodes. For this reason, if one defines $m_{i\_off} = 1 - m_{i\_on}$ (where m_{i_on} is the switch on-duty ratio of switch (i)), the relation in Equation (1) has to be satisfied. The relationships between v_1, v_2, v_3 and switch on-duty ratios are shown in Figure 2 [32].

$$m_{1\_off} + m_{2\_off} + m_{3\_off} = 1 \tag{1}$$

With reference to Figure 1b, the following relationships (2) can be derived:

$$v_1 = E_1 - R_1 \cdot i_1 - L_1 \cdot \frac{di_1}{dt}$$

$$v_2 = E_2 - R_2 \cdot i_2 - L_2 \cdot \frac{di_2}{dt} \tag{2}$$

$$v_{out} = v_1 + v_2 + v_3$$

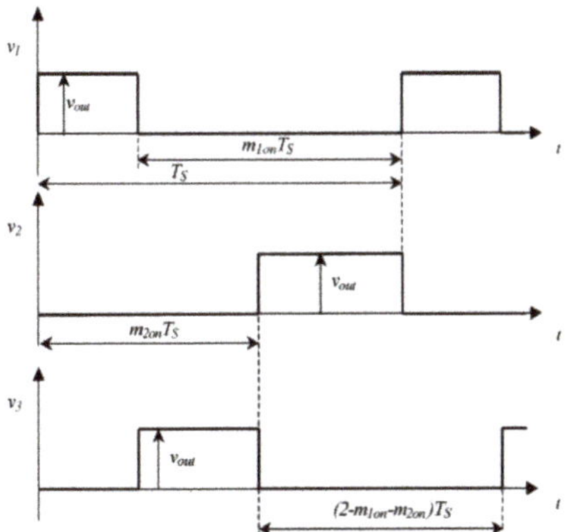

Figure 2. Relationships among v_1, v_2 and v_3 voltages and switch on-duty ratios.

The v_{out} voltage variation can be neglected during the switching period, so that $v_{out} \approx V_{out}$ and the average voltages V_1, V_2 and V_3 can be easily obtained:

$$V_1 = (1 - m_{1\_on}) \cdot V_{out}$$
$$V_2 = (1 - m_{2\_on}) \cdot V_{out} \tag{3}$$
$$V_3 = (m_{1\_on} + m_{2\_on} - 1) \cdot V_{out}$$

The averaged currents I_1 and I_2 can be expressed as in Equation (4) [32]:

$$I_1 = \frac{E_1 - (1 - m_{1\_on}) \cdot V_{out}}{R_1}$$
$$I_2 = \frac{E_2 - (1 - m_{2\_on}) \cdot V_{out}}{R_2} \tag{4}$$

According to what was previously reported, firing pulse generation logic can be defined as shown in Figure 3a and the resulting firing pulse generation signals are reported in Figure 3b [32].

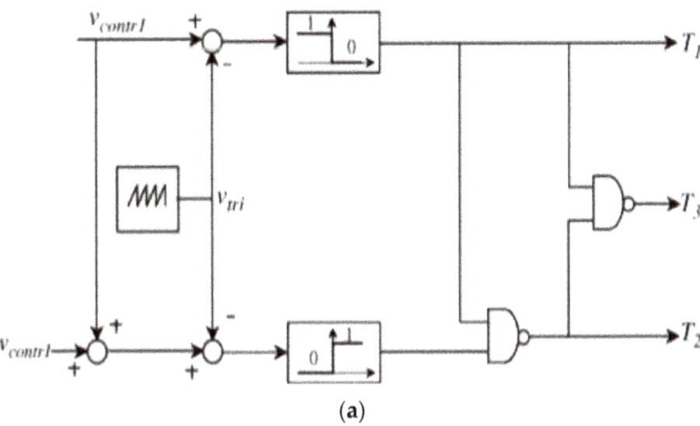

(a)

Figure 3. *Cont.*

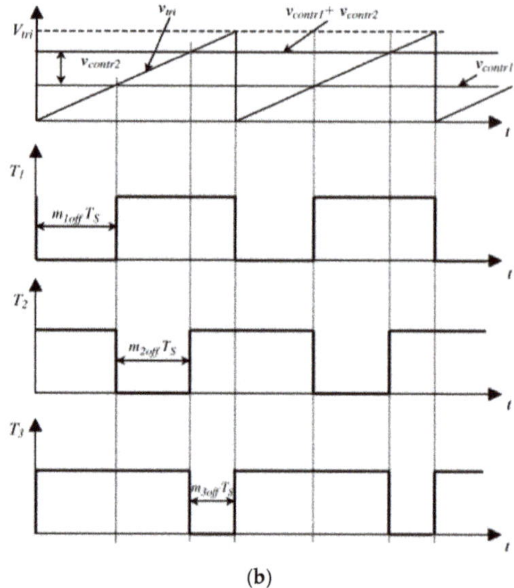

(b)

Figure 3. (**a**) Firing pulse generation logic. (**b**) Firing pulse generation signals.

2.2. MATLAB/Simulink Plecs Model

In order to establish converter losses, a converter model was created in MATLAB/Simulink Plecs environment, with a thermal model for IGBTs and diodes. Conduction losses were evaluated according to the data shown in Figures 4a and 5a, and the switching losses were evaluated according to the data shown in Figures 4b and 5b. Data were taken from the datasheet provided by the manufacturer of IGBT SKM400GA124D [40]. Indeed, this IGBT was used in the converter prototype. Converter parameters are reported in Table 1, where f_{sw} is the switching frequency; the same parameters were used for both 4SC and 3SC efficiency evaluation. The IGBTs had 1200 V rated voltage and 400 A rated current.

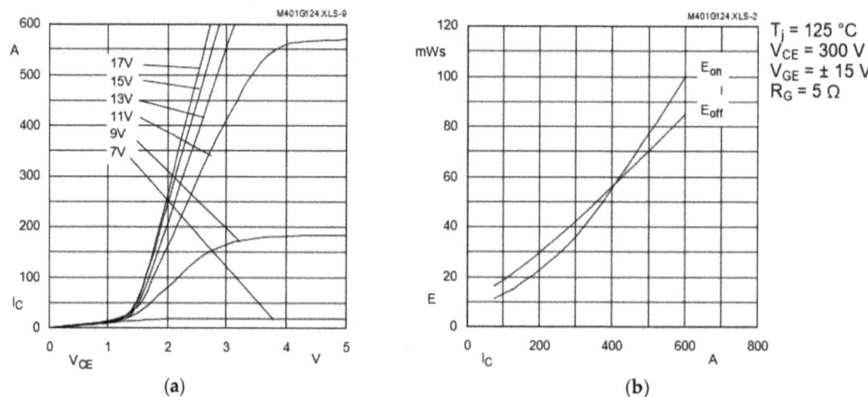

(a)   (b)

Figure 4. IGBT features. (**a**) IGBT characteristic (**b**) IGBT switching losses.

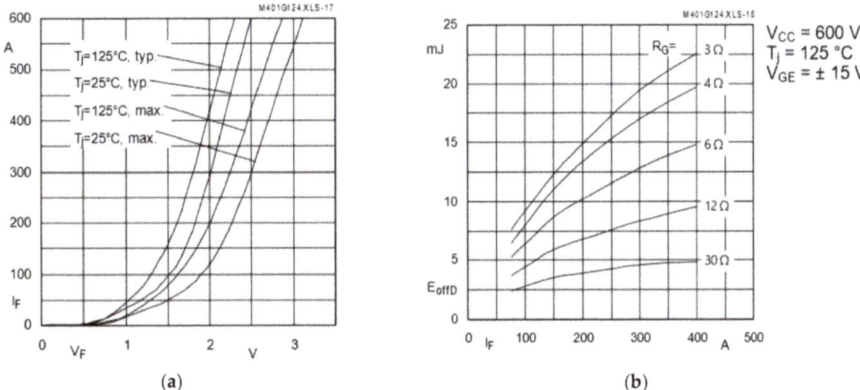

(a)   (b)

Figure 5. Diode features. (**a**) Diode characteristic (**b**) Diode turn-off losses.

Table 1. 3SC parameters.

Parameter	Value
E_1	100 V
L_1	200 µH
R_1	20 mΩ
E_2	50 V
L_2	100 µH
R_2	10 mΩ
C_{out}	10 mF
V_{out}	400 V
f_{sw}	10 kHz

Plecs model for the 3SC is reported in Figure 6. Since the simulation time was very low if compared to charging/discharging time, storages were modelled as ideal DC sources. A current source was connected to converter output in order to simulate the load, which can be both active or passive (i.e., injecting or absorbing current). IGBT and diode features of Figures 4 and 5 were implemented in the thermal model of the switches and a heat sink was connected to each of them. Losses were evaluated as the sum of all IGBT losses. Same IGBT thermal model was used in the 4SC model.

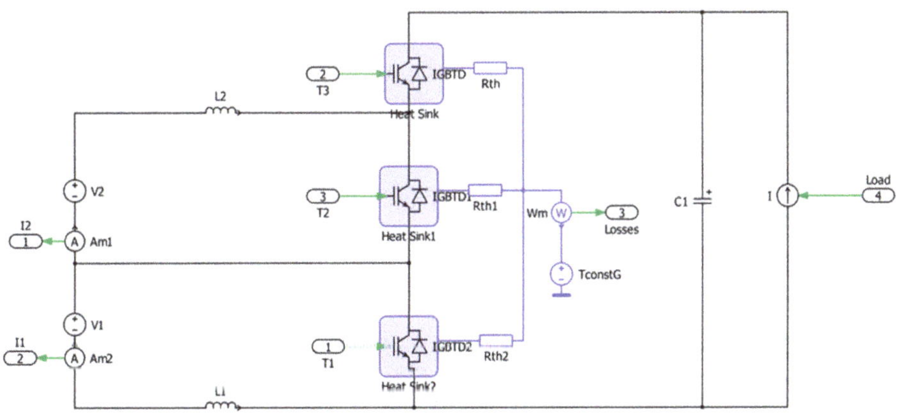

Figure 6. 3SC Plecs model.

3. Simulation Results

A comparison of conduction losses between 4SC and 3SC is shown in [32]; nevertheless, an efficiency analysis considering also switching losses is missing in the literature and therefore, is evaluated in this section. Converter losses (i.e., IGBT and diode conduction losses, IGBT switching losses, diode reverse recovery losses and inductor core losses) were evaluated in all the converters' working ranges. 4SC losses and efficiency contour maps are shown in Figure 7, whereas 3SC contour maps are shown in Figure 8. In Figure 9, losses and efficiency comparisons, i.e., differences, between 4SC and 3SC are reported. When P_1 and P_2 have opposite sign, 4SC losses are slightly lower than 3SC, however, when P_1 and P_2 have the same sign, 3SC losses are significantly lower than 4SC losses. Indeed, observing converter topology in Figure 1b, one can note that current in T1 is i1 when T3 is off and i1-i2 when T2 is off, current in T2 is i2 when T3 is off and i2-i1 when T1 is off, whereas current in T3 is i1 when T1 is off and i2 when T2 is off. For this reason, when the currents are concordant, in switches T1 and T2 flows a current which is significantly lower than the one that flows in the switches of a 4SC. Moreover, the condition in which P_1 and P_2 are concordant is the most common condition in hybrid vehicles [32]. Please note that in all the contour maps, P1 is reported on the X-axis, P2 is reported on the Y-axis, and losses (or efficiency) are reported on the Z-axis.

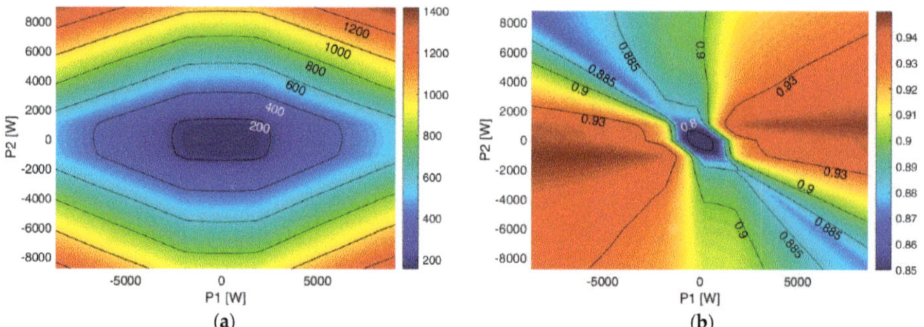

Figure 7. 4SC (**a**) Losses [W] (**b**) Efficiency.

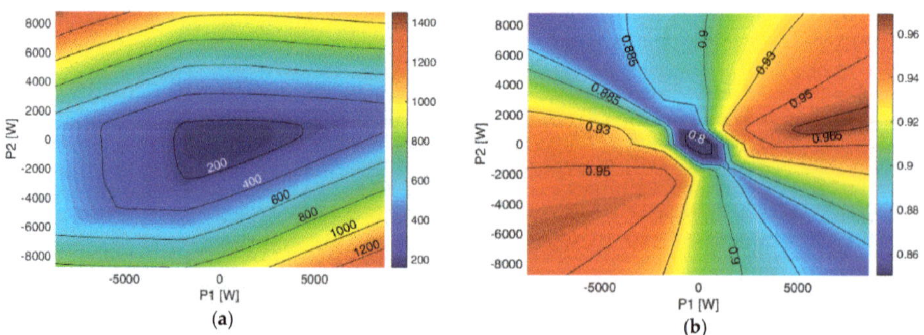

Figure 8. 3SC (**a**) Losses [W] (**b**) Efficiency.

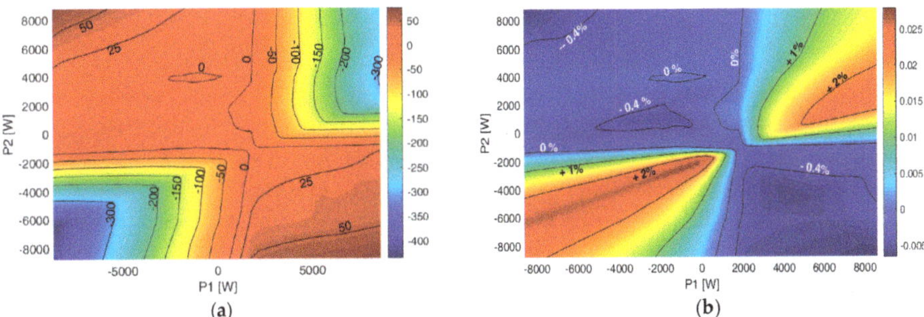

Figure 9. Comparison between 4SC and 3SC (**a**) 3SC losses—4SC losses [W] (**b**) 3SC efficiency—4SC efficiency.

4. Experimental Results

4.1. Experimental Test Bench and Validation Approach

In order to validate the simulation model, a prototype of the converter was realized, which is shown in Figures 10 and 11.

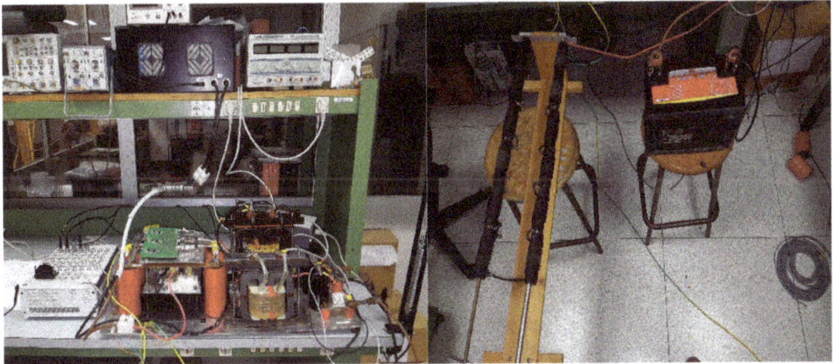

Figure 10. Experimental setup.

Figure 11. Converter prototype.

The prototype has three IGBTs SKM400GA124D, as in the simulation model, but it has different inductance and voltage values, as reported in Table 2. The aim of the prototype was to validate the simulation model, therefore, other simulations were performed with voltage and inductance values used in the converter prototype. Indeed, the converter efficiency was evaluated at first in the rated working points, in order to evaluate converter potential performance; however, since the experimental tests were performed at lower voltages and currents, simulations were performed also at the same values as in the experimental tests. In this way, it was possible to compare current waveforms and efficiency obtained via simulations with the ones obtained via experimental tests. A supercapacitor was used as E_2 (with a voltage of about 22 V) whereas a battery was used as E_1 (with a voltage of about 12 V). A 166 Ω resistor was connected to the DC-link and could be disconnected to test the converter only with power exchange between storages. Due to this configuration, the prototype could not be tested in full-buck operation (i.e., both, storages in charging condition) but the other working operations could all be tested: full-boost operation (i.e., both storages in discharging mode) and buck-boost operation (i.e., one storage in charging mode with the other one in discharging mode and vice versa).

Table 2. Prototype parameters.

Parameter	Value
IGBTs	SKM400GA124D
E_1	12 V
L_1	330 µH
E_2	22 V
L_2	950 µH
C_{out}	10 mF
V_{out}	100 V
f_{sw}	10 kHz

The aim of the experimental tests was to validate the simulation model with two different approaches:

1. Comparing the current waveforms obtained via simulations with the ones obtained via experimental tests.
2. Comparing the efficiency obtained via simulations with the one obtained via experimental tests.

Both approaches were useful for model validation. Indeed, since Vce = Vce (Ic), Vf = Vf (If), Eon = Eon (Ic), Eoff = Eoff (Ic) and Eoffd = Eoffd (If) are all monotonous increasing functions (Figures 4 and 5), a reduction in switching and/or conduction losses is inevitably caused by a reduction in current values during switching or conduction. For this reason, it is important to verify that the current waveforms obtained with simulations are in accordance with the current waveforms obtained with the experimental tests. On the other hand, it is important to verify that the efficiency obtained with the prototype converter is comparable with the one obtained with simulation, so that the efficiency increase (or decrease) moving from 4SC to 3SC can be considered sufficiently accurate. However, evaluating only this second aspect would be not sufficient; indeed, many conditions intervene in real applications, therefore, it is obvious that the efficiency evaluation with simulation cannot be very precise. Nevertheless, if the current waveforms are well simulated, the efficiency variation between 4SC and 3SC is accurate, thanks to the monotony of the previous functions. Definitively, even if the absolute converter efficiency cannot be modelled precisely with simulations, the efficiency variation, between 4SC and 3SC, can be modelled precisely if the current waveforms are correctly simulated.

4.2. Experimental Results

Three different tests were performed, which correspond to the three different working conditions. Test 1 (Table 3) was with the converter in full-boost operation (i.e., both storages in discharging mode), whereas Tests 2 and 3 ewre in buck-boost operations: in Test 2, the battery was discharging and the supercapacitor was charging; in Test 3, the battery was charging and the supercapacitor was discharging. In all three configurations, there was a load of 0.6 A on the DC-link. In Table 4, simulation losses are shown, together with the efficiency evaluated via simulations and the efficiency evaluated with the experimental tests. The efficiency was evaluated according to the equations in Table 3.

Table 3. Efficiency evaluation.

Test	Efficiency via Simulation	Efficiency via Experimental Test
1	$\eta = \dfrac{V_{out} \cdot I_{out}}{V_{out} \cdot I_{out} + Losses}$	$\eta = \dfrac{V_{out} \cdot I_{out}}{V_1 \cdot I_1 + V_2 \cdot I_2}$
2	$\eta = \dfrac{V_{out} \cdot I_{out} + |V_2 \cdot I_2|}{V_{out} \cdot I_{out} + |V_2 \cdot I_2| + Losses}$	$\eta = \dfrac{V_{out} \cdot I_{out} + |V_2 \cdot I_2|}{V_1 \cdot I_1}$
3	$\dfrac{V_{out} \cdot I_{out} + |V_1 \cdot I_1|}{V_{out} \cdot I_{out} + |V_1 \cdot I_1| + Losses}$	$\eta = \dfrac{V_{out} \cdot I_{out} + |V_1 \cdot I_1|}{V_2 \cdot I_2}$

From Table 4, it can be noticed that the thermal model in MATLAB/Simulink Plecs well represents the converter efficiency; indeed, the efficiency obtained with simulations is very close to the one obtained with experimental results, in all the tests.

Table 4. Experimental test current values and efficiency.

Test	I 1 (V1)	I 2 (V2)	I out (Vout)	Simulation Losses (of which for Conduction)	η Simulation	η Experimental
1	6 A (12 V)	0.5 A (22 V)	0.6 A (100 V)	21.5 W (1.2 W)	74%	72%
2	14 A (12 V)	−2 A (22 V)	0.6 A (100 V)	70 W (9 W)	60%	62%
3	−2 A (12 V)	5 A (22 V)	0.6 A (100 V)	27.6 W (1.5 W)	75%	76%

The 3SC efficiency is greater than 90% in the majority of the working points, as can be seen in Figure 8b, even if one has to note that converter efficiency shown in Table 4 is quite low. The cause of such a low efficiency is that, in the experimental tests, the converter did not work in conditions close to the rated working points. The evaluation of converter efficiency in the rated working points is carried out in Section 3 via simulation. The aim of the experimental test was to validate the Plecs model and, especially, the thermal model used in it. Moreover, an error of 2% between simulated and experimentally measured efficiency seems to be quite high. However, one has to consider the loss percent error. Indeed, an error of 2% on efficiency while efficiency is 96% corresponds to an error of 50% on the losses, which is not tolerable. In the results in Table 4, instead, the worst case is an error of 2% with 74% efficiency, which is equivalent to an error of about 7% on the losses, which is significantly low considering all the non-idealities of a real converter.

In Figures 12–14, simulation and experimental current waveforms in the three different tests are plotted. One can note that the simulation model well represents the converter behaviour. Indeed, very similar results were obtained from simulations and experimental tests.

Ultimately, the converter is well modelled since, on the one hand, the converter efficiency obtained via simulations is close to the measured one and also because the current waveforms are very similar in the simulations and experimental tests. All in all, the two approaches of validation described in Section 4.1 are both verified and the results obtained in Section 3 are therefore validated.

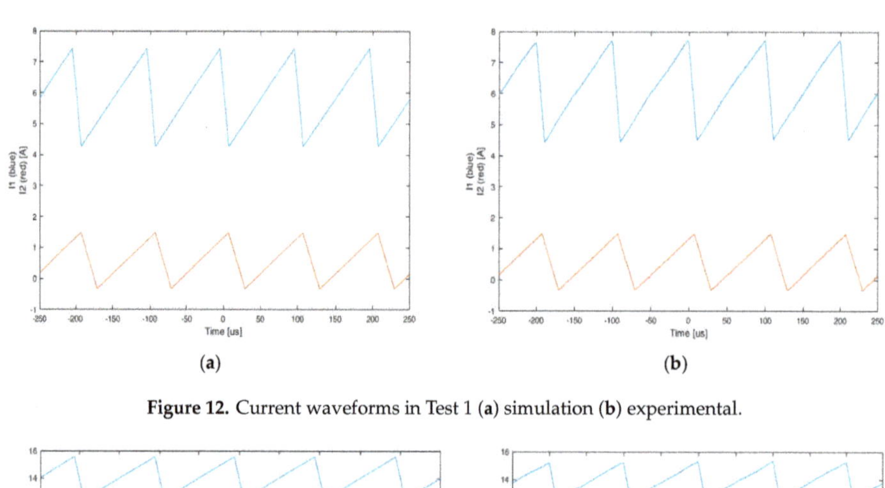

(a)   (b)

Figure 12. Current waveforms in Test 1 (**a**) simulation (**b**) experimental.

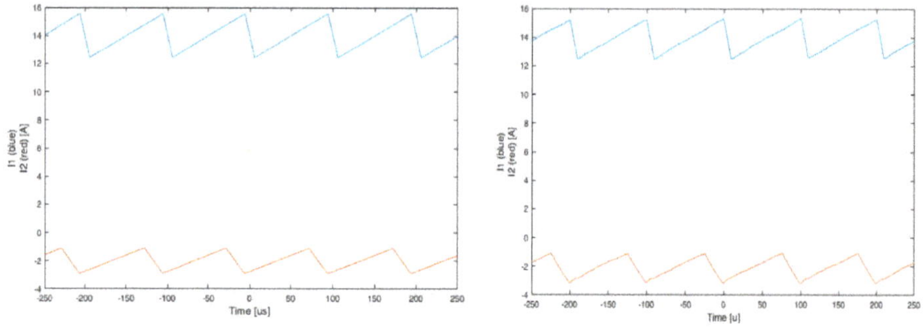

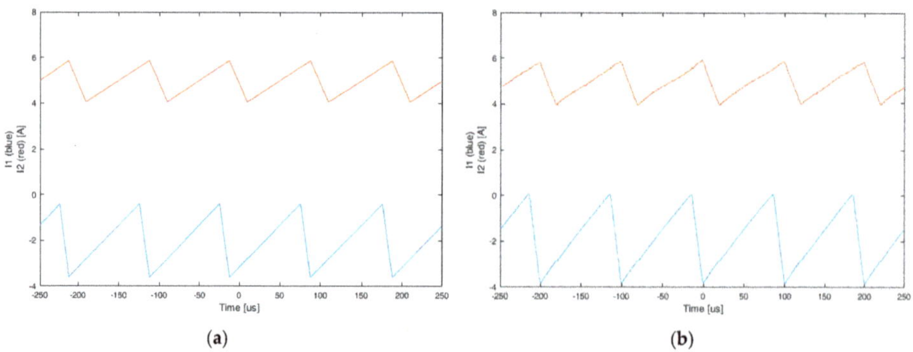

(a)   (b)

Figure 13. Current waveforms in Test 2 (**a**) simulation (**b**) experimental.

(a)   (b)

Figure 14. Current waveforms in Test 3 (**a**) simulation (**b**) experimental.

5. Conclusions

In this paper, the efficiency of an innovative three-switch DC-DC converter for hybrid vehicle applications was analysed. A refined model of the converter was created in MATLAB/Simulink Plecs environment. The model was used to evaluate the efficiency of both the traditional four-switch converter (i.e., two DC-DC converters in parallel connection) and the three-switch converter, in order to compare them. This study shows that not only does the three-switch converter use a reduced number of IGBTs, but the efficiency is significantly higher when the power flows of battery and supercapacitor are concordant (which is the most common condition).

The Plecs model was validated by comparing the results obtained via simulations with the results obtained with the experimental tests. The tests were performed on a prototype converter that has the same switches used to perform the simulations.

The experimental tests show an efficiency very close to the simulated one and moreover, the current waveforms obtained with the experimental tests are very similar to the ones obtained with simulations. The simulation model is therefore validated and the efficiency comparison between the four-switch converter and the three-switch converter can be considered very accurate.

Author Contributions: Conceptualization, M.M. and L.V.; Data curation, M.P.; Formal analysis, M.P.; Funding acquisition, M.M.; Investigation, M.P.; Methodology, L.V.; Resources, L.V.; Software, M.P.; Supervision, M.M.; Validation, L.V.; Writing—original draft, M.P.; Writing—review & editing, M.M. All authors have read and agreed to the published version of the manuscript.

Funding: This research received no external funding.

Conflicts of Interest: The authors declare no conflict of interest.

References

1. Burress, T.A.; Campbell, S.L.; Coomer, C.L.; Ayers, C.W.; Wereszczak, A.A.; Cunningham, J.P.; Marlino, L.D.; Seiber, L.E.; Lin, H.T. *Evaluation of the 2010 toyota prius hybrid synergy drive system. Power Electronics and Electric Machinery Research Facility*; Technical Report ORNL/TM2010/253; Oak Ridge National Laboratory (ORNL): Oak Ridge, TN, USA, 2011.

2. Zhang, X.; Peng, H.; Wang, H.; Ouyang, M. Hybrid lithium iron phosphate battery and lithium titanate battery systems for electric buses. *IEEE Trans. Veh. Technol.* **2018**, *67*, 956–965. [CrossRef]

3. Lanzarotto, D.; Marchesoni, M.; Passalacqua, M.; Prato, A.P.; Repetto, M. Overview of different hybrid vehicle architectures. *IFAC-PapersOnLine* **2018**, *51*, 218–222. [CrossRef]

4. Sinkaram, C.; Asirvadam, V.S.; Nor, N.B.M. Capacity study of lithium ion battery for hybrid electrical vehicle (HEV) a simulation approach. In Proceedings of the 2013 IEEE International Conference on Signal and Image Processing Applications, Melaka, Malaysia, 8–10 October 2013; pp. 112–116.

5. Xiong, R.; He, H.; Sun, F.; Zhao, K. Online estimation of peak power capability of li-ion batteries in electric vehicles by a hardware-in-loop approach. *Energies* **2012**, *5*, 1455–1469. [CrossRef]

6. Bonfiglio, A.; Lanzarotto, D.; Marchesoni, M.; Passalacqua, M.; Procopio, R.; Repetto, M. Electrical-loss analysis of power-split hybrid electric vehicles. *Energies* **2017**, *10*, 2142. [CrossRef]

7. Camara, M.B.; Gualous, H.; Gustin, F.; Berthon, A.; Dakyo, B. Dc/dc converter design for supercapacitor and battery power management in hybrid vehicle applications—polynomial control strategy. *IEEE Trans. Ind. Electron.* **2010**, *57*, 587–597. [CrossRef]

8. Golchoubian, P.; Azad, N.L. Real-time nonlinear model predictive control of a battery–supercapacitor hybrid energy storage system in electric vehicles. *IEEE Trans. Veh. Technol.* **2017**, *66*, 9678–9688. [CrossRef]

9. Liu, S.; Peng, J.; Li, L.; Gong, X.; Lu, H. A mpc based energy management strategy for battery-supercapacitor combined energy storage system of HEV. In Proceedings of the 2016 35th Chinese Control Conference (CCC), Chengdu, China, 27–29 July 2016; pp. 8727–8731.

10. Cao, J.; Emadi, A. A new battery/ultracapacitor hybrid energy storage system for electric, hybrid, and plug-in hybrid electric vehicles. *IEEE Trans. Power Electronics* **2012**, *27*, 122–132.

11. Sampietro, J.L.; Puig, V.; Costa-Castelló, R. Optimal sizing of storage elements for a vehicle based on fuel cells, supercapacitors, and batteries. *Energies* **2019**, *12*, 925. [CrossRef]

12. Cheng, L.; Wang, W.; Wei, S.; Lin, H.; Jia, Z. An improved energy management strategy for hybrid energy storage system in light rail vehicles. *Energies* **2018**, *11*, 423. [CrossRef]

13. Zhang, C.; Min, H.; Yu, Y.; Wang, D.; Luke, J.; Opila, D.; Saxena, S. Using cpe function to size capacitor storage for electric vehicles and quantifying battery degradation during different driving cycles. *Energies* **2016**, *9*, 903. [CrossRef]

14. Zhang, Q.; Deng, W. An adaptive energy management system for electric vehicles based on driving cycle identification and wavelet transform. *Energies* **2016**, *9*, 341. [CrossRef]

15. Odeim, F.; Roes, J.; Heinzel, A. Power management optimization of an experimental fuel cell/battery/supercapacitor hybrid system. *Energies* **2015**, *8*, 6302–6327. [CrossRef]

16. Marchesoni, M.; Savio, S. Reliability analysis of a fuel cell electric city car. In Proceedings of the 2005 European Conference on Power Electronics and Applications, Dresden, Germany, 11–14 September 2005; p. 10.

17. Passalacqua, M.; Lanzarotto, D.; Repetto, M.; Marchesoni, M. Advantages of using supercapacitors and silicon carbide on hybrid vehicle series architecture. *Energies* **2017**, *10*, 920. [CrossRef]

18. Zhao, Y.; Yao, J.; Zhong, Z.m.; Sun, Z.c. The research of powertrain for supercapacitor-based series hybrid bus. In Proceedings of the 2008 IEEE Vehicle Power and Propulsion Conference, Harbin, China, 3–5 September 2008; pp. 1–4.

19. Wu, W.; Partridge, J.; Bucknall, R. Development and evaluation of a degree of hybridisation identification strategy for a fuel cell supercapacitor hybrid bus. *Energies* **2019**, *12*, 142. [CrossRef]

20. Passalacqua, M.; Lanzarotto, D.; Repetto, M.; Vaccaro, L.; Bonfiglio, A.; Marchesoni, M. Fuel economy and ems for a series hybrid vehicle based on supercapacitor storage. *IEEE Trans. Power Electron.* **2019**, *34*, 9966–9977. [CrossRef]

21. Moreno, J.; Ortúzar, M.E.; Dixon, J.W. Energy-management system for a hybrid electric vehicle, using ultracapacitors and neural networks. *IEEE Trans. Ind. Electron.* **2006**, *53*, 614–623. [CrossRef]

22. Passalacqua, M.; Carpita, M.; Gavin, S.; Marchesoni, M.; Repetto, M.; Vaccaro, L.; Wasterlain, S. Supercapacitor storage sizing analysis for a series hybrid vehicle. *Energies* **2019**, *12*, 1759. [CrossRef]

23. Partridge, J.; Abouelamaimen, D.I. The role of supercapacitors in regenerative braking systems. *Energies* **2019**, *12*, 2683. [CrossRef]

24. Thounthong, P.; Chunkag, V.; Sethakul, P.; Davat, B.; Hinaje, M. Comparative study of fuel-cell vehicle hybridization with battery or supercapacitor storage device. *IEEE Trans. Veh. Technol.* **2009**, *58*, 3892–3904. [CrossRef]

25. Fadil, H.E.; Giri, F.; Guerrero, J.M.; Tahri, A. Modeling and nonlinear control of a fuel cell/supercapacitor hybrid energy storage system for electric vehicles. *IEEE Trans. Veh. Technol.* **2014**, *63*, 3011–3018. [CrossRef]

26. Passalacqua, M.; Lanzarotto, D.; Repetto, M.; Marchesoni, M. Conceptual design upgrade on hybrid pow ertrains resulting from electric improv ements. *Int. J. Transp. Dev. Integr.* **2018**, *2*, 146–154. [CrossRef]

27. Zhang, R.; Tao, J.; Zhou, H. Fuzzy optimal energy management for fuel cell and supercapacitor systems using neural network based driving pattern recognition. *IEEE Trans. Fuzzy Syst.* **2019**, *27*, 45–57. [CrossRef]

28. Lanzarotto, D.; Passalacqua, M.; Repetto, M. Energy comparison between different parallel hybrid vehicles architectures. *Int. J. Energy Prod. Manag.* **2017**, *2*, 370–380. [CrossRef]

29. Lukic, S.M.; Wirasingha, S.G.; Rodriguez, F.; Cao, J.; Emadi, A. Power management of an ultracapacitor/battery hybrid energy storage system in an HEV. In Proceedings of the 2006 IEEE Vehicle Power and Propulsion Conference, Windsor, UK, 6–8 September 2006; pp. 1–6.

30. Onar, O.C.; Khaligh, A. A novel integrated magnetic structure based dc/dc converter for hybrid battery/ultracapacitor energy storage systems. *IEEE Trans. Smart Grid* **2012**, *3*, 296–307. [CrossRef]

31. Shen, J.; Khaligh, A. A supervisory energy management control strategy in a battery/ultracapacitor hybrid energy storage system. *IEEE Trans. Transp. Electrif.* **2015**, *1*, 223–231. [CrossRef]

32. Marchesoni, M.; Vacca, C. New dc–dc converter for energy storage system interfacing in fuel cell hybrid electric vehicles. *IEEE Trans. Power Electron.* **2007**, *22*, 301–308. [CrossRef]

33. Carpaneto, M.; Ferrando, G.; Marchesoni, M.; Vacca, C. The average switch model of a new double-input dc/dc boost converter for hybrid fuel-cell vehicles. In Proceedings of the IEEE International Symposium on Industrial Electronics(ISIE 2005), Dubrovnik, Croatia, 20–23 June 2005; pp. 601–607.

34. Carpaneto, M.; Ferrando, G.; Marchesoni, M.; Savio, S. A new conversion system for the interface of generating and storage devices in hybrid fuel-cell vehicles. In Proceedings of the IEEE International Symposium on Industrial Electronics (ISIE 2005), Dubrovnik, Croatia, 20–23 June 2005; pp. 1477–1482.

35. Barabino, G.; Carpaneto, M.; Comacchio, L.; Marchesoni, M.; Novella, G. A new energy storage and conversion system for boat propulsion in protected marine areas. In Proceedings of the 2009 International Conference on Clean Electrical Power, Capri, Italy, 9–11 June 2009; pp. 363–369.

36. Chen, G.; Jin, Z.; Deng, Y.; He, X.; Qing, X. Principle and topology synthesis of integrated single-input dual-output and dual-input single-output dc–dc converters. *IEEE Trans. Ind. Electron.* **2018**, *65*, 3815–3825. [CrossRef]

37. Wai, R.; Lin, C.; Chen, B. High-efficiency dc–dc converter with two input power sources. *IEEE Trans. Power Electron.* **2012**, *27*, 1862–1875. [CrossRef]
38. Liu, Y.; Chen, Y. A systematic approach to synthesizing multi-input dc–dc converters. *IEEE Trans. Power Electron.* **2009**, *24*, 116–127. [CrossRef]
39. Ganjavi, A.; Ghoreishy, H.; Ahmad, A.A. A novel single-input dual-output three-level dc–dc converter. *IEEE Trans. Ind. Electron.* **2018**, *65*, 8101–8111. [CrossRef]
40. Semikron. Available online: https://www.semikron.com (accessed on 31 December 2019).

Article

A Power Loss Decrease Method Based on Finite Set Model Predictive Control for a Motor Emulator with Reduced Switch Count

Rui Qin, Chunhua Yang, Hongwei Tao *, Tao Peng, Chao Yang and Zhiwen Chen

School of Automation, Central South University, Changsha 410083, China; ruiqin@csu.edu.cn (R.Q.); ychh@csu.edu.cn (C.Y.); pandtao@csu.edu.cn (T.P.); chaoyang@csu.edu.cn (C.Y.); zhiwen.chen@csu.edu.cn (Z.C.)
* Correspondence: hongwei.tao@csu.edu.cn; Tel.: +86-151-1113-1734

Received: 7 October 2019; Accepted: 4 December 2019; Published: 6 December 2019

Abstract: This paper presents a power loss decrease method based on finite set model predictive control (FSMPC) with delay compensation for a motor emulator with reduced switch count. Specifically, the topology and mathematical model of the proposed motor emulator with reduced switch count are firstly built. Secondly, in light of given instructions, the normal or fault reference current of the motor emulator is set by a reference current setter. Then delay compensation is applied for the predictive current model to calculate the current residual generated by each switch control signal, and the current tracking performance under actions of two adjacent switch control signals is evaluated for each sector. Finally, a switch power loss objective function is defined, then the two adjacent switch control signals that generate the lowest switch power loss are selected for the next second instant, which minimizes the power loss of the motor emulator with ensuring satisfied current tracking performance. Simulation and experimental results show the feasibility and effectiveness of the proposed method.

Keywords: motor emulator; power loss; current tracking; finite set model predictive control

---

1. Introduction

As key power equipment, motors are widely used in various applications such as defense military, industrial production, and rail transit [1–3]. In the past few decades, the research about motor operating on fault state and condition suddenly alteration have become hot topics in high voltage and power applications [4,5]. The physical motors in the experiments of motor faults are usually damaged, which are not beneficial for performing experiments repeatedly and the cost of the experiments is high. Moreover, the drive shafts of the two physical motors in the experiments of speed with sudden alteration are connected together by adopting the output torque of one motor drive system as the load torque of another motor drive system. The speed of physical motors in the experiments is difficult to sudden change. Stator current of a motor is one of the most reliable electrical signals that reflecting specific features of motor operating on fault state and condition suddenly alteration [6,7]. Motor emulator was proposed by the British scholar H.J. Slater for the first time [8,9], in which the port characteristics of load current are the same as that of stator current of physical motor by controlling power electronic components. So various experiments that motor operating on fault state and condition sudden alteration could be conducted by motor emulator instead of physical motor. It is more secure and economical to perform these experiments by motor emulator compared with a physical motor.

The core of the motor emulator is the load current tracking stator current of motor [10,11]. During several years, motor emulators with plenty of electronic components are employed to simulate port characteristics of stator current [12,13]. In [14,15], the three-level converter of motor emulator is constituted by twelve Insulated Gate Bipolar Translators (IGBTs) and eighteen antiparallel diodes.

Other motor emulators can be found in [16,17], in which the two-level converter is composed of six IGBTs and six antiparallel diodes. In [18], linear inverter structure is applied to reduce the harmonics of load current in the motor emulator, but the cost is double the number of electronic components compare with traditional two-level inverter. The load model of motor emulator proposed in [19] is an inductor-capacitor-inductor (LCL) filter, which improves current ripple induced by high-frequency switch, but increase reactive power consumption. There are many electronic components are used in these above motor emulators, which increase volume, weight, cost and directly incur excessive switch power loss [20]. Therefore, the power loss of the motor emulator caused by electronic components has become an important issue to be considered.

The load current tracking control of the motor emulator is mostly realized by PI modulation in the present motor emulators. In [21], the dynamic mathematical model of asynchronous motor is established to realize real-time calculation of stator current and tracking control of load current. A wind generator emulator is proposed in [22], which simulates static characteristics of stator current at different wind speeds by controlling a converter. But the steady-state error between stator current and load current is hardly eliminated by PI modulation. A current tracking method based on FSMPC is presented in [23], which improves the accuracy and rapidity of current tracking when compare to PI modulation. In [24], an optimized current control method with delay compensation is proposed to overcome harmonic current in distributed generation converters, and reference current at the next instant is tracked by controlling one switch control signal to generate one basic space voltage vector during every sampling period. However, these presented motor emulators only pursuit current tracking performance as the control object, while power loss of motor emulators is severely neglected. For motor emulators, especially applying to high voltage and power systems, various crucial factors such as on-off time and frequency of switch, can affect switch power loss, which directly linked with service behavior and life of switch. To the best of our knowledge, there is a lack of work about the power loss control method of motor emulator.

Motivated by the above discussion, a power loss decrease method for a reduced switch count motor emulator is proposed in this paper, which is realized by FSMPC with delay compensation. At the first, a topology of the motor emulator is proposed, in which the switch count of converter is reduced. Secondly, a power loss decrease method based on FSMPC is presented, an objective function is designed to select two adjacent switch control signals that generating lowest switch power loss on the premise of ensuring current tracking performance. Finally, the delay compensation is applied to improve current tracking accuracy, which is realized by calculating the current residual between predicted load current and reference current at the next second instant.

The rest of the paper is organized as follows. The topology and modeling of the motor emulator with reduced switch count are illustrated in Section 2. The power loss decrease method based on FSMPC with delay compensation is elaborated upon in Section 3. Section 4 is dedicated to present experimental results and discussion of switch power loss and performance of current tracking of motor emulator. Finally, the conclusions of this paper are reviewed in Section 5.

2. Motor Emulator with Reduced Switch Count

2.1. Topology of Motor Emulator

The topology of the motor emulator with reduced switch count is presented in Figure 1. It consists of the three-phase four-switch converter, coupled load network, motor model, the reference current setter and predictive controller. In three-phase four-switch converter, b and c phase are constructed of four active switches and four antiparallel diodes, and a phase is composed of two dc-link capacitors. The coupled load network consists of three-phase resistance-inductance load, where each phase includes a coupled resistance R and a coupled inductance L. The left and right ports of coupled load networks are connected to three-phase midpoint of three-phase four-switch converter and voltage source. The motor model is a squirrel cage asynchronous motor model, which provides a normal three-phase stator current signal. According to given instruction, reference current is generated by reference current setter. When normal instruction is given, a three-phase stator current is adopted directly as reference current signal. When fault instruction is given, fault reference current signal is generated by conditioning three-phase stator current signal and specific fault signal, which is detailed in [25]. By calculating the current residual between reference current signal and predictive value of load current, the best applicable switch control signal is selected by predictive controller as switch control signal of three-phase four-switch converter for the next second instant.

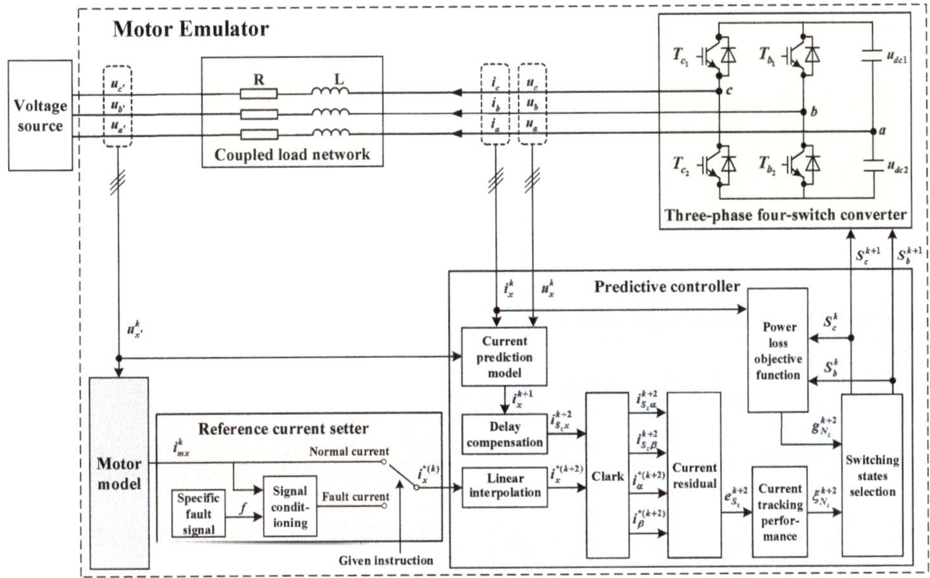

Figure 1. The topology of motor emulator with reduced switch count.

2.2. Modeling of Motor Emulator

Three-phase output voltages of three-phase four-switch converter can be expressed by switching state as:

$$\begin{bmatrix} u_a \\ u_b \\ u_c \end{bmatrix} = u_{dc} \begin{bmatrix} 0.5 \\ S_b \\ S_c \end{bmatrix} \tag{1}$$

where u_a, u_b and u_c are output voltage values of a, b and c phases of three-phase four-switch converter, u_{dc} is input voltage of three-phase four-switch converter, which is the sum of voltages of split capacitors u_{dc1} and u_{dc2} at the dc side. S_b and S_c are switching states of b and c phases in the three-phase four-switch converter, each of them has two logical states 0 and 1 [26,27]. When $S_X = 0$, T_{X_1} is off and T_{X_2} is on. When $S_X = 1$, T_{X_1} is on and T_{X_2} is off , $X = b, c$.

In addition, the switch control signal of three-phase four-switch converter can be defined as S_z, which can be given by $S_z = S_b S_c$, and z represents the selected number of S_z. The basic space voltage vector corresponding to switch control signal S_z is defined as V_z, which is synthesized by three-phase output voltages u_a, u_b and u_c in three-phase abc static coordinate frame. The basic space voltage vector V_z is described as (2) and the spatial distribution on the abc static coordinate frame is shown Figure 2. The relationship of switch control signal S_z, three-phase output voltages u_a, u_b and u_c, and basic space voltage vectors V_z are listed in Table 1.

$$V_z = u_a + u_b + u_c \tag{2}$$

Table 1. Switch control signals, three-phase output voltages and basic space voltage vectors.

Selected Number	Switch Control Signals	Three-Phase Output Voltages			Basic Space Voltage Vectors
z	$S_z = S_b S_c$	u_a	u_b	u_c	V_z
1	00	$\frac{1}{2}u_{dc}e^{j0}$	0	0	$\frac{1}{2}u_{dc}e^{j0}$
2	01	$\frac{1}{2}u_{dc}e^{j0}$	0	$u_{dc}e^{j\frac{4\pi}{3}}$	$\frac{\sqrt{3}}{2}u_{dc}e^{j\frac{3\pi}{2}}$
3	11	$\frac{1}{2}u_{dc}e^{j0}$	$u_{dc}e^{j\frac{2\pi}{3}}$	$u_{dc}e^{j\frac{4\pi}{3}}$	$\frac{1}{2}u_{dc}e^{j\pi}$
4	10	$\frac{1}{2}u_{dc}e^{j0}$	$u_{dc}e^{j\frac{2\pi}{3}}$	0	$\frac{\sqrt{3}}{2}u_{dc}e^{j\frac{\pi}{2}}$

In *abc* static coordinate frame, every work period of the three-phase four-switch converter is divided into four sectors by four basic space voltage vectors. The sector formed by two adjacent basic space voltage vectors V_z and V_{z+1} is named as N_z, where $(z, z+1)$ can be $(1,2)$, $(2,3)$, $(3,4)$ and $(4,1)$, as shown in Figure 2.

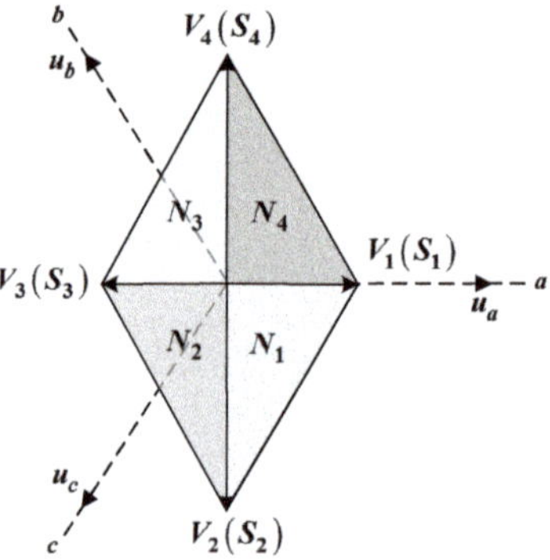

Figure 2. Spatial distribution diagram of basic space voltage vectors and sectors.

The voltage state equation of the motor emulator can be expressed as:

$$u_x = u_{x'} + Ri_x + L\frac{di_x}{dt}, \quad x = a,b,c; \; x' = a',b',c'. \tag{3}$$

where u_x is output voltage of x phase of three-phase four-switch converter, $u_{x'}$ is output voltage of x' phase of voltage source. i_x is load current of x phase of coupled load network. R and L are the resistance and inductance of coupled load network respectively.

Then, predictive current at the $(k+1)$-th instant can be obtained by discretizing Equation (3):

$$i_x^{k+1} = (1 - \frac{RT_s}{L})i_x^k + \frac{T_s}{L}(u_x^k - u_{x'}^k) \tag{4}$$

where k is sampling instant and $k = 1, 2, 3 \cdots$. T_s is the sampling period. i_x^{k+1} is predictive current at the $(k+1)$-th instant. i_x^k is sampling of i_x at the k-th instant. u_x^k and $u_{x'}^k$ are samplings of u_x and $u_{x'}$ at the k-th instant respectively.

3. Power Loss Decrease Method Based on FSMPC with Delay Compensation

3.1. Current Tracking Performance

The theory of ideal FSMPC is shown in Figure 3a. Reference stator current i_x^* is obtained and actual load current i_x^k are sampled at the k-th instant, and the optimal switch control signal is determined during the k-th sampling period and applied to the system at the k-th instant, then load current will reach the expected value at the $(k + 1)$-th instant. However, digital process system needs time to perform algorithm, as shown in Figure 3b. The sampling is accomplished at the k-th instant, but the optimal switch control signal is applied to system after t_d delay, which results in error between actual load current and expected value at the $(k + 1)$-th instant. Thus, the accuracy and rapidity of current tracking is partly decreased, especially for motor operating on fault state and condition suddenly alteration. Therefore, the delay compensation is adopted to regulate action time of the optimal switch control signal, as shown in Figure 3c. Sampling is completed at the k-th instant and calculating the optimal switch control signal during the the k-th sampling period, and applied to the system at the $(k + 1)$-th instant, then load current will reach the expected value at the $(k + 2)$-th instant. The delay compensation not only makes up for computation time of the algorithm, but also effectively raises the performance of current tracking for motor operating on fault state and condition suddenly alteration.

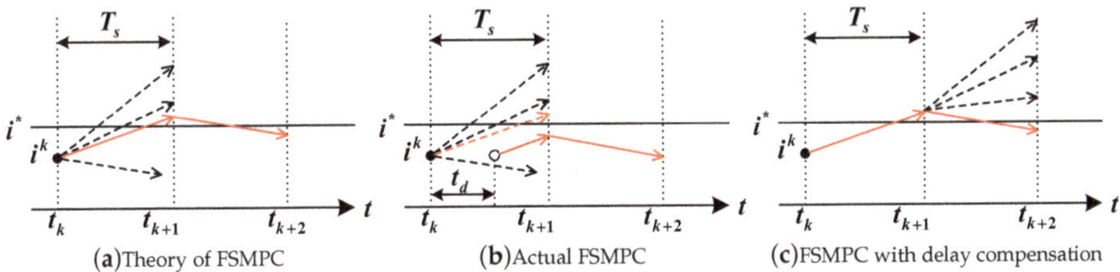

Figure 3. Principle of delay compensation.

According to principle of delay compensation [24], all switch control signals are used to calculate the predictive current at the $(k + 2)$-th instant:

$$i_{S_z x}^{k+2} = (1 - \tfrac{RT_s}{L})i_x^{k+1} + \tfrac{T_s}{L}(u_{S_z x}^{k+1} - u_{x'}^k), \qquad z = 1,2,3,4. \tag{5}$$

where $i_{S_z x}^{k+2}$ is predictive current at the $(k + 2)$-th instant corresponding to switch control signal S_z. $u_{S_z x}^{k+1}$ predictive voltage at the $(k + 1)$-th instant corresponding to switch control signal S_z. Considering that the change of voltage source during one sampling period is not obvious, the sampling of $u_{x'}$ at the $(k + 1)$th instant is approximatively equal to $u_{x'}^k$.

After Clark coordinate transformation, $i_{S_z x}^{k+2}$ can be converted as:

$$\begin{bmatrix} i_{S_z \alpha}^{k+2} \\ i_{S_z \beta}^{k+2} \end{bmatrix} = \sqrt{\frac{2}{3}} \begin{bmatrix} 1 & -\frac{1}{2} & -\frac{1}{2} \\ 0 & \frac{\sqrt{3}}{2} & -\frac{\sqrt{3}}{2} \end{bmatrix} \begin{bmatrix} i_{S_z a}^{k+2} \\ i_{S_z b}^{k+2} \\ i_{S_z c}^{k+2} \end{bmatrix} \tag{6}$$

where $i_{S_z \alpha}^{k+2}$ and $i_{S_z \beta}^{k+2}$ are the values of predictive current $i_{S_z x}^{k+2}$ in $\alpha\beta$ stationary frame.

Then, the current residual function at the $(k + 2)$-th instant can be defined as:

$$e_{S_z}^{k+2} = |i_{S_z \alpha}^{k+2} - i_\alpha^{*(k+2)}|^2 + |i_{S_z \beta}^{k+2} - i_\beta^{*(k+2)}|^2 \tag{7}$$

where $e_{S_z}^{k+2}$ is current residual at the $(k + 2)$-th instant corresponding to switch control signal S_z. $i_\alpha^{*(k+2)}$ and $i_\beta^{*(k+2)}$ are the reference currents at the $(k + 2)$-th instant in $\alpha\beta$ stationary frame. The $i_\alpha^{*(k+2)}$ and $i_\beta^{*(k+2)}$ can be constructed by the reference currents at the $(k-1)$th, the k-th and the $(k + 1)$-th instants according to linear interpolation theorem:

$$\begin{cases} i_\alpha^{*(k+2)} = 3i_\alpha^{*(k+1)} - 3i_\alpha^{*(k)} + i_\alpha^{*(k-1)} \\ i_\beta^{*(k+2)} = 3i_\beta^{*(k+1)} - 3i_\beta^{*(k)} + i_\beta^{*(k-1)} \end{cases} \tag{8}$$

where, when k is equal to 1, the values of $i_\alpha^{*(0)}$ and $i_\beta^{*(0)}$ are set to equal the values of $i_\alpha^{*(1)}$ and $i_\beta^{*(1)}$ respectively.

In the traditional method of current tracking, only one basic space voltage vector V_z is generated by controlling a switch control signal S_z during every sampling period. When the traditional strategy is adopted for the current tracking control of the motor emulator, the error between load current and stator current is larger and current tracking accuracy is lower because of the finiteness and space distribution fixity of basic space voltage vectors. In the method of current tracking in this paper, there are two adjacent basic space voltage vectors V_z and V_{z+1} are generated by controlling two adjacent switch control signals S_z and S_{z+1} during every sampling period. By allocating the action time proportion of the two space voltage vectors during one period, the error between load current and stator current can be minimized.

Since the sum of action time of two adjacent switch control signals S_z and S_{z+1} is constant T_s, and the action time of each switch control signal is inversely proportional to the current residual generated by it, larger current residual leads to smaller action time. Thus the action time of two adjacent switch control signals S_z and S_{z+1} for the sector N_z are derived as:

$$
\begin{cases}
d_{N_z S_z}^{k+2} = \dfrac{e_{S_{z+1}}^{k+2}}{e_{S_z}^{k+2}+e_{S_{z+1}}^{k+2}} T_s \\[2mm]
d_{N_z S_{z+1}}^{k+2} = \dfrac{e_{S_z}^{k+2}}{e_{S_z}^{k+2}+e_{S_{z+1}}^{k+2}} T_s \\[2mm]
T_s = d_{N_z S_z}^{k+2} + d_{N_z S_{z+1}}^{k+2}
\end{cases}
\tag{9}
$$

where $d_{N_z S_z}^{k+2}$ and $d_{N_z S_{z+1}}^{k+2}$ are action times two switch control signals S_z and S_{z+1} of sector N_z for the $(k+2)$-th instant.

The current tracking performance function for the sector N_z at the $(k+2)$-th instant can be designed as:

$$
\zeta_{N_z}^{k+2} = 1 - \frac{T_s}{4 I_{mx}^*} \left(\frac{e_{S_{z+1}}^{k+2}}{d_{N_z S_z}^{k+2}} + \frac{e_{S_z}^{k+2}}{d_{N_z S_{z+1}}^{k+2}} \right)
\tag{10}
$$

where $\zeta_{N_z}^{k+2}$ expresses current tracking performance at the $(k+2)$-th instant under the action of the two switch control signals S_z and S_{z+1} for sector N_z. I_{mx}^* is peak value of three-phase reference current i_x^*. The value of $\zeta_{N_z}^{k+2}$ is between 0 to 1. When $\zeta_{N_z}^{k+2} = 1$, current tracking performance is optimal, $\zeta_{N_z}^{k+2}$ is dropping gradually as the current tracking performance decreases.

3.2. Power Loss Minimization

The switch power loss generated by switch control signal S_z at the $(k+2)$-th instant is defined as $P_{S_z}^{k+2}$, which can be given by:

$$
P_{S_z}^{k+2} = P_{S_z b_1}^{k+2} + P_{S_z b_2}^{k+2} + P_{S_z c_1}^{k+2} + P_{S_z c_2}^{k+2}
\tag{11}
$$

where, $P_{S_z b_1}^{k+2}$, $P_{S_z b_2}^{k+2}$, $P_{S_z c_1}^{k+2}$ and $P_{S_z c_2}^{k+2}$ are switch power losses of switches T_{b_1}, T_{b_2}, T_{c_1} and T_{c_2} at the $(k+2)$-th instant under the action of switch control signal S_z, they can be expressed as follows:

$$
\begin{cases}
P_{S_z X_1}^{k+2} = \delta_X^k [S_{X_1}^k (|i_X^k| \times u_{S_z X_1}^{k+1} \times T_s) + L_{S_z X_1}^{k+1}] \\[1mm]
P_{S_z X_2}^{k+2} = \delta_X^k [S_{X_2}^k (|i_X^k| \times u_{S_z X_2}^{k+1} \times T_s) + L_{S_z X_2}^{k+1}]
\end{cases}
, \quad X = b, c.
\tag{12}
$$

where, i_X^k is sampling of i_X at the k-th instant. δ_X^k is flag of i_X^k, if $i_X^k \geqslant 0$, $\delta_X^k = 1$; if $i_X^k < 0$, $\delta_X^k = 0$. $S_{X_1}^k$ and $S_{X_2}^k$ are pulse signals of T_{X_1} and T_{X_2} at the k-th instant, $S_{X_1}^k = 1$ when T_{X_1} on, and $S_{X_1}^k = 0$ when T_{X_1} off, which also beseem to $S_{X_2}^k$ and T_{X_2}. $u_{S_z X_1}^{k+1}$ and $u_{S_z X_2}^{k+1}$ are on-state voltage drops of T_{X_1} and T_{X_2}

at the $(k + 1)$-th instant corresponding to switch control signal S_z, they are assumed to be constant values, which can be found in [25]. $L_{S_zX_1}^{k+1}$ and $L_{S_zX_2}^{k+1}$ are on-off losses of T_{X_1} and T_{X_2} at the $(k + 1)$-th instant corresponding to switch control signal S_z, they can be given by:

$$L_{S_zX_m}^{k+1} = \begin{cases} E_{on}, & S_{zX_m}^{k+1} - S_{X_m}^k = 1 \\ 0, & S_{zX_m}^{k+1} - S_{X_m}^k = 0 \quad , \quad X = b,c; \; m = 1,2. \\ E_{off}, & S_{zX_m}^{k+1} - S_{X_m}^k = -1 \end{cases} \tag{13}$$

where, E_{on} and E_{off} are on and off power loss of switch, they are assumed to be constant values, which can be found in [25]. $S_{zX_m}^{k+1}$ is pulse signal of T_{X_m} at the $(k + 1)$-th instant corresponding to switch control signal S_z.

The switch power loss objective function is designed as:

$$g_{N_z}^{k+2} = \frac{P_{S_z}^{k+2} + P_{S_{z+1}}^{k+2}}{8} \tag{14}$$

where, $g_{N_z}^{k+2}$ is average switch power loss at the $(k + 2)$-th instant under the action of the two adjacent switch control signals S_z and S_{z+1} for sector N_z. During the the k-th sampling period, two adjacent switch control signals of each sectors are orderly used to calculate value of switch power loss objective function $g_{N_z}^{k+2}$.

The switching states of b and c phases can be obtained according to Equation (15):

$$g[S_b^{k+1}, S_c^{k+1}] = \min_{\substack{g_{N_z}^{k+2} \in (0.95,1] \\ z = 1,2,3,4}} \{g_{N_z}^{k+2}\} \tag{15}$$

where, S_b^{k+1} and S_c^{k+1} are switching states of b and c phases in the three-phase four-switch converter at the $(k + 1)$-th instant. On the condition that current tracking performance greater than 95%, the switches that generating lowest switch power loss are selected to operate during every sampling period, then switch power loss can be reduced.

The flowchart of proposed power loss decrease method based on FSMPC with delay compensation is shown in Figure 4.

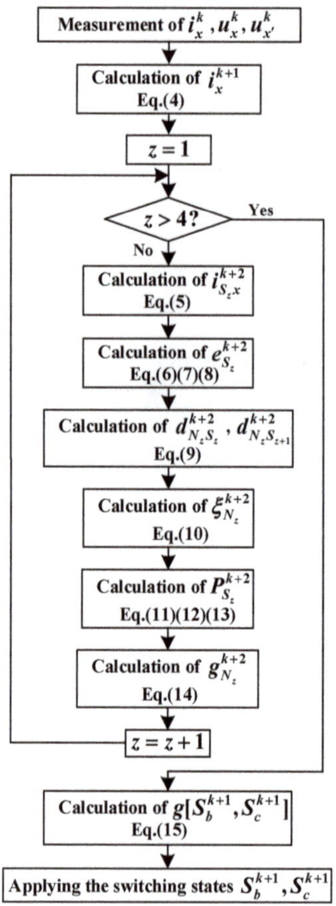

Figure 4. Flowchart of the FSMPC with delay compensation.

4. Experimental Results

In this section, the power loss decrease method for motor emulator based on reducing switch count and FSMPC with delay compensation is verified on a hardware-in-the-loop platform as shown in Figure 5. The corresponding component parameters are indicated in Table 2. The platform consists of a physical controller, real-time simulator, and PC. Physical controller adopts TMS320F28335 control chip with high processing capacity and rich interface resources to realize real-time control of voltage source. The real-time simulator includes DS1007CPU board and DS5203FPGA board, the former is used for real-time calculation of reference current setter and predictive controller, and the latter is used for real-time simulation of motor model, coupled load network and three-phase four-switch converter. The PC collects real-time data from model by real-time simulation software, and monitors running state of observation point.

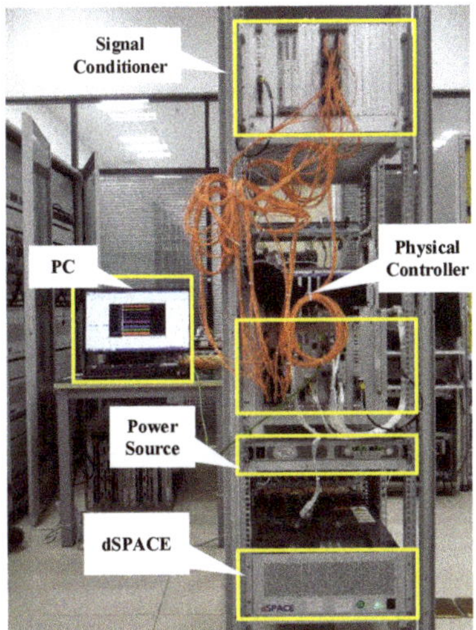

Figure 5. Experimental platform.

Table 2. Parameters of the experimental platform.

	Parameters	High Power Motor	Low Power Motor
Motor model	Rated power	300 kW	100 W
	Stator inductance	0.142 H	0.021 H
	Rotor inductance	0.130 H	0.018 H
	Stator resistor	0.144 Ω	0.014 Ω
	Rotor resistor	0.142 HΩ	0.012 Ω
Coupled load network	Load resistor	1 Ω	0.05 Ω
	Load inductance	0.004 H	0.00013 H
Frequency	Sampling frequency	20 kHz	20 kHz
	Switching frequency	10 kHz	10 kHz

4.1. Motor Emulators with Different Switch Counts

On the condition that high power motor is operating with reference load current set 60 A/20 Hz and low power motor is operating with reference load current set 7 A/30 Hz, respectively. The comparison of current tracking performance and switch power loss for traditional six-switch motor emulator and proposed four-switch motor emulator without power loss minimization control are displayed in this section.

For the high power motor, the experimental waveforms of a-phase reference current, a-phase load current of six-switch motor emulator, and a-phase load current of four-switch motor emulator are shown in Figure 6a. Based on these conditions, the experimental waveforms of average three-phase current residual of between reference current and load current of six-switch and four-switch motor emulator are shown in Figure 6b, respectively. The experimental waveforms of average three-phase current THD (total harmonic distortion) of reference current, load current of six-switch and four-switch motor emulator are shown in Figure 6c, respectively. The switch power loss during entire simulation period of six-switch and four-switch motor emulator are shown in Table 3. Similarly, the experimental waveforms for the low power motor are shown in Figure 7a–c and Table 4.

Table 3. Switch power losses of traditional six-switch and proposed four-switch motor emulator for high power motor.

	Switch Power Loss (KJ)								σ (%)
	p_{a1}	p_{a2}	p_{b1}	p_{b2}	p_{c1}	p_{c2}	p_{sum}	p_{ave}	
Traditional six-switch	21.09	18.62	19.16	20.43	20.27	19.75	119.32	19.89	23.8
Proposed four-switch	—	—	22.40	23.36	23.25	21.83	90.84	22.71	

Table 4. Switch power losses of traditional six-switch and proposed four-switch motor emulator for low power motor.

	Switch Power Loss (J)								σ (%)
	p_{a1}	p_{a2}	p_{b1}	p_{b2}	p_{c1}	p_{c2}	p_{sum}	p_{ave}	
Traditional six-switch	34.74	35.22	35.46	34.87	35.91	34.05	210.25	35.04	20.7
Proposed four-switch	—	—	41.75	42.16	41.84	40.97	166.72	41.68	

As shown in Figure 6a, comparing to the zero-crossing time of reference current, the zero-crossing time of six-switch motor emulator is delayed by about 30 us, which is expressed as t_n^T, and the zero-crossing time of four-switch motor emulator is delayed by about 35 us, which is denoted as t_n^P. From Figure 6b, average three-phase current residual of six-switch motor emulator is about 1.1A, current tracking accuracy is 98.2% according to (16). Average three-phase current residual of four-switch motor emulator is about 1.3 A, current tracking accuracy is 97.8% according to (17). In Figure 6c, the average THD of three-phase reference current is 1.30%, and average THD of three-phase load current of six-switch and four-switch motor emulator are about 1.7% and 1.80%, respectively. From Table 3, comparing to a traditional six-switch motor emulator, although the average switch power loss has increased, the sum of switch power loss decreases 23.81%, according to (18).

$$\varepsilon^T = \frac{i^R - \Delta i^{RT}}{i^R} \times 100\%$$ (16)

where ε^T is current tracking accuracy of traditional six-switch motor emulator, i^R is reference current, Δi^{RT} is average three-phase current residual between reference current and load current of traditional six-switch motor emulator.

$$\varepsilon^P = \frac{i^R - \Delta i^{RP}}{i^R} \times 100\%$$ (17)

where ε^P is current tracking accuracy of proposed four-switch motor emulator, i^R is reference current, Δi^{RP} is average three-phase current residual between reference current and load current of proposed four-switch motor emulator.

$$\sigma = \frac{P_{sum}^T - P_{sum}^P}{P_{sum}^T} \times 100\%$$ (18)

where σ is decrease percent of total switch power loss, P_{sum}^T is sum of switch power loss of traditional six-switch motor emulator, P_{sum}^P is sum of switch power loss of proposed four-switch motor emulator.

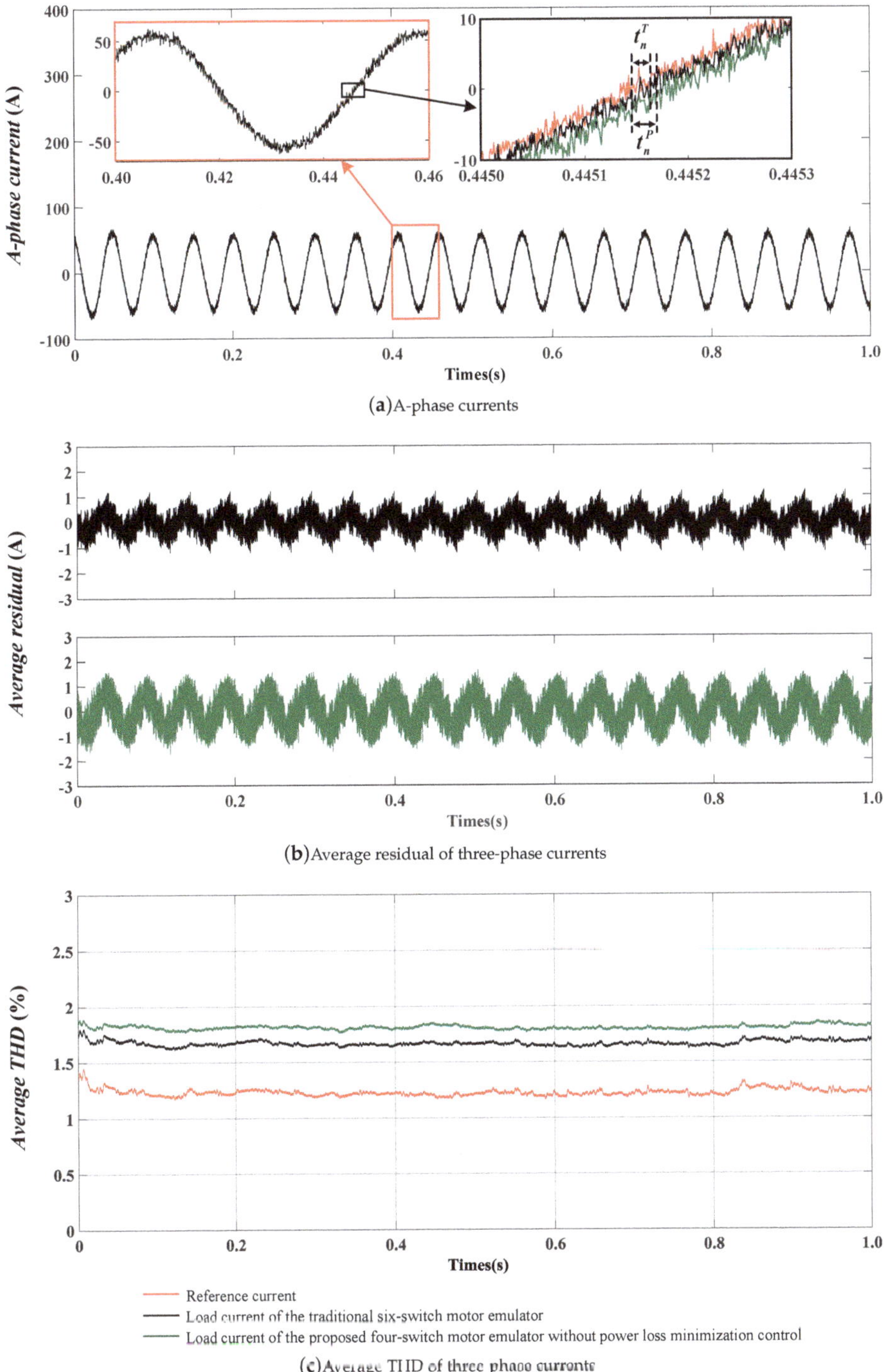

Figure 6. Waveforms of traditional six-switch and proposed four-switch motor emulator for high power motor.

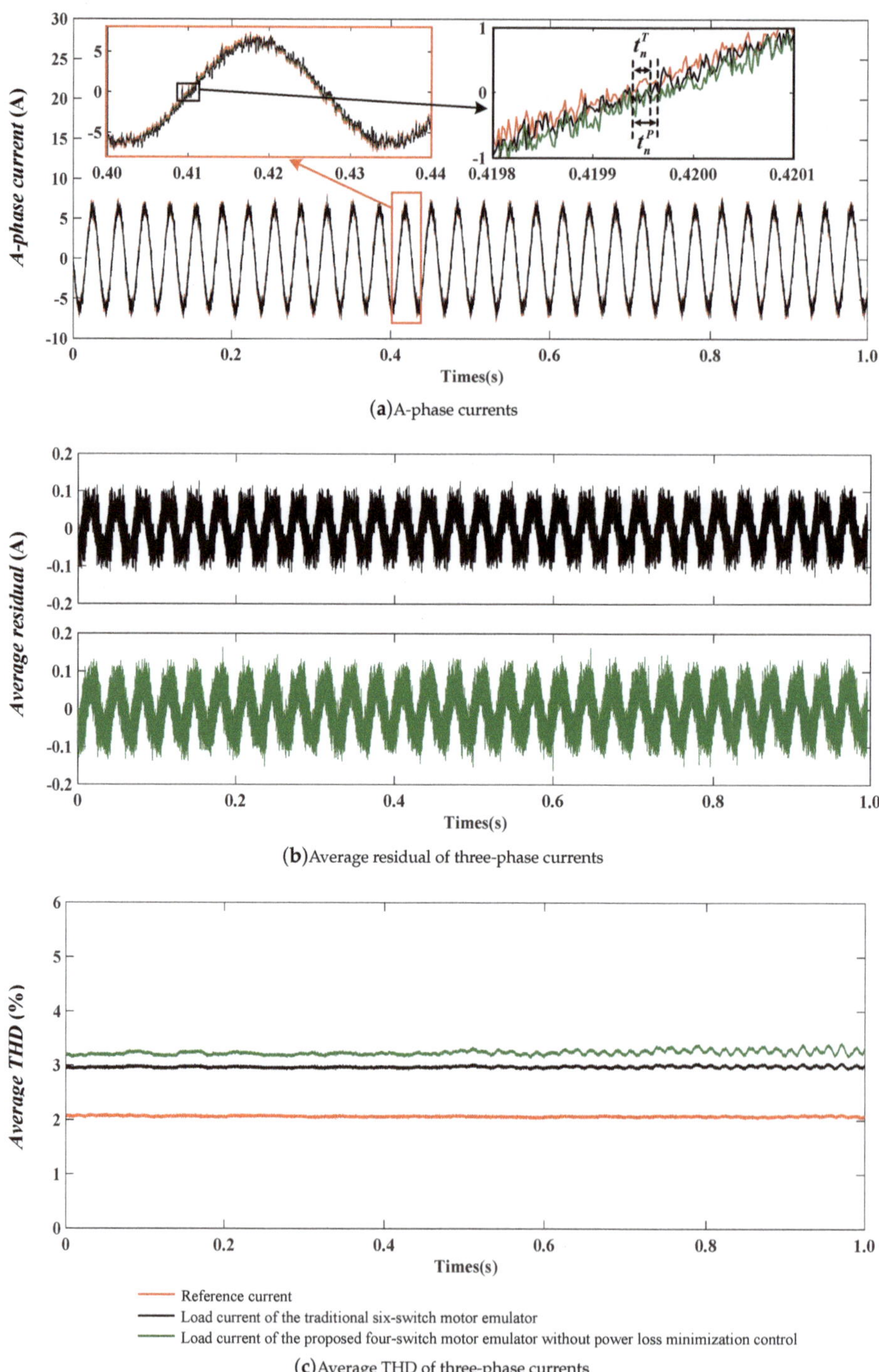

(**a**)A-phase currents

(**b**)Average residual of three-phase currents

(**c**)Average THD of three-phase currents

Figure 7. Waveforms of traditional six-switch and proposed four-switch motor emulator for low power motor.

As shown in Figure 7a, comparing to the zero-crossing time of reference current, the zero-crossing time of six-switch motor emulator is delayed by about 30 us, which is expressed as t_n^T, and the zero-crossing time of four-switch motor emulator is delayed by about 35 us, which is denoted as t_n^P. From Figure 7b, average three-phase current residual of six-switch motor emulator is about 0.1 A, current tracking accuracy is 98.6%. Average three-phase current residual of four-switch motor emulator is about 0.12 A, current tracking accuracy is 98.3%. In Figure 7c, the average THD of three-phase reference current is 2.1%, and average THD of three-phase load current of six-switch and four-switch motor emulator are about 2.9% and 3.3%, respectively. From Table 4, comparing to traditional six-switch motor emulator, although the average switch power loss has increased, the sum of switch power loss decreases 20.7%. The experimental results have shown that the proposed four-switch motor emulator decrease effectively switch power loss on the premise of ensuring current tracking effect of motor emulator for the high and low power motor.

4.2. Current Tracking Performance Based on FSMPC and FSMPC with Delay Compensation

On the condition that high power motor is operating with reference load current set 60 A/20 Hz and low power motor is operating with reference load current set 7 A/30 Hz, respectively. The comparison of current tracking performance and switch power loss based on FSMPC and FSMPC with delay compensation on the condition without power loss minimization control are displayed in this section.

For the high power motor, the experimental waveforms of *a*-phase reference current, *a*-phase load current based on FSMPC, and *a*-phase load current based on FSMPC with delay compensation are shown in Figure 8a. With these conditions, the experimental waveforms of average three-phase current residual of between reference current and load current based on FSMPC and FSMPC with delay compensation are shown in Figure 8b, respectively. The experimental waveforms of average three-phase current THD of reference current, load current based on FSMPC and FSMPC with delay compensation are shown in Figure 8c, respectively. The switch power loss results during entire simulation period based on FSMPC and FSMPC with delay compensation are shown in Table 5. Similarly, the experimental waveforms for the low power motor are shown in Figure 9a–c and Table 6. Comparison of based on FSMPC and FSMPC with delay compensation for high and low power motor are shown in Table 7.

Table 5. Switch power losses based on FSMPC and FSMPC with delay compensation for high power motor.

	Switch Power Loss (KJ)						σ (%)
	p_{b1}	p_{b2}	p_{c1}	p_{c2}	p_{sum}	p_{ave}	
FSMPC	22.40	23.36	23.25	21.83	90.84	22.71	0.13
FSMPC with delay compensation	21.84	23.47	23.69	21.72	90.72	22.68	

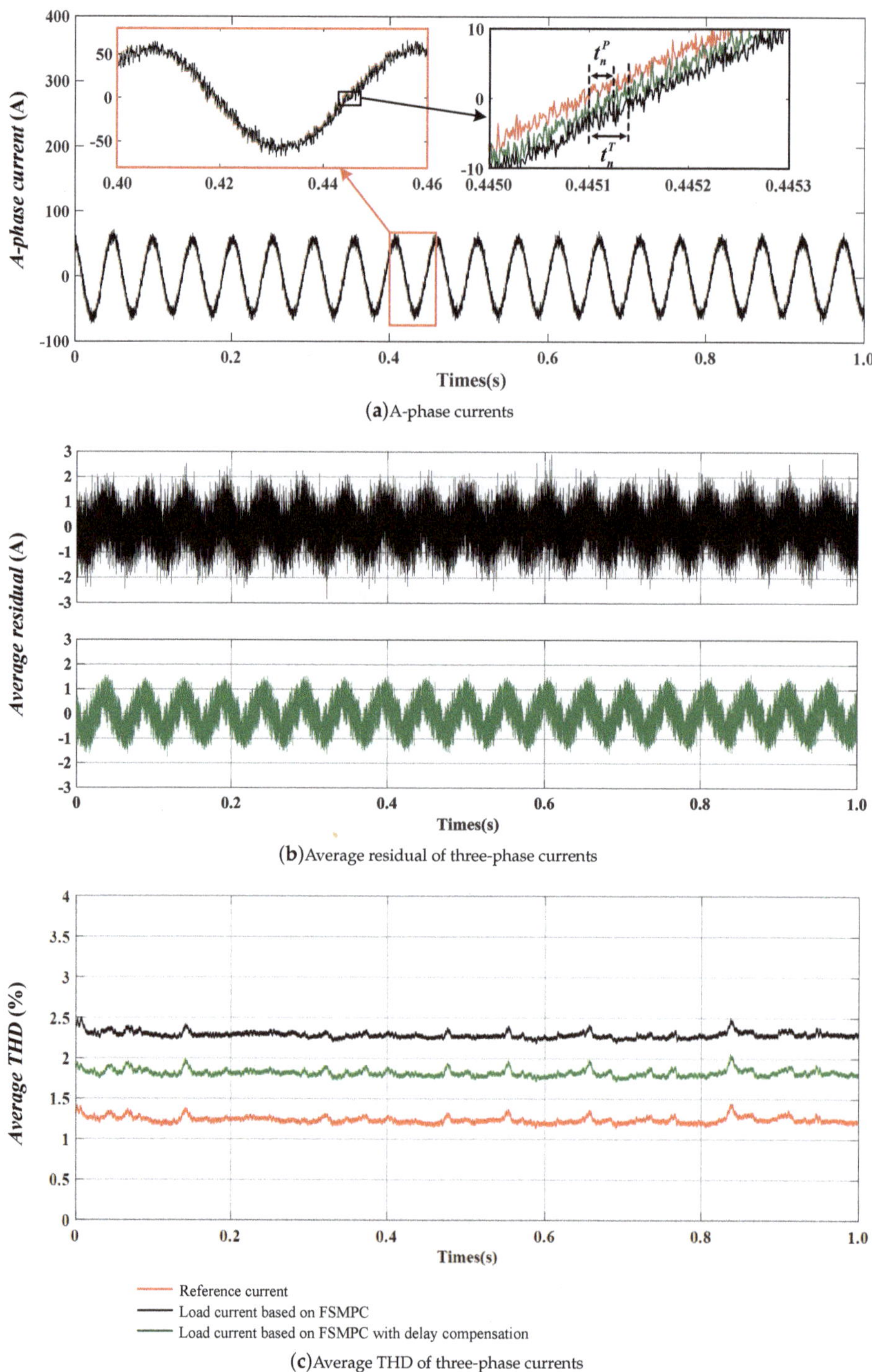

(**a**)A-phase currents

(**b**)Average residual of three-phase currents

(**c**)Average THD of three-phase currents

Figure 8. Waveforms based on FSMPC and FSMPC with delay compensation for high power motor.

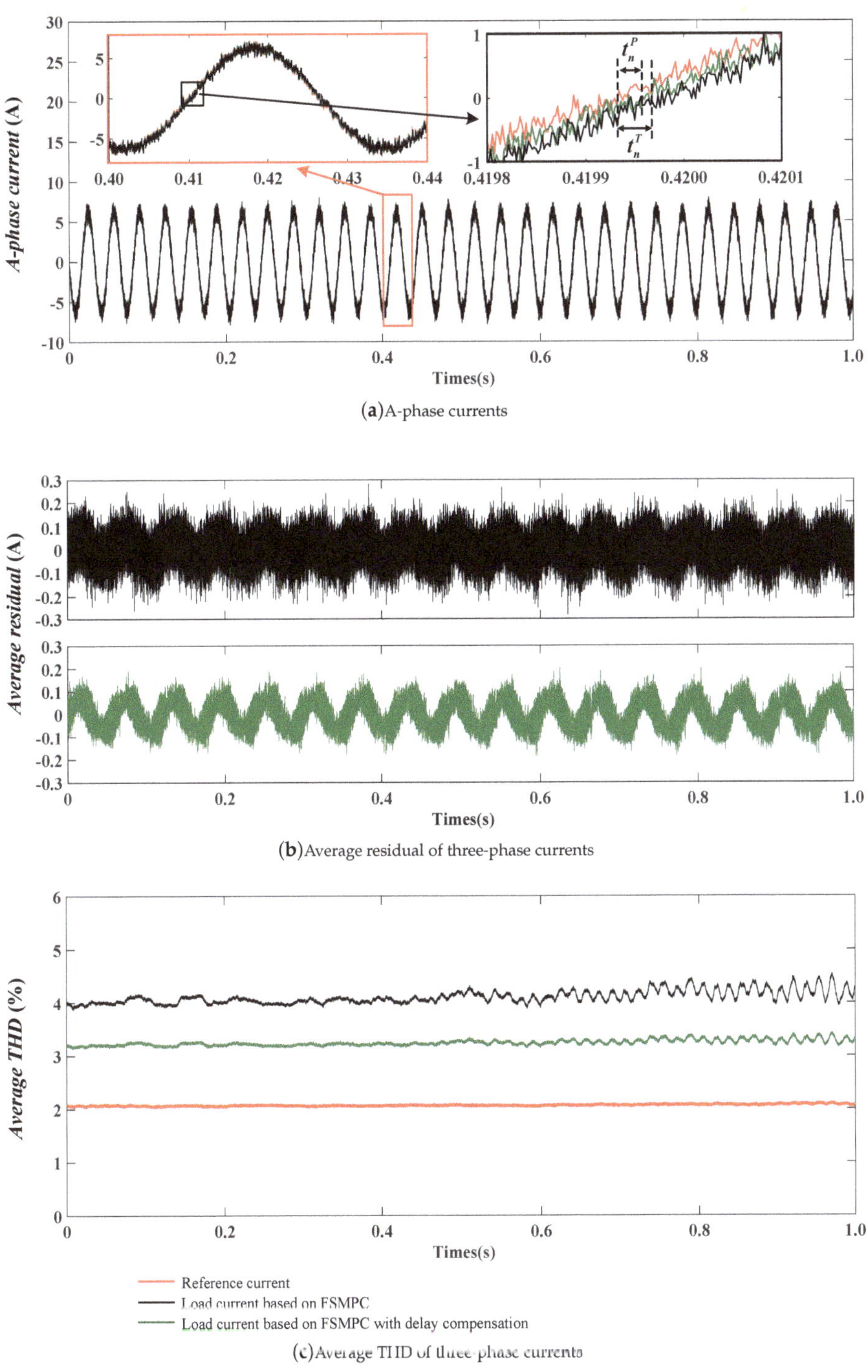

(**a**)A-phase currents

(**b**)Average residual of three-phase currents

(**c**)Average THD of three-phase currents

Figure 9. Waveforms based on FSMPC and FSMPC with delay compensation for low power motor.

As shown in Figure 8a, comparing to the zero-crossing time of reference current, the zero-crossing time of load current based on FSMPC is delayed by about 45 us, which is expressed as t_n^T, and the zero-crossing time of load current based on FSMPC with delay compensation is delayed by about 35 us, which is denoted as t_n^P. From Figure 8b, average three-phase current residual based on FSMPC is about 1.9 A, current tracking accuracy is 96.8%. Average three-phase current residual based on FSMPC with delay compensation is about 1.3 A, current tracking accuracy is 97.8%. In Figure 8c, the average THD of three-phase reference current is 1.30%, and average THD of three-phase load current based on FSMPC and FSMPC with delay compensation are about 2.4% and 1.80%, respectively. From the Table 5, the sum and average switch power loss based on FSMPC and FSMPC with delay compensation are almost equal.

Table 6. Switch power losses based on FSMPC and FSMPC with delay compensation for low power motor.

	Switch Power Loss (J)						σ (%)
	p_{b1}	p_{b2}	p_{c1}	p_{c2}	p_{sum}	p_{ave}	
FSMPC	41.75	42.16	41.84	40.97	166.72	41.68	0.22
FSMPC with delay compensation	41.80	41.94	41.57	41.06	163.37	41.59	

As shown in Figure 9a, comparing to the zero-crossing time of reference current, the zero-crossing time of load current based on FSMPC is delayed by about 45 us, which is expressed as t_n^T, and the zero-crossing time of load current based on FSMPC with delay compensation is delayed by about 35 us, which is denoted as t_n^P. From Figure 9b, average three-phase current residual based on FSMPC is about 0.17 A, current tracking accuracy is 97.5%. Average three-phase current residual based on FSMPC with delay compensation is about 0.12 A, current tracking accuracy is 98.3%. In Figure 9c, the average THD of three-phase reference current is 2.1%, and average THD of three-phase load current based on FSMPC and FSMPC with delay compensation are about 3.2% and 4.3%, respectively. From the Table 6, the sum and average switch power loss based on FSMPC and FSMPC with delay compensation are almost equal. The experimental results have shown that, for the high and low power motor, the residual between load current and reference current is reduced and current tracking accuracy is improved by applying FSMPC with delay compensation, when compared with FSMPC.

Table 7. Comparison of based on FSMPC and FSMPC with delay compensation for high and low power motor.

	High Power Motor		Low Power Motor	
	FSMPC	**FSMPC with Delay Compensation**	**FSMPC**	**FSMPC with Delay Compensation**
Tracking accuracy of load current	96.8%	97.8%	97.5%	98.3%
Zero-crossing delay time	45 us	35 us	45 us	35 us
THD	2.4%	1.8%	4.3%	3.2%
Decrease percent of switch power loss	0.13%		0.22%	
Load inductance	4 mH		0.13 mH	
Switching frequency	10 kHz			
Sampling frequency	20 kHz			

4.3. Motor Rotor Broken Bar Fault

When the motor is operating under rotor broken bar fault with the reference current for high power motor set 60 A/20 Hz and low power motor set 7 A/30 Hz respectively, and fault level set 20%. The comparison of current tracking performance and switch power loss for proposed motor emulator without and with power loss minimum control are displayed in this section.

For the high power motor, the experimental waveforms of a-phase reference current, a-phase load current of proposed motor emulator without and with power loss minimization control are shown in Figure 10a. Based on these conditions, the experimental waveforms of average three-phase current residual of between reference current and load current of without and with power loss minimization control are shown in Figure 10b, respectively. The experimental waveforms of average three-phase current THD of reference current, load current with or without power loss minimization control are shown in Figure 10c, respectively. The switch power loss results during entire simulation period of proposed motor emulator without and with power loss minimization control are shown in Table 8. Similarly, the experimental waveforms for the low power motor are shown in Figure 11a–c and Table 9.

Table 8. Switch power losses of motor rotor broken bar fault for high power motor.

	Switch Power Loss (KJ)						σ (%)
	p_{b1}	p_{b2}	p_{c1}	p_{c2}	p_{sum}	p_{ave}	
Without	22.91	23.84	23.75	22.36	92.86	23.22	22.74
With	17.52	18.37	18.65	17.22	71.76	17.94	

As shown in Figure 10a, when the motor is operating under rotor broken bar fault, comparing to the zero-crossing time of reference current, the zero-crossing time of load current without power loss minimization control is delayed by about 40 us, which is expressed as t_n^P, and the zero-crossing time of load current with power loss minimization control is delayed by about 50 us, which is denoted as t_y^P. From Figure 10b, average three-phase current residual of without and with power loss minimization control are about 2.0 A and 2.2 A, and current tracking accuracy are 96.6% and 96.3%. In Figure 10c, the average THD of three-phase reference current is 19%, and average THD of three-phase load current of without and with power loss minimization control is 22% and 24%, respectively. From Table 8, consideration of power loss, the sum and average switch power loss of the latter have both decreased by 22.74% than the former.

Table 9. Switch power losses of motor rotor broken bar fault for low power motor.

	Switch Power Loss (J)						σ (%)
	p_{b1}	p_{b2}	p_{c1}	p_{c2}	p_{sum}	p_{ave}	
Without	42.29	42.61	41.77	42.85	169.52	42.38	21.47
With	32.96	33.72	33.36	33.04	133.08	33.27	

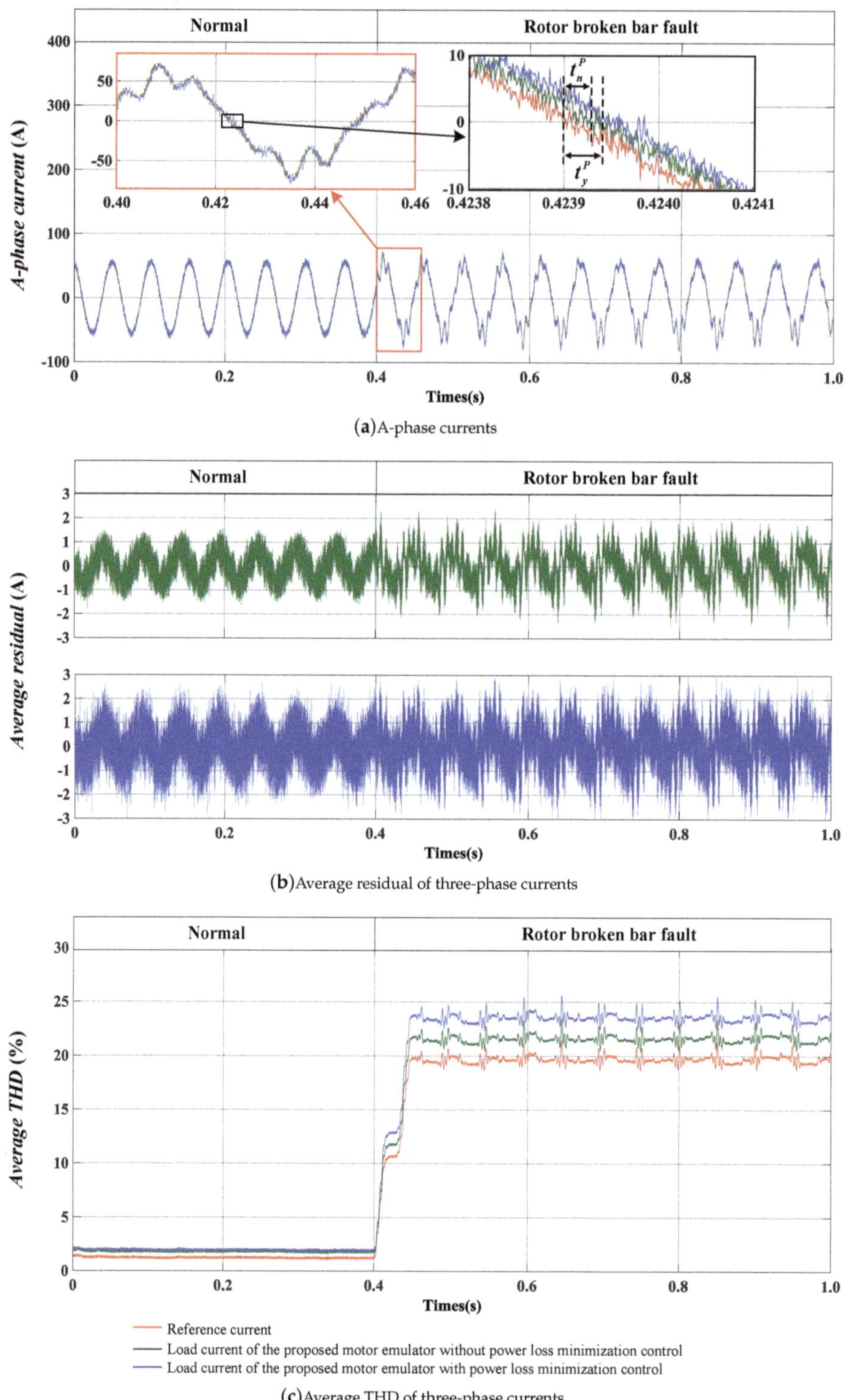

(**a**) A-phase currents

(**b**) Average residual of three-phase currents

(**c**) Average THD of three-phase currents

Figure 10. Waveforms of rotor broken bar fault for high power motor.

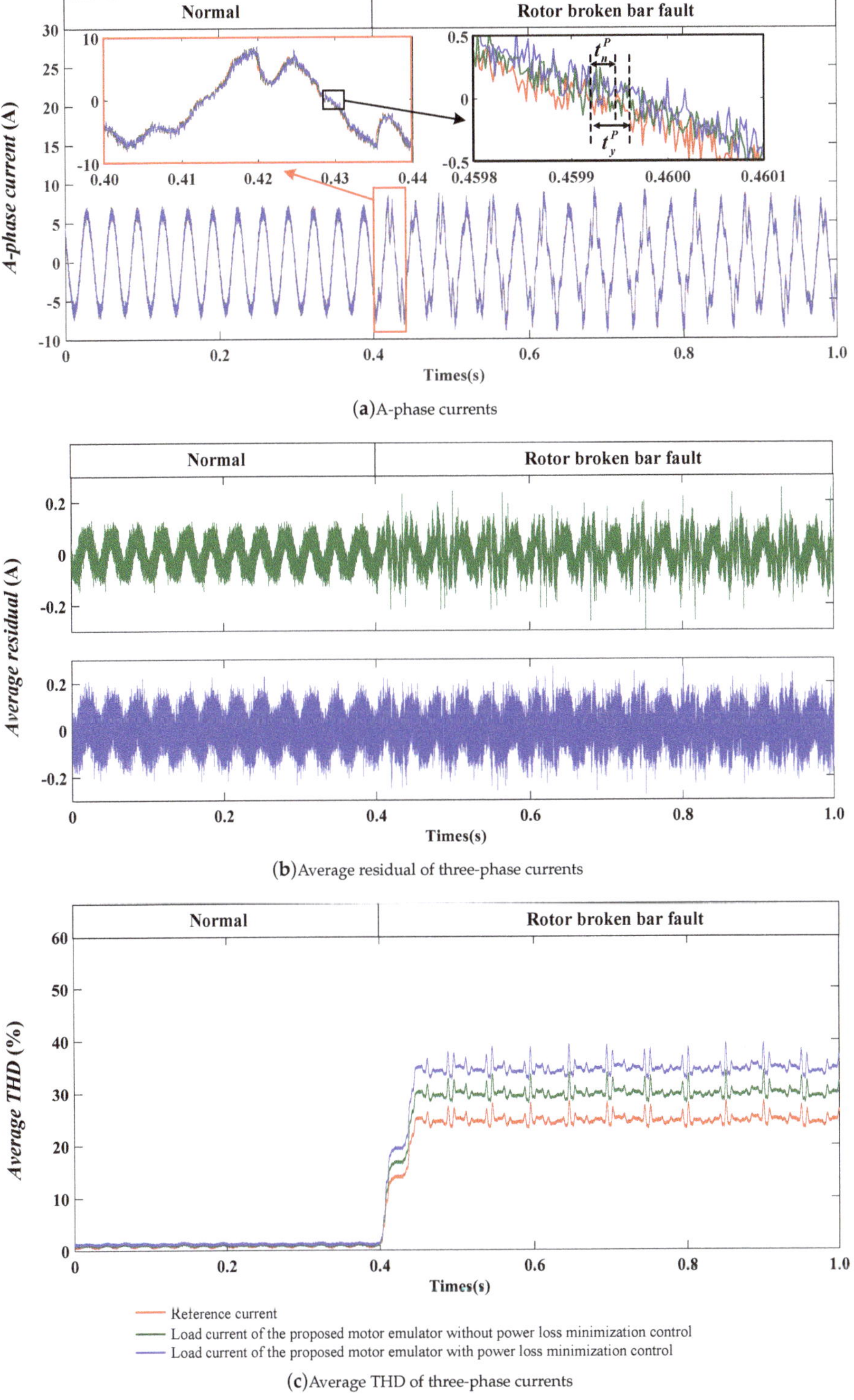

Figure 11. Waveforms of rotor broken bar fault for low power motor.

As shown in Figure 11a, when the motor is operating under rotor broken bar fault, comparing to the zero-crossing time of reference current, the zero-crossing time of load current without power loss minimization control is delayed by about 40 us, which is expressed as t_n^P, and the zero-crossing time of load current with power loss minimization control is delayed by about 50 us, which is denoted as t_y^P. From Figure 11b, average three-phase current residual of without and with power loss minimization control are about 0.15 A and 0.18 A, and current tracking accuracy are 97.8% and 97.4%. In Figure 11c, the average THD of three-phase reference current is 25%, and average THD of three-phase load current of without and with power loss minimization control is 30% and 35%, respectively. From Table 9, consideration of power loss, the sum and average switch power loss of the latter have both decreased by 21.47% than the former, respectively. The experimental results have shown that, for the high and low power motor, the proposed power loss decrease method can reduce effectively switch power loss on the premise of ensuring the current tracking effect of the motor emulator when the motor is operating under rotor broken bar fault.

4.4. Motor Speed Suddenly Alteration

When the motor is operating under speed suddenly alteration with speed set 1250 rpm to 2500 rpm for high power motor and speed set 1000 rpm to 1800 rpm for low power motor respectively, as show in Figures 12a and 13a. The comparison of current tracking performance and switch power loss for proposed motor emulator without and with power loss minimization control are displayed in this section.

For the high power motor, the experimental waveforms of a-phase reference current, a-phase load current of proposed motor emulator without and with power loss minimization control are shown in Figure 12b. Based on these conditions, the experimental waveforms of average three-phase current residual of between reference current and load current of without and with power loss minimization control are shown in Figure 12c, respectively. The experimental waveforms of average three-phase current THD of reference current, load current of proposed motor emulator without and with power loss minimization control are shown in Figure 12d, respectively. The switch power loss results during the entire simulation period of the proposed motor emulator without and with power loss minimization control are shown in Table 10. Similarly, the experimental waveforms for the low power motor are shown in Figure 13b–d and Table 11.

Table 10. Switch power losses of speed suddenly alteration for high power motor.

	Switch Power Loss (KJ)						σ (%)
	p_{b1}	p_{b2}	p_{c1}	p_{c2}	p_{sum}	p_{ave}	
Without	22.76	23.95	23.47	22.36	92.54	23.13	20.83
With	17.72	18.96	18.75	17.83	73.26	18.32	

As shown in Figure 12b, when the motor is operating under speed sudden alteration, comparing to the zero-crossing time of reference current, the zero-crossing time of load current without power loss minimization control is delayed by about 40 us, which is expressed as t_n^P, and the zero-crossing time of load current with power loss minimization control is delayed by about 60 us, which is denoted as t_y^P. From Figure 12c, average three-phase current residual of without and with power loss minimization control are about 2.0 A and 2.2 A, and current tracking accuracy are 96.6% and 96.3%. In Figure 12d, the average THD of three-phase reference current is 1.7%, and average THD of three-phase load current of without and with power loss minimization control is 1.9% and 2.0%, respectively. From Table 10, consideration of power loss, the sum and average switch power loss of the latter have both decreased by 20.83% than the former.

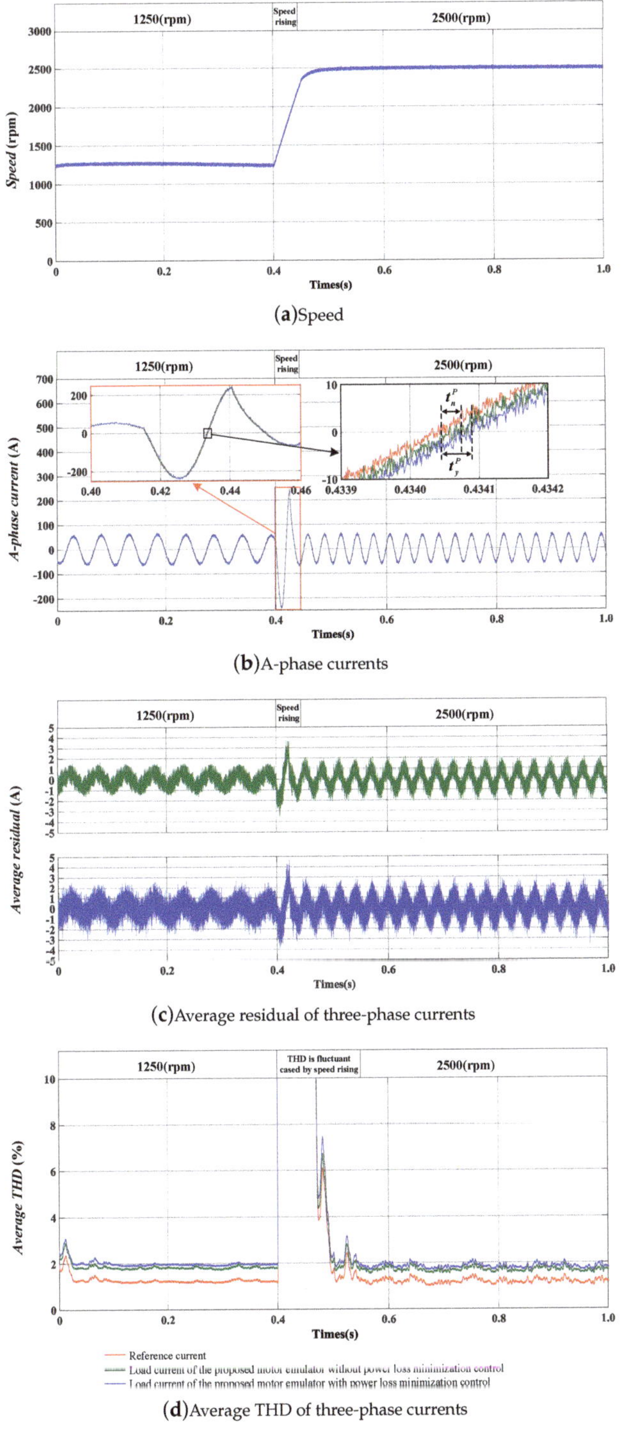

(**a**)Speed

(**b**)A-phase currents

(**c**)Average residual of three-phase currents

(**d**)Average THD of three-phase currents

Figure 12. Waveforms of speed suddenly alteration for high power motor.

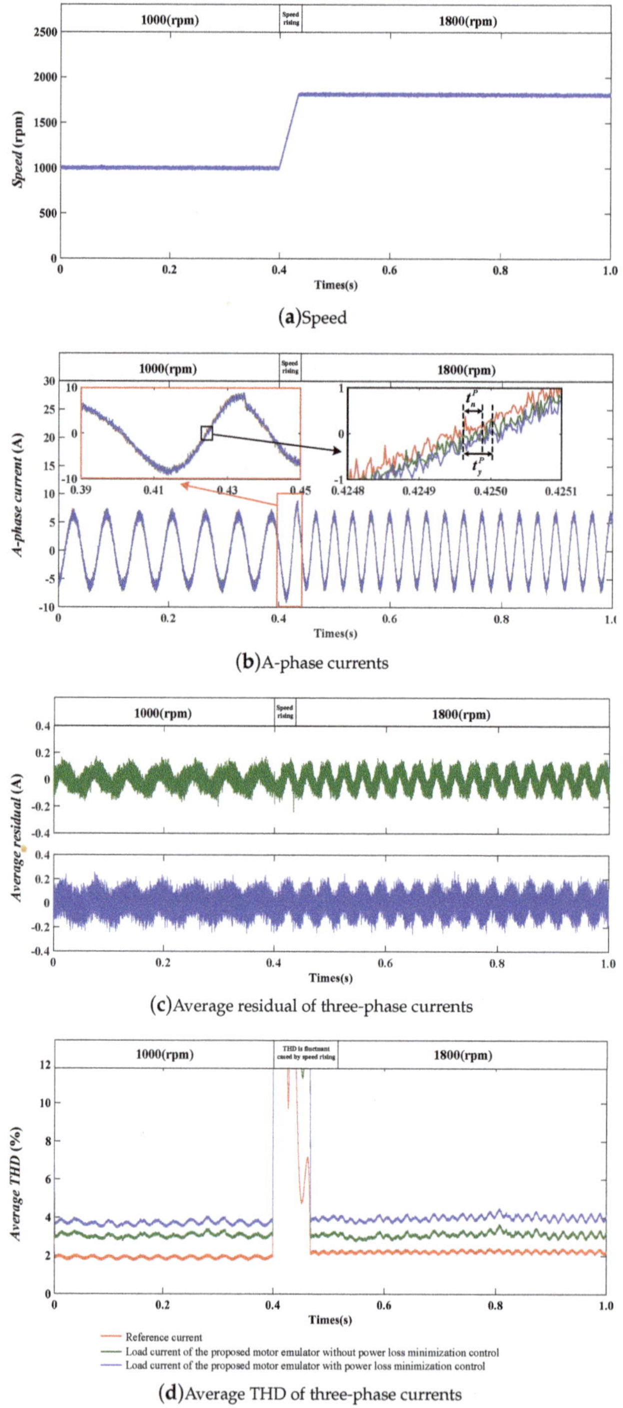

Figure 13. Waveforms of speed suddenly alteration for low power motor.

Table 11. Switch power losses of speed suddenly alteration for low power motor.

	Switch Power Loss (J)						σ (%)
	p_{b1}	p_{b2}	p_{c1}	p_{c2}	p_{sum}	p_{ave}	
Without	45.64	46.32	45.27	45.70	182.94	45.73	20.88
With	36.31	35.97	35.86	36.58	144.72	36.18	

As shown in Figure 13b, when the motor is operating under speed suddenly alteration, comparing to the zero-crossing time of reference current, the zero-crossing time of load current without power loss minimization control is delayed by about 40 us, which is expressed as t_n^P, and the zero-crossing time of load current with power loss minimization control is delayed by about 60 us, which is denoted as t_y^P. From Figure 13c, average three-phase current residual of without and with power loss minimization control are about 0.16 A and 0.19 A, and current tracking accuracy are 97.7% and 97.3%. In Figure 13d, the average THD of three-phase reference current is 2.0%, and average THD of three-phase load current of without and with power loss minimization control is 3.2% and 4.0%, respectively. From Table 11, consideration of power loss, the sum and average switch power loss of the latter have both decreased by 20.88% than the former, respectively. The experimental results show that, for the high and low power motor, the proposed power loss decrease method can reduce effectively switch power loss on the premise of ensuring the current tracking effect of the motor emulator, when the motor is operating under speed suddenly alteration.

5. Conclusions

A power loss decrease method based on FSMPC with delay compensation for a reduced switch count motor emulator is proposed in this paper. In the proposed motor emulator topology, converter consists of four active switches and two capacitors. Within the power loss decrease method based on FSMPC with delay compensation, an objective function is designed to select the two adjacent switch control signals that generating lowest switch power loss while keeping satisfied current tracking performance. The simulation and experiment results show the feasibility and effectiveness of the proposed method which achieving minimum power loss and ensuring current tracking performance greater than 95%. Besides, they also testify that the current can track stator current accurately and rapidly when the motor operating on the cases, namely in the normal state, or the fault state or the speed suddenly alteration. A real-time platform of a motor emulator for the presented method has been built to provide a reliable environment and offers more authentic data for motor fault injection, diagnosis, and tolerance research.

Author Contributions: conceptualization, R.Q. and C.Y. (Chunhua Yang); methodology, R.Q. and H.T.; software, R.Q.; validation, R.Q. and T.P.; formal analysis, C.Y. (Chunhua Yang); investigation, C.Y. (Chao Yang); resources, C.Y. (Chunhua Yang); data curation, H.T.; writing—original draft preparation, R.Q.; writing—review and editing, C.Y. (Chunhua Yang) and Z.C.; visualization, R.Q.; supervision, Z.C.; project administration, T.P.; funding acquisition, C.Y. (Chao Yang).

Acknowledgments: This research was supported by the National Natural Science Foundation of China (No. 61490702, 61773407, 61621062 and 61803390), by the Key Laboratory of Energy Saving Control and Safety Monitoring for Rail Transportation (No. 2017TP1002), by the Program of the Joint Pre-research Foundation of the Chinese Ministry of Education (No. 6141A02022110), by the General Program of the Equipment Pre-research Field Foundation of China (No. 61400030501), by the Hunan Provincial Innovation Foundation for Postgraduate (No. CX20190064), by the Fundamental Research Funds for the Central Universities of Central South University (No. 2018zzts592).

Conflicts of Interest: The authors declare no conflict of interest.

References

1. Chen, Z.; Li, X.; Yang, C.; Peng, T.; Yang, C.; Karimi, H.R.; Gui, W. A data-driven ground fault detection and isolation method for main circuit in railway electrical traction system. *ISA Trans.* **2019**, *87*, 264–271. [CrossRef] [PubMed]
2. Yang, X.; Yang, C.; Peng, T.; Chen, Z.; Liu, B.; Gui, W. Hardware-in-the-Loop Fault Injection for Traction Control System. *IEEE J. Emerg. Sel. Top. Power Electron.* **2018**, *6*, 696–706. [CrossRef]
3. Chen, Z.; Ding, S.; Peng, T. Fault Detection for Non-Gaussian Processes Using Generalized Canonical Correlation Analysis and Randomized Algorithms. *IEEE Trans. Ind. Electron.* **2018**, *65*, 1559–1567. [CrossRef]
4. Chen, Z.; Yang, C.; Peng, T.; Dan H.; Li, C.; Gui, W. A Cumulative Canonical Correlation Analysis-Based Sensor Precision Degradation Detection Method. *IEEE Trans. Ind. Electron.* **2019**, *66*, 6321–6330. [CrossRef]
5. Glowacz, A. Acoustic-Based Fault Diagnosis of Commutator Motor. *Electronics* **2018**, *7*, 299. [CrossRef]
6. Primavera, S.; Rella, G.; Maddaleno, F. One-cycle controlled three-phase electronic load. *IET Power Electron.* **2012**, *5*, 827–832. [CrossRef]
7. Tsang, K.M.; Chan, W.L. Fast Acting Regenerative DC Electronic Load Based on a SEPIC Converter. *IEEE Trans. Power Electron.* **2012**, *27*, 269–275. [CrossRef]
8. Jack, A.G.; Atkinson, D.J.; Slater, H.J. Real-time emulation for power equipment development. Part 1: Real-time simulation. *IEE Proc.-Electr. Power Appl.* **1998**, *145*, 92–97. [CrossRef]
9. Slater, H.J.; Atkinson, D.J.; Jack, A.G. Real-time emulation for power equipment development. Part 2: The virtual machine. *IEE Proc.-Electr. Power Appl.* **1998**, *145*, 153–158. [CrossRef]
10. Roodsari, B.N.; Nowicki, E.P. Analysis and Experimental Investigation of the Improved Distributed Electronic Load Controller. *IEEE Trans. Energy Convers.* **2018**, *33*, 905–914. [CrossRef]
11. Singh, B; Murthy, S.; Gupta, S. Transient analysis of self-excited induction generator with electronic load controller (ELC) supplying static and dynamic loads *IEEE Trans. Ind. Appl.* **2005**, *41*, 1194–1204. [CrossRef]
12. Wang, J.; Yang, L.; Ma, Y. Static and Dynamic Power System Load Emulation in a Converter-Based Reconfigurable Power Grid Emulator. *IEEE Trans. Power Electron.* **2016**, *31*, 3239–3251. [CrossRef]
13. Geng, Z.; Gu, D.; Hong, T. Programmable Electronic AC Load Based on a Hybrid Multilevel Voltage Source Inverter. *IEEE Trans. Ind. Appl.* **2018**, *54*, 5512–5522. [CrossRef]
14. Heerdt, J.A.; Coutinho, D.F.; Mussa, S.A. Control Strategy for Current Harmonic Programmed AC Active Electronic Power Loads. *IEEE Trans. Ind. Electron.* **2014**, *61*, 3810–3822. [CrossRef]
15. Kesler, M.; Ozdemir, E.; Kisacikoglu, M.C. Power Converter-Based Three-Phase Nonlinear Load Emulator for a Hardware Testbed System. *IEEE Trans. Power Electron.* **2014**, *29*, 5806–5812. [CrossRef]
16. Grubic, S.; Amlang, B.; Schumacher, W. A High-Performance Electronic Hardware-in-the-Loop Drive-Load Simulation Using a Linear Inverter (LinVerter). *IEEE Trans. Ind. Electron.* **2010**, *57*, 1208–1216. [CrossRef]
17. Rao, Y.S.; Chandorkar, M.C. Real-Time Electrical Load Emulator Using Optimal Feedback Control Technique. *IEEE Trans. Ind. Electron.* **2010**, *57*, 1217–1225.
18. Marguc, J.; Truntic, M.; Rodic, M. FPGA Based Real-Time Emulation System for Power Electronics Converters. *Energies* **2019**, *12*, 969. [CrossRef]
19. Moussa, I.; Khedher, A.; Bouallegue, A. Design of a Low-Cost PV Emulator Applied for PVECS. *Electronics* **2019**, *8*, 232. [CrossRef]
20. Sadigh, A.; Dargahi, V.; Corzine, K. Investigation of Conduction and Switching Power Losses in Modified Stacked Multicell Converters. *IEEE Trans. Ind. Electron.* **2016**, *63*, 7780–7791. [CrossRef]
21. Chen, P.; Hu, K.; Lin, Y. Development of a Prime Mover Emulator Using a Permanent-Magnet Synchronous Motor Drive. *IEEE Trans. Power Electron.* **2018**, *33*, 6114–6125. [CrossRef]
22. Kojabadi, H.M.; Chang, L.C.; Boutot, T. Development of a Novel Wind Turbine Simulator for Wind Energy Conversion Systems Using an Inverter-Controlled Induction Motor. *IEEE Trans. Energy Convers.* **2004**, *19*, 547–552. [CrossRef]
23. Peng, T.; Liu, B.; Yang, C.; Qin, R.; Yang, C. Fault current tracking method of motor simulator based on finite set model prediction. *J. Cent. South Univ.* **2019**, *50*, 1098–1104.
24. Naderi, Y.; Hosseini, S.; Zadeh, S.; Mohammadi, B.; Savaghebi, M.; Guerrero, J. An optimized direct control method applied to multilevel inverter for microgrid power quality enhancement. *Int. J. Electr. Power Energy Syst.* **2018**, *170*, 496–506. [CrossRef]

25. Yang, C.; Yang, C.; Peng, T.; Yang X.; Gui, W. A Fault-Injection Strategy for Traction Drive Control System. *IEEE Trans. Ind. Electron.* **2017**, *64*, 5719–5727. [CrossRef]
26. Chen, X.; Wu, W.; Gao, N.; Liu, J.; Henry S.; Frede B. Finite Control Set Model Predictive Control for an LCL-Filtered Grid-Tied Inverter with Full Status Estimations under Unbalanced Grid Voltage. *Energies* **2019**, *12*, 2691. [CrossRef]
27. Hu, S.; Liu, G.; Jin, N.; Guo, L. Constant-Frequency Model Predictive Direct Power Control for Fault-Tolerant Bidirectional Voltage-Source Converter with Balanced Capacitor Voltage. *Energies* **2018**, *11*, 2692.

MDPI

St. Alban-Anlage 66

4052 Basel

Switzerland

Tel. +41 61 683 77 34

Fax +41 61 302 89 18

www.mdpi.com

Energies Editorial Office

E-mail: energies@mdpi.com

www.mdpi.com/journal/energies